W9-ATX-533

ANNUAL REVIEW OF
INFORMATION SCIENCE
AND TECHNOLOGY

Volume 22, 1987

ISSN: 0066-4200
CODEN: ARISBC

ANNUAL REVIEW OF INFORMATION SCIENCE AND TECHNOLOGY

Volume 22, 1987

Edited by

Martha E. Williams

University of Illinois
Urbana, Illinois, USA

Published on behalf of the
American Society for Information Science

1987

ELSEVIER SCIENCE PUBLISHERS
AMSTERDAM · NEW YORK · OXFORD · TOKYO

LSL
Ref
Z
699
A 1
A 65
v. 22

ISBN: 0 444 70302 0
ISSN: 0066 - 4200
CODEN: ARISBC

Published by:
ELSEVIER SCIENCE PUBLISHERS B.V.
P.O. Box 1991
1000 BZ Amsterdam
The Netherlands

for
American Society for Information Science
1424 Sixteenth Street N.W.
Washington, DC 20036, U.S.A.

Sole distributors for the U.S.A. and Canada:
ELSEVIER SCIENCE PUBLISHING COMPANY, INC.
52 Vanderbilt Avenue
New York, NY 10017
U.S.A.

LC No. 66-25096

ARIST Production Staff, for ASIS:
Charles & Linda Holder, Graphic Compositors
PRINTED IN THE NETHERLANDS

Contents

III
Applications

Preface

This is the 22nd volume of the *Annual Review of Information Science and Technology* (*ARIST*) produced for the American Society for Information Science (ASIS) and published by Elsevier Science Publishers B.V. ASIS initiated the series in 1966 with the publication of Volume 1 under the editorship of Carlos A. Cuadra, who continued as Editor through Volume 10. Martha E. Williams has served as Editor starting with Volume 11. ASIS is the owner of *ARIST*, maintains the editorial control, and has the sole rights to the series.

Through the years several organizations have been responsible for publishing and marketing *ARIST*. Volumes 1 and 2 were published by Interscience Publishers, a division of John Wiley & Sons. Volumes 3 through 6 were published by Encyclopaedia Britannica, Inc. Volumes 7 through 11 were published by ASIS itself, Volumes 12 through 21 were published by Knowledge Industry Publications, Inc. and with Volume 22 Elsevier Science Publishers B.V., Amsterdam, The Netherlands assumed the role of publisher of *ARIST* for ASIS.

Policy. ARIST is an annual publication that reviews numerous topics within the broad field of information science and technology. No single topic is treated on an annual basis; it is the publication of the book that occurs annually. Inasmuch as the field is dynamic, the contents (chapters) of the various *ARIST* volumes must change to reflect this dynamism. *ARIST* chapters are scholarly reviews of specific topics as substantiated by the published literature. Some material may be included, even though not backed up by literature, if it is needed to provide a balanced and complete picture of the state of the art for the subject of the chapter. The time period covered varies from chapter to chapter, depending on whether the topic has been treated previously by *ARIST* and, if so, on the length of the interval from the last treatment to the current one. Thus, reviews may cover a one-year or a multiyear period. The reviews aim to be critical in that they provide the author's expert opinion regarding developments and activities within the chapter's subject area. The review guides the reader to or from specific publications. Chapters aim to be scholarly, thorough within the scope defined by the chapter author, up to date, well written, and readable by an audience that goes beyond the author's immediate peer group to researchers and practitioners in information science and technology, in general, and ASIS members, in particular.

Purpose. The purpose of *ARIST* is to describe and to appraise activities and trends in the field of information science and technology. Material presented should be substantiated by references to the literature. *ARIST* provides an annual review of topics in the field. One volume is produced each year. A master plan for the series encompasses the entire field in all its aspects, and topics for each volume are selected from the plan on the basis of timeliness and an assessment of reader interest.

References cited in text and bibliography. The format for referring to bibliographic citations within the text involves use of the cited author's name instead of reference numbers. The cited author's surname is printed in upper case letters. The reader, wishing to find the bibliographic references, can readily locate the appropriate reference in the bibliography (alphabetically arranged by first author's last name). A single author appears as SMITH; coauthors as SMITH & JONES; and multiple authors as SMITH ET AL. If multiple papers by the same author are cited, the distinction is made by indicating the year of publication after the last name (e.g., SMITH, 1986), and if a further distinction is required for multiple papers within the same year, a lower case alpha character follows the year (e.g., SMITH, 1986a). Except for the fact that all authors in multi-authored papers are included in bibliographic references, the same basic conventions are used in the chapter bibliographies. Thus, the reader can easily locate in the bibliography any references discussed in the text.

Because of the emphasis placed on the requirement for chapter authors to discuss the key papers and significant developments reported in the literature, and because *ARIST* readers have expressed their liking for comprehensive bibliographies associated with the chapters, more references may be listed in the bibliographies than are discussed in the text.

The format used for references in the bibliographies is based on the *American National Standard for Bibliographic References*, ANSI Z39.29. We have followed the ANSI guidelines with respect to the sequence of bibliographic data elements and the punctuation used to separate the elements. Adoption of this convention should facilitate conversion of the references to machine-readable form as need arises. Journal article references follow the ANSI guide as closely as possible. Conference papers and microform publications follow an *ARIST* adaptation of the format.

Structure of the volume. In accordance with the *ARIST* master plan, this volume's nine chapters fit within a basic framework: I. Planning Information Systems and Services; II. Basic Techniques and Technologies; and III. Applications. Chapter titles are provided in the Table of Contents, and an Introduction to each section highlights the events, trends, and evaluations given by the chapter authors. An Index to the entire volume is provided to help the user locate material relevant to the subject content, authors, and organizations cited in the book. An explanation of the guidelines employed in the Index is provided in the Introduction to the Index.

Acknowledgments. Appreciation should be expressed to many individuals and organizations for their roles in creating this volume. First and foremost are the authors of the individual chapters who have generously contributed their time and efforts in searching, reviewing, and evaluating the large body of literature on which their chapters are based. The *ARIST* Advisory Committee Members and *ARIST* Reviewers provided valuable feedback and constructive criticism of the content. Major contributions toward the production of this publication were made by Mary W. Rakow, Copy Editor, Elaine Tisch, Bibliographic Editor, and Debora Shaw, Index Editor. Appreciation is expressed to all of the members of the *ARIST* technical support staff who are listed on the Acknowledgments page.

Martha E. Williams

Acknowledgments

The American Society for Information Science and the Editor wish to acknowledge the contributions of the three principals on the editorial staff and the technical support staff.

Mary W. Rakow, Copy Editor

Debora Shaw, Index Editor

Elaine Tisch, Bibliographic Editor

Technical Support Staff

El-Siddig At-Taras, Technical Advisor

Laurence Lannom, Technical Advisor

Laurel Preece, Technical Advisor

Scott E. Preece, Technical Advisor

Linda C. Smith, Technical Advisor

Sheila Carnder, Editorial Assistant

Margery Johnson, Clerical Support

Linda (O'Brien) Holder, Compositor

Advisory Committee for *ARIST*

Toni Carbo Bearman

Everett H. Brenner

José-Marie Griffiths

Donald Hawkins

Mary Ellen Jacob

Ben-Ami Lipetz

Thomas H. Martin

M. Lynne Neufeld

Elliot Siegel

Contributors

Bryce L. Allen
School of Library and
    Information Science
University of Western Ontario
London, Ontario
Canada N6G 1H1

Nicholas J. Belkin
School of Communication,
    Information, and Library Studies
Rutgers University
4 Huntington Street
New Brunswick, NJ 08903

W. Bruce Croft
Department of Computer and
    Information Science
University of Massachusetts at
    Amherst
Amherst, MA 01003

Patricia B. Culkin
Colorado Alliance of Research
    Libraries
CARL
777 Grant, Suite 304
Denver, CO 80203

Robyn C. Frank
U.S. Department of Agriculture
National Agricultural Library
Room 304
Beltsville, MD 20705

Mark T. Kinnucan
School of Library and Information
    Science
University of Western Ontario
London, Ontario
Canada N6G 1H1

Jounghyoun Lee
Engineering Library
221 Engineering Hall
University of Illinois at
    Urbana-Champaign
1308 W. Green Street
Urbana, IL 61801

Lois F. Lunin
Herner and Company
1700 North Moore Street
Suite 700
Arlington, VA 22209

William H. Mischo
Engineering Library
221 Engineering Hall
University of Illinois at
    Urbana-Champaign
1308 W. Green Street
Urbana, IL 61801

Michael J. Nelson
School of Library and Information
    Science
University of Western Ontario
London, Ontario
Canada N6G 1H1

Aatto J. Repo
Technical Research Centre of Finland
Technical Information Service
Vourimiehentie 5, SF-02150
Espoo, Finland

Ward Shaw
Colorado Alliance of Research
    Libraries
CARL
777 Grant, Suite 304
Denver, CO 80203

Linda C. Smith
410 David Kinley Hall
University of Illinois at Urbana-
    Champaign
1407 W. Gregory Drive
Urbana, IL 61801

Amy J. Warner
School of Library and Information
    Studies
University of Wisconsin at Madison
Helen C. White Building-Room 4255
600 N. Park Street
Madison, WI 53706

Chapter Reviewers

Henriette D. Avram

Harold Bamford

Toni Carbo Bearman

David Becker

Wesley T. Brandhorst

Everett H. Brenner

Walter Carlson

Martha Evens

Margaret T. Fischer

José-Marie Griffiths

Glynn Harmon

Donald Hawkins

Mary Ellen Jacob

Karen B. Levitan

Ben-Ami Lipetz

Donald Masys

M. Lynne Neufeld

Nancy Roderer

Gerard Salton

Peter B. Schipma

Elliot Siegel

Donald E. Walker

Edward Weiss

Herbert S. White

I

Planning Information Systems and Services

Section I includes a chapter on "Economics of Information" by Aatto J. Repo of the Technical Research Centre of Finland. Repo describes and analyzes studies on the economics of information during the past two decades from the point of view of the library and information service community. The literature covered is largely that which has been published in the United States and Great Britain. Some of the basic concepts and issues of importance are described and defined briefly. Repo traces the history of the phrase "economics of information" and explains it in terms of information supply and demand and effectiveness of information services, value of information to the user, and productivity of information work. The chapter deals primarily with empirical studies on the economics of information, and these are presented in nine categories: 1) costs of information products and services; 2) price of information; 3) evaluation of effectiveness and efficiency of library and information services; 4) cost-benefit analysis of information transfer; 5) value of information (case studies); 6) information service as a value-added process; 7) economics of information retrieval; 8) macroeconomic studies of information and productivity; and 9) economics of information processing. Repo comments on areas where there is a need for future research. These include: basic information on international and national information markets; awareness of costs and benefits of services within the library and information service community; and case studies of the value of information within organization.

Repo points out that most of the research into the economics of information is conceptually vague and that consequently the results are also often vague. While studies of the economics of information have frequently used classical economic theories, some features that are specific to information have emerged as a result of the studies. These special features are: 1) information products cannot be replaced with other products if the information contents are not identical, 2) the benefits of information depend on the user's ability to exploit them, 3) information does not deteriorate, 4) information is not a constant, 5) information is an abstraction, 6) total production cost of information is seldom included in its market price, and 7) because the benefit of information is tied to use, it is difficult to measure.

1 Economics of Information

AATTO J. REPO
Technical Research Centre of Finland

INTRODUCTION

General Background

This review describes the past 20 years of research on the economics of information and has been written from the viewpoint of library and information services. The literature covered has been published mostly in the United States and in Great Britain.

Research into the economics of information is conceptually vague and even though much research has been done, the results are still vague. Even though this report presents empirical research, it is necessary to introduce the basic concepts and ideas of the economics of information. Opinions are included among definitions because even the important concepts are somewhat hazy and there is disagreement even about the starting points of the research. A brief glossary of terms used in the economics of information is given by ROBERTS and includes valuable examples of the definitions used in the literature.

ARIST has reviewed this subject in four previous chapters (GRIFFITHS; HINDLE & RAPER; LAMBERTON; MICK). Economists who have studied the economics of information include MACHLUP (1962; 1980) and FLOWERDEW & WHITEHEAD. In information science the subject has been treated recently by KING ET AL. (1982; 1984), TAYLOR (1982a; 1982b; 1984a), CRONIN (1986), and CRONIN & GUDIM. LANCASTER (1977) evaluated the research done as of 1977 and BLAGDEN (1980a) reviewed research in the United Kingdom.

Annual Review of Information Science and Technology (*ARIST*), Volume 22, 1987
Martha E. Williams, Editor
Published for the American Society for Information Science (ASIS)
by Elsevier Science Publishers B.V.

The review of earlier studies on cost-benefit analysis in information science by FLOWERDEW & WHITEHEAD is recommended. It has now been supplemented by the seminal report of MARTYN & FLOWERDEW. WILLS & OLDMAN supplement the previously mentioned British viewpoints from the perspective of special libraries (see also BLAGDEN, 1980a).

For the U.S. studies, the collection of articles in KING ET AL. (1983) contains the latest writings about the costs and pricing of information products and services and the value of information. VAN HOUSE recently surveyed the research on the economics of information within libraries, and VARLEJS has covered this topic as it concerns public libraries.

Economics and Information

The phrase "economics of information" began to appear in the terminology of economists in the 1960s. In information science it appeared primarily in connection with evaluation studies. After the discussions on the costs and the effectiveness of information services, the value of information for the user and the productivity of information work have become issues of the 1980s. The economics of information can best be described by naming the central areas of interest in the information market (CASPER; VAN HOUSE): 1) information supply (economy of information production, pricing information products), 2) information demand (acquisition and use of information), 3) special problems of information as a product or resource (e.g., pricing of use, public funding of information production).

A study of the economics of information presupposes a definition of information itself. Many information scientists define information in terms of its method of presentation. For example, "information is stored knowledge," according to MARTYN & FLOWERDEW. Others define it as a process in which data become information, knowledge, and even wisdom (CLEVELAND; HORTON, 1979; TAYLOR, 1986). Data are raw material of information collected from nature or tradition (historical information), which is transformed, processed, collected, transferred, made available, and so forth. Information becomes knowledge when someone applies it to something useful. Wisdom can be seen as processed, integrated information (e.g., CLEVELAND). Information is thus seen as a kind of product, even as a resource (HORTON, 1979), but information is a unique resource (CLEVELAND; CRONIN, 1985; FLOWERDEW & WHITEHEAD; HORTON, 1982). The special features of information are now examined separately from the theoretical and practical viewpoint.

Recent discussion has focused on information in the context of a profit-oriented market. This, of course, makes the whole concept structure very shaky. CLEVELAND, who is a political scientist, characterizes information in the following way: 1) information is human- there is information only through human observation; 2) information is expandable; "The more we use it the more profitable it becomes"; one basic limit is the biological age of human individuals and groups; 3) information is compressible; the increasing amount of information can be controlled by centralization and integration and by compressing it so that it is useful in different surroundings; 4) information is substitutable- it can replace other resources like money, manpower, and raw materials; for instance, the accumulation of information related to automation replaces several million workers annually; 5) information is transferrable; the speed and facility of information transfer are a considerable factor in developing and shaping communities; 6) information is diffusive—it tends to leak out and spread despite our efforts to protect individuals and innovations; 7) information is shareable; goods can be exchanged, but in information exchange the giver still "retains" what he has given away.

At this point the above features of information must be left only as comments criticizing simple resource thinking. The study of the economics of information, in concrete investigations, has often used classical economic theories. Some special features of information have, however, emerged during empirical measurements and studies: 1) information products cannot be replaced with other information products if the information contents are not identical; 2) information products add value, but their benefit also depends on the ability of the user to exploit them; 3) information does not deteriorate by use; only time makes information similar to consumer goods in some cases (e.g., information on stock exchange rates; otherwise information is more like an investment good); 4) information is not a constant- i.e., generally it cannot be quantified. The model of SHANNON & WEAVER is only suitable for examining the amounts of information transferred, not for defining the value of the information content; 5) information is an abstraction- i.e., it is produced, disseminated, stored, and used through different devices and services. This feature causes plenty of confusion, for example, when one estimates the value of information indirectly by what someone is willing to pay for it; 6) new information is produced mainly with public funds (especially basic research), but the total production costs are rarely included in its market price; and 7) the real benefit of information is difficult to measure because it is tied to its use, which is unpredictable. These characteristics of information make studies of the economics of information problematic. The economics of information is, however, becoming an important research object in information science with expanding information investments and information work and economic pressures for effectiveness of information activities.

Research on the value of information is one of the key areas in the economics of information. These studies must be connected closely with the studies of how information is used. However, recent articles that criticize user studies suggest that such studies are not very common (see, for example, CAPLAN, DERR, and JARVELIN & REPO).

On the other hand, some have tried to define information products. While WILLS & OLDMAN argue strongly against an idea of information as a product, others like TAYLOR (1982b; 1984a) try to characterize information products. According to Taylor an information product (or service) must meet the following conditions: 1) it must have a noticeable role in the activity studied; it has to be measurable and comparable with other information products; 2) it has to be stable enough so that its production costs can be measured; 3) it must have such a form that it can be identified for other uses.

Such a definition of an information product (and service) ought to be available because it allows us to examine information in a general and situation-independent way. The definition also includes information content as part of the definition of information product. In the production and processing of pieces of information in an abstract information system the definition seems usable, but it does not offer a way to study the use of information. In practice, we are forced to examine separately, information products, information services and information systems on the one hand, and the use of information and its benefit and value, on the other.

Approaches and Methods

As previously stated, questions concerning the economics of information are closely connected to the evaluation of information services and libraries. The basic evaluation theory is simple: the goals are described, and then we measure how well the particular activity (or activities) helps us to realize the goals. Economic analysis has a central position in this evaluation. Cost-benefit analysis is needed in decision making in general as well as in the evaluation of previous decisions and in seeking the partial systems important for the activity. Cost-benefit analysis aids decision making when we seek answers to the following questions (FLOWERDEW & WHITEHEAD; WILKINSON): 1) how much should the organization invest in information products and services (e.g., compared with competitors); 2) how much of the organization's running costs should be invested in information services and how much in other activities; 3) is the size of the present information budget adequate, or should it be changed; 4) how are the resources allocated among different information services; 5) are the investments in particular information services successful, or should they be changed; 6) how useful are the individual information service projects; 7) how are the information services priced; 8) how should information production and dissemination be supported or taxed; 9) is pricing suitable for management and investment plans; 10) what kind of measures are needed to monitor the performance of the information services.

Flowerdew and Whitehead restrict themselves to cost-benefit analysis of the measures in terms of money. This approach, which is characteristic of economics, has been criticized because benefits cannot always be measured in terms of money. Thus, some researchers (e.g., OLDMAN, WHITE) deny the usefulness of cost-benefit analysis in measuring the value of information in the library environment. Stronger but questionable criticism is presented by Machlup (in VARLEJS) who notes that it is futile to try to measure the influence of information services on the users, the value of information or its social benefit. LANCASTER (1971), too, presents similar thoughts. Although these latter thoughts have been picked from connections, where the writers have criticized badly made and faulty generalizations in cost-benefit analyses of information, they indicate that some leading scientists both in information science and economics doubt the possibility of measuring the value-in-use of information. But we can at least name those benefits that cannot be measured with money. This view is used here because the difficulty of explaining the benefits of information in terms of money has led the analyses in practice toward this extended view of cost-benefit analysis. Perhaps one should speak of the value of information rather than the cost-benefit of information to avoid confusion. In presenting the empirical studies we further specify costs, effectiveness, and benefit. By combining these concepts we get five levels on which to examine the economics of information (FLOWERDEW & WHITE-HEAD; LANCASTER, 1977; WILKINSON).

Costs. The definition of costs is often the first task in valuing an information activity. In defining the costs for information production, dissemination, and so forth, time has to be included as a cost. Allocating overhead costs to the service or product evaluated also can cause problems. The empirical studies of costs have often been too vague and sketchy to be of much benefit for decision making in practice. The calculations of costs have usually been limited to those associated with the production of information products. Less study has been devoted to the costs of information use, and the few studies made are limited to the costs of the use of certain information-acquisition channels.

In connection with the costs of information and its dissemination, the public expenditure connected to the production of information has often been discussed. Comparison of public expenditure with private costs in acquiring information is difficult and has not been studied much (for a broader presentation of costs see ROBERTS).

Effectiveness. An evaluation of the effectiveness of a service or product presupposes a comparison of the goals with the results. In most evaluations of information services and products, large sets of data must be handled and measured—i.e., both "hard" data (amounts, costs) and "soft" (opinions, views).

Examination of cost/effectiveness gives, in principle, good possibilities for evaluation of the effectiveness of alternative plans and allocation of resources. For the ideal evaluation of effectiveness we need a measurement of the value of each product and service, and then the possible sets are compared. The benefits and costs that cannot be directly given in monetary terms can be assigned relative weight coefficients. In proper comparisons, statistical data are collected from the users on standpoints, use, etc., as a background for evaluating weight coefficients, but still weighting is a problem.

Efficiency. While effectiveness studies compare goals and results, efficiency studies concentrate on how well the activities as such are performed. These studies have been very popular in evaluations of library and information services.

In measuring the efficiency of work processes, alternative processes and their costs are studied. Although this type of investigation is important (e.g., for the management of information services), when effectiveness and benefit studies are missing, interpretation of results continuously causes problems. This is evident in the compilation of national statistics. When it has been difficult to get statistics on the actual use of libraries, the performance of a library has been described in terms of circulation statistics and interpreted as a measure of the library's effectiveness. Naturally those statistics only tell us how busy some librarians have been and not much about how well the library has served its clients.

Benefits. The benefits from the use of information and the measurement difficulties have always been priority areas in the economics of information. Recently, these investigations on benefits have often been called studies of the value of information (e.g., GRIFFITHS). There is much speculative literature on the definition of information value. Still, the definitions converge at least in that information value has to be defined from the viewpoint of use or users.

To trace the benefit of information services and products, TAYLOR (1984a) has suggested: 1) statistical methods to collect user experiences, and 2) the so-called critical success factors approach (CSF), wherein the most important tasks (of the managers) and the related information needs are clarified by interviews. Previously, user interviews and surveys (e.g., on willingness to pay and time savings) and demand analyses have been used to determine the benefits. "Hard" data have been produced by calculating the actual payments for information products and by costing the time spent in identifying, acquiring, and reading (KING ET AL., 1982; 1984). There are many measurement problems and even problems connected to the basic approaches applied in the studies, which are mainly due to the special characteristics of information discussed earlier.

Value (or cost-benefit analysis). Cost-benefit analysis examines the benefits and costs of an activity. All the previously described levels can thus be used to describe the prevailing situation or alternative plans. A thorough analysis records entities that can be measured monetarily as well as those that can't. ROTHENBERG, for instance, writes about the starting points and the methodical basis of cost-benefit analysis.

Although cost-benefit analysis has been much criticized in evaluating information systems and services, analyses have been made and certainly will be made. The need for cost-benefit analysis is due to emphasis on the economic significance of information in organizations and communities. A good cost-benefit analysis should consider problems arising from the fact that it is often easier to measure the cost than the benefit; otherwise the cheapest solutions (and probably the most useless) are wrongly emphasized in the analyses.

The methods of research into the economics of information and associated problems have also been analyzed (BLAGDEN, 1980b; CRONIN, 1982; FLOWERDEW & WHITEHEAD; LANCASTER, 1977; MARTYN & FLOWERDEW). The problems of defining information have led the appraisers to stress concrete approaches (particularly Lancaster) and a more accurate definition of information markets (e.g., MARTYN & FLOWERDEW).

MARTYN & FLOWERDEW criticize studies performed on the following grounds: 1) the goals of the studies are vague; few investigations have concrete goals; 2) the goals are sometimes "impossible"; Martyn refers here particularly to efforts to define the value of information, and he suggests case studies in this regard; 3) the studies are based on inadequate data; 4) unfamiliarity with the field; the economists do not know the special problems of information science and vice-versa, especially; a possible improvement would be cooperation between economists and information scientists; 5) too much emphasis is given to the "scientific" quality of the research results at the expense of the practical benefit of the studies; 6) the methodological defects are obvious—they are to a large extent due to the special characteristics of information; 7) the field is artificially restricted to the information services of science and technology. Even though similar lists can be made from several fields of research, the problems of a new research field are clearly emphasized here as well as the special characteristics of information as an economic entity.

All the methodological approaches typical to social sciences—experimental, statistical, case study, comparative, and descriptive study—are available to study the economics of information. The most important data-collecting modes are collection of facts and user interviews and questionnaires. In research performed all the above approaches have been used, but because exact descriptions and definitions in the field cause problems, the best results are achieved by an analysis of case studies (see, e.g., MARTYN & FLOWERDEW).

EMPIRICAL STUDIES OF THE ECONOMICS OF INFORMATION

General

This section presents the main research topics of the economics of information and studies completed in each field. The review is based on the literature of information science. Several studies contain material that makes it possible to present them in several contexts. However, each study is presented in the one category thought most appropriate for its main focus.

The studies are grouped according to the research material available. They do not necessarily indicate the activity in the whole field. We try only to specify presentation and, to some degree, to specify the development of the research in the economics of information during the past two decades. It has not been possible to acquire all research reports. Thus, most of the material consists of articles.

Costs of Information Products and Services

Information products and services incur costs. The costs are generally stated in monetary terms, but in some cases certain disadvantages are considered as costs. Such costs can then be defined as a lost benefit. The costs of information products and services have been treated in many review articles (HINDLE & RAPER; KING ET AL., 1983; LANDAU, 1969; MICK). Some textbooks also treat the topic (ATHERTON; LANCASTER, 1977; ROBERTS).

All the reviews and textbooks worry about the quality of the cost measurements and research. Research methodology is considered particularly inadequate as well as documentation. Research has concentrated on costs of individual services and information services as units. Comparability of the results of badly documented investigations is rather poor. Even some of the reviews are only vaguely organized (e.g., MICK) or collect isolated examples of studies done (e.g., in KING ET AL., 1983). It is also interesting to notice that Landau, nearly 20 years ago, cited the above-mentioned problems associated with performing cost measurements that still prevail today.

LANCASTER (1977) analyzed research of the costs of libraries and information services thoroughly. The priority fields in which cost analyses have been made are acquisition, cataloging, online information search and information production. In acquisition the influence of literature prices and the costs of handling have been studied. In cataloging, most of the studies done have compared manual and automated cataloging; for instance, Pierce and Taylor (in KING ET AL., 1983) measured the costs of one manual and two computerized cataloging systems (OCLC and BALLOTS).

One popular research topic during the past decade has been the costs of information searching (also discussed later). Shirley (in MICK) cites many different costs arising from an information search. However, in several investigations the scope of the costs is limited so that the research results are of limited value. For instance, Bement (in MICK) compares the costs of the results of two information searches from three online services (BRS, DIALOG, and SDC) by calculating only the connect time and the costs arising from the number of references retrieved. Thus, the costs for the preparation of the search and for offline output are not considered, and the causes for the significant differences in cost among systems are not understood.

The costs of producing information have also been studied. For instance, Senders et al. (in MICK) justify electronic solutions by presenting in detail the costs of a scientific publishing scheme. In 1976 King (in MICK) collected statistical data of production costs of information. Altogether, research in this field has been scanty because it is difficult to collect data about the costs of producing information.

It is also instructive to look at cost analyses from the viewpoint of their purposes. From the articles collected by KING ET AL. (1983) we can pick five different types of analyses.

Direct cost analyses. In direct cost analyses the costs of the activity are collected once or at regular intervals. As an example of one-time collection of costs we can take the 1973 study of Vickers (in KING ET AL., 1983). This investigation collected cost data from 18 European online systems. The aim was to study how automation affects costs. The investigation showed that costs are affected by the arrangements of information system management, salaries and productivity of the personnel more than by automation. Talavage et al. (in MICK) collected similar descriptive data about costs, appropriations reserved for the activity, etc., of approximately 500 information services.

Cost data are collected at regular intervals mostly for management. Helmkamp (in KING ET AL., 1983) planned an automated cost calculation system for a technical information service. The costs were specified per service, and the overhead costs were allocated among different services. The calculation system produced cost reports at regular intervals. MOISSE presents the cost calculation system of information activities at the Battelle's Geneva research center, which also includes the information services required for research. Costs for time spent in reading and using the information are considered along with information acquisition and seeking costs.

Theoretical models. Cooper (in KING ET AL., 1983) provides an example of a theoretical model for cost definition. In 1972 he created a mathematical model for minimizing the costs of an online information service. The model is based on minimizing the system costs plus user costs. The article, however, does not include data on how the model could be applied to studying concrete systems.

Planning cost analyses. In cost analyses done to support planning, data on the cost of alternative solutions are generally produced. For instance in 1979 Kraft and Liesener (in KING ET AL., 1983) described an operations-research approach connected to planning a school library. Cost data are produced to support the evaluation of alternatives in planning. (The researchers claim that they are making cost-benefit analyses although they are only recording the costs of alternatives.)

Independent cost comparisons. Cost comparisons can also be done separately from the planning process. Pierce and Taylor (in KING ET AL., 1983) built a model for comparison of manual and automated cataloging. Data on the one-time, fixed, operating and use-independent costs of the different alternatives were collected for the model. The average costs of cataloging a single item were calculated for each alternative. This study is a good example of modeling, in which the key variables are first collected and their interactions and relations with the environment are then defined for comparisons.

Prognostication cost analyses. In 1979 Barwise (in KING ET AL., 1983) predicted, with the aid of cost analysis, the price development of an online information service to the year 1985. The prognosis is based on an inquiry sent to the providers of databases. It is essential to the model that it consider the variable external factors; for example, computers and communications become cheaper all the time, but the use costs of databases obviously do not decrease because when the significance of online use of databases increases in the profit formulation compared with printed services for the database providers, the use costs will probably increase considerably at the end of the prognosis period. Now we can see that the prognosis somewhat exaggerated the increase in use costs, but the basic idea of the prognosis was appropriate.

The possibilities of using cost analyses are thus wide. The criticism of the quality of research can be rejected in part by recognizing that the most important research has been done to support practical management. The reporting of the research methods has been of secondary interest. Lately interest has shifted from mere cost analyses to other, wider questions. Nevertheless, we should remember that cost definition is an important part of these more comprehensive and, as such, more important studies.

Price of Information

Information has a price because its production, storage, dissemination, and use incur costs. The question is, who pays. The production and dissemination of information are usually subsidiary activities, whose costs are not always charged to the user. This information has not been priced and is sometimes considered to be free.

With more emphasis on the economics of information, librarians, especially those in public libraries, have started to talk about the price of information and fees for some services. The pricing of information is also being studied (KING ET AL., 1983; MICK), even though the literature reports it much less than the general studies mentioned above.

The price of information hardly ever arises from information itself and its value-in-use but from transfer or equipment costs and sometimes also from commercial profit. For instance, it is rather comical to pay for the references from an information search according to how long one has been in contact with a computer. The partial public-good character of information makes pricing difficult, though its dissemination is to a great extent commercial. In principle, pricing can have two bases: average costs + profit or marginal costs + profit.

The use of average costs and profit as a basis for pricing is hampered by the problems of calculation of total costs and prediction of demand. Several economists have therefore considered the best solution to be the calculation of the price of information from the marginal costs of dissemination. It is said that in this way also the use of information can be maximized. In research on the pricing of information we can find at least three areas of interest, enumerated below.

Economic examination of charging. McKenzie (in KING ET AL., 1983) presented in 1979 the economists' basic ideas on library services. According to him, more effective library services require the pricing of services and competition between libraries. Gell (in KING ET AL., 1983) presented similar thoughts; she considers that the services should be at least partly charged for if they are to become effective and because public funding for information activities is likely to decrease (CASPER (in KING ET AL., 1983) used regression analysis when she tested changes in the demand for library services when prices change. Her most important observation was that the influence of price depended strongly on user income. Other variables affecting use included knowledge about services, their accessibility, and competitive services.

Examination of the relationship between the price and use of information. The previously mentioned study by Casper could also be included in this group. Cooper and DeWath (in KING ET AL., 1983) studied information searches before and after invoicing was introduced. They noted that invoicing improved the quality of searches, and that the number of searches dropped only by a little more than ten percent. When King (in KING ET AL., 1983) talked about the necessity of marketing information, he stated that price is only one factor influencing the decision on information acquisition. According to him information providers will have a problem when the same information is disseminated in paper form as well as in electronic form and these products compete with each other. He emphasizes the need to monitor the situation to find a suitable (and necessary) mix in product supply. HUNTER, too, considers the pricing problems of online databanks by emphasizing cooperation between the libraries and system producers in the sensible development of pricing.

Pricing from the perspective of the information producer. Berg and Braunstein (both in KING ET AL., 1983) have studied the pricing problems of a publisher of a scientific journal. Berg drafted a model that named the factors that influence the demand and supply of the journal, including functions (e.g., between costs and sales, number of pages, and production costs). Braunstein considered particularly the simultaneous maximization of the publisher's profit and public benefit. Both studies include empirical tests of the models presented. BAUMOL ET AL. have drafted a guide for the definition of prices and costs for organizations disseminating information. It is based on several empirical studies on the costs and pricing of publishing activity. It states that it is impossible to include in the journal price the total costs of information production and that pricing according to the customer groups is needed; also the author page price and subscription prices must be optimized simultaneously.

Effectiveness and Efficiency of Information Services

Research on the effectiveness of an information service is typical evaluation research and has been the subject of several studies in the library and information service field. LANCASTER (1977) presents a number of such studies. Only some recent articles that explicitly deal with cost-effectiveness analysis are reviewed here. The investigation by WOLFE ET AL., however, is deferred until later because its problems are primarily connected to cost-benefit analysis.

The effectiveness (and efficiency) of an information service can be studied in several different ways. Approaches used often include collection of statistics about the use of all or some services provided and analysis of the statistics, as well as comparison between alternative services (for instance, own service vs. service bought from outside the company from the viewpoint of the information service of the organization) and market studies of the services (BLICK, 1977a).

Aslib has been performing these studies in British organizations. VICKERS suggests the following approach for improving the effectiveness of an organization: 1) the information needs in the organization are clarified with personnel interviews; 2) the information sources are then examined; 3) the goals for the information service are formulated with the management; 4) the services needed and changes in existing services are planned; and 5) the services are effectively organized. Here one should pay special attention to the results of the services. Too often stress is laid on the efficiency factors and so the focus has been on information acquisition and processing.

Another somewhat different approach to evaluating effectiveness within an organization has been inspired by "programmed budgeting" (D. MASON). The five phases of this "planning-programming-budgeting system" are: 1) clarification of the goals of information service; 2) description of activities;

3) analysis of activity modes; 4) cost calculation (to clarify the unit costs of each service); and 5) evaluation of cost-effectiveness.

These and similar approaches give general frames for the studies, but they seldom offer any special tool for measuring information. It is essential to the adoption of different approaches in information service that the special features of this activity are recognized. It can be catastrophic if outsiders begin to improve an information service on the basis of plain cost data. Thus, those who run information services must do the effectiveness analysis themselves or at least participate in these studies.

The development of the effectiveness of a library has recently been subject to an extensive study of the use of the library (BREMBER & LEGGATE). Considerable versatility in collecting basic data is demonstrated in this study. Data were collected with structured user interviews, questionnaires, feedback forms connected to the services, observations, follow-up of disseminated references, and loan files.

Also the effectiveness of the individual services of information units has been subject to several studies. BLICK (1977a; 1977b) has collected comparison data for decision making on organizing a current-awareness service in the drug industry. Three different service modes were examined: 1) current-awareness service as general follow-up: "Do we do it ourselves or do we buy it as a service"; 2) arrangement of the current-awareness service on a certain topic; 3) "Do we use our own data base or outside services in information searches."

The different measures of each service (time, costs, availability, etc.) were collected, and weight coefficients were given to the alternatives. The best alternative was obtained by calculating the sums of the weighted characteristics.

Several studies have also been devoted to the effectiveness of information search systems. For instance, WILLIAMS collected empirical data about the effectiveness of an information search when the search is done by those who need the information, when an information scientist seeks it for a user, or when both search it together. The best result was obtained by searching together, but the searches made by an information scientist were the cheapest. The weights on the relations between benefit and costs remained open in this study.

Cost-Benefit Analysis of the Dissemination of Information

Much has been written about cost-benefit analysis in the library and information service area during the past decades. In reading the articles one gets the feeling that almost everything associated with the economics of information has sometimes been called cost-benefit analysis. There are also many interesting empirical studies. Some are discussed below.

This type of research has a long tradition, and there are several good review studies available. For instance, FLOWERDEW & WHITEHEAD and later MARTYN & FLOWERDEW have presented economic evaluations criticizing the knowledge of and skill in cost-benefit analysis in information science. GRIFFITHS reviews several studies on the value of information.

WILLS & OLDMAN analyze the studies more strongly from a library per-spective. In addition, several collections of articles and papers repeat the con-tents of the publications mentioned: for instance BLAGDEN (1980a; 1980b) and HANNABUSS stress cost-benefit analysis from the point of view of the information user.

These studies are presented in five groups. The grouping does not necessarily do justice to the full extent of the research. The studies by King Research, Inc. are presented at the end because they summarize previous knowledge in the field.

Pseudo-cost-benefit analyses. Several studies claim to have used cost-benefit analysis but in truth did not. Often what has really been measured is some cost savings. Another issue that makes the grouping of cost-benefit studies difficult is that some reviewers (e.g., FLOWERDEW & WHITEHEAD) include in their analysis only those benefits and costs that can be measured in monetary terms. However, here we are including non-monetary benefits and costs in the measurements.

An example of pseudo-cost-benefit analyses is the type of study done to compare the costs of alternatives when an organization is planning a new information system. In Finland such (cost) analyses have been made, e.g., in administration when information systems have been designed. Such investiga-tions, at their worst, are little more than after-the-event calculations which are intended to support decisions already taken.

Studies on resource allocation. WILLS & OLDMAN evaluated certain studies done in the late 1960s, in which library budgets were studied by operations-research methods. The basic idea was that past decisions on resource allocation showed the value of each library service (see, for example, HAWGOOD & MORLEY). The weakness of these studies is that they imagine that the resource allocations that were made really provided the best services. Wills and Oldman also studied this question of resource allocation, but their most important analyses were based on the viewpoint of the library users.

Demand analyses. An example of demand analyses comes from public libraries. Newhouse and Alexander (in WILLS & OLDMAN) asked library users how willing they were to lend or buy books when the prices increased. The aim was to determine the levels at which the borrowing of books would benefit the library users most. The problem of the study was that demand was also influenced by variables other than price change (e.g., location of the library, knowledge about services). Still, this kind of study offers significant practical information for library acquisitions. MASON & SASSONE and WOLFE ET AL. have also used demand analysis as part of their studies. Because of their broader scope, these studies are reviewed below.

"Direct" CBAs. The "direct" cost-benefit analyses (CBAs) of information are most often based on time savings or willingness to pay or a combination of both. An often cited "classic" study is the comprehensive one by WOLFE ET AL. on the effectiveness of scientific information services. This study, based on "classical" economic theories and questions, was proposed to both the suppliers and users of information services concerning: 1) time savings when secondary sources are used in seeking information, 2) additional time, if no

secondary sources are available, 3) need for extra salary in this case, and 4) dependence on secondary services.

FLOWERDEW & WHITEHEAD, who considered the study important and interesting, thought that the definition of the optimality of scientist's time use presented a problem. The interests of scientist and employer can conflict; for example, an extensive current-awareness service in one's research area is important for the career of the scientist, but it may hamper the productivity of the work from the point of view of the employer at least in the short term. This is just one point where cost-benefit analyses are powerless because of problems in defining the value of information. The analysis does give information on direct benefits and direct costs to individuals, but another, broader level—the information environment and its "social" ramifications- is essential in evaluating information services.

WILLS & OLDMAN, too, doubt the ability of scientists to read only what is of prime importance for ongoing research. They brought the investigation of WOLFE ET AL. back to an examination of the effectiveness of secondary services. FLOWERDEW & WHITEHEAD refer also to several other studies on time savings, where monetary values for time savings are calculated. These studies do not seem to provide enough versatile data about the benefit gained from information.

An example of time-saving studies in practical surroundings can be taken from KRAMER. According to him, a librarian has to be able to answer questions about the benefit of library services. In Kramer's investigation library users were asked: "how much money or time did the information search save you or your research group," and "how long would it have taken you to acquire the information without the aid of the library." The replies yielded concrete data about the importance of libraries to be presented to management.

Another much-used use indicator has been willingness to pay. FLOWERDEW & WHITEHEAD and WILLS & OLDMAN present several such studies. For instance, DAMMERS studied current-awareness service such that, in addition to follow-up of time use, users were asked what they were willing to pay for the service. The problem with such a survey is to know how realistic the replies of the users are when they do not pay for the services themselves, and even if they do have to pay, the question becomes hypothetical.

HAWGOOD & MORLEY asked the users of a current-awareness service both about their own willingness to pay and about how much they thought their university would be willing to pay. Then the benefit of the information both for the individual and the organization was taken into account but only as evaluated by the individual users. It was also interesting to observe that data about the willingness to pay given by the individuals and organizations were not interdependent. FLOWERDEW & WHITEHEAD estimate, however, that such investigations, when carefully made, can be applied to a definition of the value of information services.

ANDERSON & MEADE, at the National Oceanic and Atmospheric Administration (NOAA), have done cost-benefit analyses to evaluate the information services of NOAA's development programs. The cost data were gathered from development projects and formed the basis for the in-house pricing of information services. The benefit was calculated from data on willingness to pay, which were considered to give the lower limit for benefit. In connection with willingness to pay, data on the present charges were also collected by inquiries and telephone interviews. The conclusion was that even in this restricted form, with regard to measurement of benefit, the analyses were useful in comparing between alternative information services. Still, the study was considered to be too arduous for continuous use.

Besides the investigation of WOLFE ET AL., another study based on the classical ideas of economics is that of MASON & SASSONE. They analyzed the supply-and-demand situation for a service of the Information Analysis Center. The costs were calculated from invested resources and the benefit came from the users' willingness to pay, which was plotted as a demand curve (social demand is estimated to be somewhat greater). The information user could either do his own search (own supply, no handbook) or reduce his costs by using the service (own supply, handbook). The total benefit of the information service is described graphically by the area below the demand curve up to the maximum amount of service acquisition. The area below the supply curve describes the total costs. The net benefit is the difference between these two.

Further, the value of a service is obtained as a difference between the net benefit of self-service and the net benefit of information service. The quantitative benefit is the saving in time when an information service instead of self-service is used. When the cost of time is known, the value of the service can be measured, and this value can be used to put a price on the service. The model suggests information about the benefits of using an information service when the user has two alternatives and when he works economically (i.e., effectively). The model does not consider the costs connected to information use. An additional problem is the collection of the empirical data needed for the model.

Braunstein (in KING ET AL., 1983) correspondingly used an economic approach when he analyzed the benefit and costs of library use from the user's viewpoint. The analysis was very rough because Braunstein included only the costs the user incurred in going into the library and the costs the library paid to supply the desired reading. He also examined the delay that other borrowers suffered when the book was out and the benefit the borrower gained when he did not have to buy the book.

CBAs based on the need and use of information. In several of the above analyses the use of information or information services was seen as a starting point for research. In the studies presented here the information needs and use have a more central role. A Swedish medical company has made an interesting study of information needs and cost-benefit analysis, which is supposed to support management in practice. (Unfortunately, only a short presentation of the research is available (LJUNGBERG & TULLGREN)).

WILLS & CHRISTOPHER studied the acquisition of information in connection with market research. They used cost-benefit analysis to define the optimum method for seeking information. This theoretical article predicted significant empirical applications for the approach, but the model has not been used since, which must be due to problems in data collection. The frequently mentioned analyses of WILLS & OLDMAN were library-use oriented. They studied, for instance, how often the customers visited the library, their information-acquisition habits, the use of a library, and influences of use. The analyses were qualitative, and Wills and Oldman saw them as a preliminary study for quantitative studies. This comprehensive inquiry, primarily of student use of libraries, has, however, remained only a criticism of rough cost-benefit analysis; it did not produce any significant new results.

Perhaps the most interesting of the latest empirical studies are those made at King Research, Inc. These researchers have completed comprehensive investigations on the value of information: "Value of the Energy Data Base" (KING ET AL., 1982), "The Use and Value of Defense Technical Information Center Products and Services" (RODERER ET AL.), and "The Value of Libraries as an Intermediary Information Service" (KING ET AL., 1984). All the reports are interesting, but perhaps the most significant is the first, which has the best description of starting points and methods. It also includes a broad review of earlier research in the area. Here we present the approach, research methods, and some results of the studies.

In these studies the value of information is defined in terms of willingness to pay (measured by what has actually been paid for the information, both in terms of time spent and monies paid), the added cost of using alternative sources for the information, and the benefits derived from using the information. In connection with the energy database, the following information was obtained: the price (in time, salaries and money) paid for information searches, articles and reports; the price (in time and salaries) paid for reading the materials; and the consequences of reading in terms of time and other savings associated with the readers' work. In addition to this user perspective on the value of information, the benefit or value of the database and certain primary information was calculated from the viewpoint of information intermediaries, organizations and funders.

The report discusses the problems of measuring value. Perhaps the most significant observation is that the value of information must be kept separate from the value of the information product or service. The value of information (content) is related to its use; whereas the value of the information product (or service) is measured in terms of the contribution it makes to the amount of use. The ultimate value of a product is determined by amount of use, the benefits derived from that use, and the benefits that would be lost if the product (or service) were not available.

Data from several sources were used in the study: information on database use, on previous investigations, and from two new inquiries. The questions included writing activity, journal-reading habits, use of articles, mode of ordering copies, and the use of bibliographic and source databases. Questions on the use of the Department of Energy's own reports included how the reports were acquired (from distribution, by asking from own information service, through the National Technical Information Service (NTIS), etc.). Further, using telephone inquiries some of those who used reports were traced from the libraries and were asked: the report in question; the form of the report; how much time was spent in acquisition and reading; how the report was found; how was it acquired; for what use; the time and other savings. More general questions were also asked about information acquisition and use. The questions were tested by sampling, and altogether thousands of inquiry and interview forms were handled.

In the investigation, total savings were calculated for the 60,000 scientists at the Department of Energy, where the annual savings in research time from reading was $13 billion. The creation of information required $5.3 billion, and dissemination and use, $500 million. These figures show the importance of exploiting research done elsewhere, and the investigation produced, in addition, much interesting material, e.g., data about the relations between information acquisition and reading. The average price paid for reading an article was $33.50. Of this $9.40 was used in acquisition and handling, $4.50 in search, $2.80 in acquisition for use, and $16.80 in reading. Reading thus represents 50% of the costs (in reports about 70%). So, if reading can be even slightly better directed by a better search, resulting in more optimal selection of readings, the benefit gained can far exceed the relatively large costs incurred by the search itself.

A similar investigation was made for the Department of Defense (RODERER ET AL). Using the approaches and methods reported above, it clarified the use and value of the information services of the department's technical information service. For instance, they studied the use of technical reports through on-demand use of the report database, through systematic distribution of reports, and through the use of some external (primarily economic) databases.

The aim of the third investigation (KING ET AL., 1984) was to clarify the contribution that libraries and information analysis services have on the value of information again from the user's viewpoint. The study yielded data about library effectiveness (speed and quality of the service), efficiency (how much certain services were used), and cost-effectiveness (cost per use). It looked at two performance attributes of online searching- turnaround time of the search and relevance of the output. These were studied relative to the value of the search to the user (in terms of the user's time) through a conjoint measurement technique. Publication delays, updating of the database and delays in receiving references were also studied.

Most users were very satisfied with the library services. The costs for online services caused most of the dissatisfaction. Interesting data were obtained when the purposes of reading were explored. Scientists at the Department of Energy's research institutes listed, in order of importance, the following purposes: professional self-education, ensuring ongoing research, developing the methods of ongoing research, supporting reporting, research proposals, preparation of lectures, and planning, budgeting, and management. The investigation gave measures similar to those used in the previous studies. The result of the conjoint measurement was that relevance was found to be more important to the user than turnaround time.

Although the studies of King Research, Inc. obviously are the most interesting of the empirical investigations, they are hampered by a certain difficulty with measuring higher order effects. KING ET AL. (1982; 1984) defend their approach with references to the special nature of information and with the claim that their analyses give concrete evidence for decision making in practice, for determining how best to configure information service alternatives. CRONIN (1985) has presented a more detailed critique on the results of the study on the energy database.

The Value of Information in Light of Examples

Every use of information is unique; one can't measure the absolute value of a chunk of information. It is necessary to measure value through samples. In practice, whoever is responsible for an information service should always have examples on hand to describe its benefits. It may be profitable to show examples, for instance, to the manager of the organization when he wonders about the increasing costs of information activities.

Only a few studies on the value of information use examples. This may be explained by the fact that it is difficult to generalize from examples and that it is difficult to acquire reliable information about the benefits even in specific situations. In the following studies, examples have been analyzed, or they have at least been used to support other analysis. LJUNGBERG also stated that it is difficult to find investigations about the monetary value of information. He found three examples from the Canadian Institute of Scientific and Technical Information. In these examples, information that was acquired considerably reduced the heating costs and rationalized the work processes, introducing 50% savings. The advice given by the institute increased the profits of its clients by over $500,000 in one year. In Astra Ab (LJUNG-BERG) the value of information was analyzed by using the same methods and principles as those used to evaluate new products (the article does not specify the methods).

The studies by King Research, Inc. (KING ET AL., 1982; 1984; RODERER ET AL.) also give examples of the value of information, which are described as follows (RODERER ET AL.): 1) considered from the perspective of scientists and engineers, value is measured in terms of what the users actually paid in their time and money to acquire and read the information; 2) another perspective is how the reading and use of information affect the users' work; 3) a third perspective involves how the work of scientists and engineers affects the objectives of their organizations. It is speculated that ultimately information has some effect on the operational effectiveness of the users' organizations, the balance of payments, quality of life, training and research in the future, and so on.

In practice several variables are difficult to describe even with examples. However, in the study by KING ET AL. (1982) monetary savings in reading and acquiring information was recorded: 75% of the 148 readers of technical reports in the field of energy saved time and/or money by reading a certain report. The average saving from reading the reports was $1,280; the variation was $0 to $1.5 million. The most common reasons for the savings were the avoidance of repeated investigations and of the trouble of seeking information.

In the investigation for the Department of Defense (RODERER ET AL.) scientists and engineers funded by the agency were surveyed. Information was obtained on specific readings of reports. These data provided estimates of the value of information in reports. They were also asked whether the report acquired was beneficial in terms of their objective and how much research time did reading it save. A 19-page enclosure to the research report lists the detailed replies, but the main points are:

- The information service alerted them to other scientists operating in the field and prompted them to begin international contacts.
- Owing to the character of the data some information could be acquired only through the department's own information service. It was hard to put a dollar value on such information.
- A comprehensive information search of the subject opened new viewpoints for the scientist and deepened an understanding of the problem.
- The information service gave the cooperative partners an idea of what the Department of Defense was interested in.
- Time was saved in writing and research because the tests needed and other procedures had already been done by other workers.

The scientists at the Department of Defense reported time and/or money savings for reports read, and the average savings (in the scientist's time as a result of having read a report) was $4,700. Although these reported consequences of readings in the investigations of King Research are rather difficult to use, it seems that only with them can the appraiser of an information service get a practical touch to his evaluations. The special benefit of such a broad set of examples is that they give quantified data of the evaluations of the service users. Still, caution is needed when using examples of monetary values for purposes of generalization, especially when the data vary widely.

The above studies are based on the information users' own evaluations of benefit. Hyami and Peterson (in FLOWERDEW & WHITEHEAD) give an example of a more objective analysis. They use two models used in agriculture to optimize the production and stocking of products. It was stated that the authorities who collect U.S. agricultural statistics can, with some cost additions, extend the samples of production and stocking statistics. The models showed that when the errors in the statistics could be decreased from 2.5% to 2.0% by enlarging the sample size, the cost-benefit ratio of the information obtained increased to 600:1.

Even though such an investigation is possible only in some special cases, it demonstrates that sometimes information can be defined by a very accurate and monetary value. CRONIN (1984) represents the other extreme. He speculated on the costs, benefit, and value of presenting his paper at a conference. Even though this speculation may easily be interpreted as a joke designed to ease the conference atmosphere, it shows the problems we face when we try to measure the value of information with money in more complex surroundings. The value of information was studied indirectly by MARTYN using questionnaires about late-discovered information in connection with research projects. A more extensive use of information services at the beginning of research could decrease the amount of overlapping research, but Martyn warned against making conclusions that were too far-reaching based on data collected by questionnaires from 266 scientists.

One viewpoint about the value of information is the negative value of missing information or ignorance. BRITTAIN has made an interesting investigation in the field of medicine that touches on this problem. This theoretical study speculates, for instance, on the possibility of defining ignorance in different fields. The definition shall be the starting point for the evaluation of the consequences of ignorance. Brittain explains that, for instance, on exotic diseases and their care there is in Great Britain enough agreement to define ignorance, but that, on the other hand, agreement on the care of cancer is too scanty for us to talk about ignorance accurately enough.

BARRET also gives many examples of the costs of not having information— e.g., in connection with savings in a tower-building program and the costs of a manufacturing failure. He emphasizes the importance of examples and case studies in demonstrating the value of information. Still, it is not just historical examples we need, but dynamic, situation-oriented, up-to-date examples that will help decision makers understand the value of information.

A familiar example of the consequences of a lack of information is the chemical accident, where the right measurements require a knowledge of information about the properties of the chemicals. Not even in these cases, or, for instance in poisonings, can we talk in terms of money only. There is the question, for instance, of the safety or even lives of people. MOISSE gives an example of monetary value: a five-year successful product development was wasted when a significant patent was not found when research began. The patent was missed because it had been indexed incorrectly. The company doing the development work lost $500,000, the cost of the research work. Another of Moisse's examples cites the negligence in following up new patent applications that caused a loss of $400,000. Similar examples have been collected by BLAGDEN (1980a).

INFORMATION SERVICE AS A VALUE-ADDED PROCESS

The value-added process approach is an interesting approach to the examination of the benefit and effectiveness of information services. The approach has been used in evaluation of office work (MITCHELL) and more widely in evaluation of the productivity of information work (STRASSMANN). TAYLOR (1982a; 1984b; 1986) has used the approach particularly in evaluating information services.

Mitchell defines the value-added process or system as any process supporting the results of the employee's work in the office. He points out that this approach is better, in rationalizing office work, than the old approach, which is based on avoidance of costs. Value-added thinking examines the costs and benefit of the operation simultaneously in order to direct resources to the most productive operations.

TAYLOR (1982a) stresses that the value-added process approach also provides an efficient opportunity for evaluating information systems. The systems are examined from a user viewpoint to counterbalance the much-used information production-oriented approaches.

Not much empirical research has been done that uses the above-described approach. The most significant is probably a study led by TAYLOR (1984b) into the evaluation of indexing and abstracting systems. The writing of abstracts and indexes was analyzed in 13 organizations with the aim of defining the activities that benefit the users in information searching. It was found that the services increased 20 of 23 defined values, which include availability (elimination of unnecessary information), scanning (easy use), coverage, cost savings, and physical availability. Though the study gave interesting data on indexing functions, some of the recorded values seemed somewhat academic.

The scientists themselves considered the analysis beneficial in the organization of information gathering, analysis, and interpretation of abstracts. For those who were indexing and writing abstracts the results are certainly beneficial because they describe the factors in the process that are important for the user. The investigation stated that indexers often do not know where or how the results of their work are used. This is, of course, a considerable problem in large indexing and abstracting organizations.

Scientists encountered problems when they tried to combine individual added values to the costs incurred by their production. In addition to those causes that are due to the general nature or starting points of the model, the problems were found to be caused, for instance, by large size differences between the processes investigated. Some of those responsible for the systems claimed also that it is impossible to define the individual costs because the interconnections between subprocesses are so complex.

CRONIN (1984) also refers to value-added services when he examines the pricing of information services; the value added can be a basis for pricing. Cronin refers to a study in which the lack of a database caused scientists' work to be worth an average £5,000. On the other hand, the input of one complete record in the database cost also about £5,000. According to the number of users and the size of the database, we can now calculate the value of the database.

Too few empirical studies have been done to make the significance of the approach for the evaluation of the benefit of information services clear. The strength of the value-added thinking seems, however, to be in that there is at least preliminary emphasis on information use, which is not the case with many production-oriented approaches.

Economics of Online Searching

We have already mentioned a few studies of information seeking. Then we presented the influences of information search fees, pricing of information searches, and the effectiveness of the information search systems. A separate presentation is justified by the extent of the research in this central interest area in information science during the past decade. Basic textbooks of information searching also often refer to economic questions and more widely to evaluation of the systems. For instance, HALL refers to the costs of information searching and to analyzing the differences between manual and automated searches. HENRY ET AL. again present some effectiveness studies of the role of an intermediary in online searching.

LANCASTER (1971; 1977) has possibly had the most effect on the research methods in this field. According to him, the effectiveness of an information search can be described by coverage, recall, precision, access time, and user time. As examples of benefits, Lancaster mentions money and time saving, increased productivity, elimination of double work, and improved operation. His empirical studies are connected to the MEDLARS database. Lancaster's thoughts have been exploited, for instance, by KABI, in his studies of the effectiveness and costs of Britain's Chemical Information Service's searches.

The cost-effectiveness of online and manual searches was compared by ELCHESEN, who studied 40 searches from seven abstract publications and corresponding SDC (System Development Corp.) databases. In this study, the searching process, information sources, and the behavior of intermediaries were also compared. The online searches proved to be the fastest, cheapest, and most effective. In some cases, the old-fashioned, manual search was more accurate and yielded the most recent references.

From cost comparisons and studies on transfer to online use, studies have proceeded to an examination of search benefits as such. MARKEE reports on time savings obtained from university library searches. The users were asked to roughly indicate the savings from search results. Markee admits that time savings is rough in benefit measurement, but it gives, however, additional information for evaluation besides mere cost data.

COLLETTE & PRICE gathered earlier cost-benefit information on information searches from the user viewpoint. The investigations gathered cost data, but benefit data rested on the users' rough (classified in advance) information on time and money saving. The cost-benefit ratio of information searches was 2.8/1. Despite the scientists' assurances, the results seem quite abstract. The abundant analyses of the economy and benefit of information searches do not seem very convincing. Too many investigations describe single variables only. No one, except perhaps Lancaster, has had the patience to present the results in sufficient detail so that generalizations could be made.

MOREHEAD ET AL. offer an interesting approach. Their studies on the information search start with a description of information value. From a statistical, multidimensional concept of information value, the scientists get a common research framework for evaluating information searching and the resulting information. The value of information is determined through definition of need (subject, framework, the writer's aim, and availability are the dimensions studied).

Macroeconomic Studies on Information and Productivity

In information science, macroeconomic questions have not been discussed much. MARTYN & FLOWERDEW present some studies that try to clarify the macroeconomic picture. The approach of Machlup to the economy of scientific publication of tens of man-years is considered problematic by MARTYN & FLOWERDEW because of the lack of basic statistical data. On the other hand, they consider that Porat extended the scope of information to include all the information activities of a society. His calculations have raised much discussion about the definitions used. Despite criticisms, however, these economists have often been referred to when "information society," information professions, and so forth have been discussed.

On the basis of Machlup's and Porat's statistics, COOPER presents estimates of the continuous growth of the information economy. He sees libraries as small partial factors in this growth but stresses that the knowledge developed in librarianship can be used in this widening field. Productivity studies of information activities are reviewed by CRONIN & GUDIM.

An interesting empirical example of macroeconomic studies in information science is provided in the broad U.S. investigation on mathematical modeling of the use of information systems (HAYES & BORKO; HAYES & ERICKSON). One partial study used the Cobb-Douglas econometric model. This model describes production as a function of labor and capital. The scientists divided the capital investments into capital, information (information investments), and other outside acquisitions of resources, and the model was used to see if there was an optimum for information investments.

Fifty industries in the United States were tested for the years 1967 and 1972 using the model. The model tested information purchases as a function of production. Porat's broad definition of information activities (information production, dissemination that includes training, insurance, advertising, information products, and some public services, like post) was used in defining information investments. It was found that information is used much less than would be optimal. This study has generated interest, but the use of econometric theories does not seem convincing, and some doubts can be cast on the interpretation of the results.

The above investigation is also hampered by the problems typical of the macroeconomy— viz., the reliability of the basic data, difficulties in separating information services from other acquisitions, and the connections of the acquisition of information services to labor (for example, recruiting of personnel is also information acquisition).

Over the past 20 years both the character and state of information services as an enterprise have also been examined. For instance, the economic potential of disseminating information electronically (particularly databases) has been widely discussed. The EUROPEAN INFORMATION PROVIDERS' ASSOCIATION (EURIPA) (1981; 1983) has sponsored two conferences on this topic: "The Electronic Information Marketplace" (1981) and "Making Money Out of Information" (1983). Statistics on activity in this field in Europe were presented and the potential for market research, the pricing of information, and particularly the responsibilities of the private and public sectors in disseminating information were covered.

The business side of database production has been studied by LANDAU (1985). He concludes that database production is a growing business and is already influencing the development of a country's economy. Landau and others involved in providing information seem to emphasize that information is a product. This can be explained by the requirements of selling, but then we easily focus on certain "product-like" groups of information (e.g., current facts). However, much information is sold at marginal prices and often the profits of the information providers are based on the free or almost free acquisition of information produced earlier, in many cases at the expense of taxpayers.

The Economics of Information and Automated Data Processing

In writing this review, we came across some articles that treat the economics of information in the field of automated data processing (ADP). Since the investigation presented here is based on publications in information science, the survey is not at all complete for this part. This marginal area is discussed here because of the similarities between information systems and ADP and because the information systems that form the research target of information science have become increasingly computerized (AGRAWAL & ZUNDE).

Some theoretical articles have appeared recently that treat the economics of information for ADP systems. YOVITS & FOULK model the use and value of information in decision making on the basis of extensive theoretical discussions. The simulation models they developed are based on problematic hypotheses. For example, they consider decision making to be comprised of distinct phases, where only directly exploitable information is defined as valuable.

KING & EPSTEIN analyzed the value of information connected to management information systems (MIS). They present statistical multidimensional definitions of the value of information. Models are created for alternative modes of operation in decision making, where the values are described with concepts such as profitability, relevance, understandability, significance, sufficiency, and practicality. CHRISTIE presents corresponding discussions on the multidimensionality of the value of information. A particular problem with these models is the gathering of reliable empirical information in practice. Their real significance remains a mere general description of the value of information.

Theoretical speculations on the costs and pricing of information have also been presented in connection with computer services. Cotton (in KING ET AL., 1983) examined in detail different pricing strategies (profit maximization, cost covering, value-based, client prioritizing, etc.) for computer services. Similar thoughts, emanating from the market situation, have also been presented by KLEIJNEN. These studies are almost analogous to the studies in information science.

A practical example of a cost-benefit analysis in ADP is given by DRAPER, who analyzed the costs and benefits of database management in the U.S. federal administration. The information control systems did not directly save money at first, so the benefit was primarily obtained from improved activity and productivity. Savings came later, for example, in programming and maintenance.

A brief idea of how to handle economic questions in ADP is given in the collection of papers on the economics of information processing by GOLD-BERG & LORIN (1982a; 1982b). The introduction states that economic studies are new to this field and speculates about definition problems associated with the use, value, costs, and so forth, of information. This two-volume work consists of seven parts: 1) organization and data processing, 2) company economic models for the data-processing industry, 4) economics in evaluation of information systems, 5) economy of data-processing management, 6) systems and development of applications, and 7) measurement of software projects.

The first two topics are connected to the use and organization of ADP services in companies. The third discusses the character of service demand and the product properties of the services. The fourth is connected to the methods for defining costs and particularly benefits of ADP. Identification of planning, maintenance, and operation costs are examined in connection with management of information services. The two last parts are devoted to measuring application costs.

These studies correspond closely to similar ones in information science (cf., for example, KING ET AL., 1983). In ADP studies, however, the emphasis is more clearly on production and techniques, and cost calculations are dominant. In the articles collected by GOLDBERG & LORIN (1982a; 1982b), the emphasis is on the fact that the studies have practical significance for the management of companies.

FUTURE RESEARCH

Discussions about the direction of future studies have yielded interesting findings in Britain. As a conclusion to their analysis of research thus far, FLOWERDEW & WHITEHEAD presented several researchable topics. The ideas they presented have not been realized, although the British Library Research & Development Department actively tried to sponsor such research in the late 1970s. One of the trends FLOWERDEW & WHITEHEAD desired has been realized, specifically, during the past few years there have been more cost-benefit analyses and examinations of information value among cost and efficiency analyses. The reasons why extensive research of this type was not sponsored in Britain are related to the difficulty of the research task, the severe criticism of the studies done so far, and the fact that the application of the new information technology has taken the most significant part of the research interest during the past years. The report of MARTYN & FLOWERDEW evaluates the situation anew, and most of the problems that were present a decade ago still seem to await solution. For instance, the British statistics associated with information still have not been organized despite several efforts.

CRONIN & GUDIM insist on more research on the connections between information and productivity. They introduce a matrix of seven research approaches (multivariate analysis, econometric modeling, case studies, matched case studies, cost-benefit analyses, national economic comparisons, and tracer studies of the links between basic R&D and technological innovation), and of three research areas (information technology, information systems and services, and information). The third dimension is the public/private sector. The matrix is "filled" by some examples from the earlier studies, and it is hoped that the matrix could be a useful basis for future research on information and productivity.

For those who start economic investigations now it is possible and beneficial to avoid the defects and errors of previous research by conducting studies in the field. The difficulties associated with compiling information statistics are due to the character of information. While the statistics are under development and hundreds of mechanized information systems are developed and studied, more information scientists could find the most fruitful research targets both at the organizational and individual level in the case studies of information use. Future research will show generalizations can be made from the more theoretical approaches.

Case-studies are needed to deepen our understanding of the economics of information. There seems to be a need for a dual approach (REPO, 1986b): 1) information products, services, systems, and channels need to be studied by using basic economic thinking; studies on information production, information markets, and the accounting and budgeting of information activities— better known as the information resource management (IRM) approach—are needed; 2) the value of information can be studied only through its use, which means that the economics of information cannot be described without studying the users of information, the use of information, and the effects of that use.

The economic importance of information activities for individuals, organizations, and societies is increasing. Economists have not been able to offer much help despite their substantial efforts (see, e.g., REPO, 1986a). There is no doubt that information scientists have also tried to answer the need by completing numerous studies during the past few decades. There is a need to analyze these studies and to develop a sounder basis for empirical research; otherwise the efforts do not increase our common understanding of the phenomenon. Case studies, especially those done in organizational settings, are most needed when theories are under development.

CONCLUSIONS

This review has analyzed research into the economics of information from the information science viewpoint. Much research has been done during the past 20 years. Numerous theoretical discussions as well as empirical studies have been performed.

After having analyzed empirical studies, it is easy to agree with MARTYN that the studies on the economics of information should be "simple, easy to understand (both by professionals and organization management), relatively cheap, and easy to apply to local circumstances." It has to be possible to demonstrate the value of information in different contexts for individuals and groups and in organizations as well as society. The first steps should be taken by increasing the economic consciousness of librarians and information specialists.

BIBLIOGRAPHY

AGRAWAL, J. C.; ZUNDE, P. 1985. Empirical Foundations of Information and Software Science. New York, NY: Plenum Press; 1985. 415p. ISBN: 0-306-42091-0.

ANDERSON, ROBERT C.; MEADE, NORMAN F. 1979. Cost-Benefit Analysis of Selected Environmental and Data Information Service Programs. Washington, DC: National Oceanic and Atmospheric Administration; 1979. 57p. (NOAA Technical Memorandum EDIS 25; PB-297189).

ATHERTON, PAULINE. 1977. Handbook for Information Systems and Services. Paris, France: United Nations Educational, Scientific and Cultural Organization (UNESCO); 1977. 244p. ISBN: 92-3-101457-9.

BARRET, A. J. 1985. The Costs of Not Having Refined Information. In: The Value of Information as an Integral Part of Aerospace and Defence R&D Programmes: Proceedings of Technical Information Panel Specialists' Meeting; 1985 September 4-5; Cheltenham, England. Springfield, VA: Advisory Group for Aerospace Research and Development (AGARD); 1985. 100p.

BAUMOL, WILLIAM J.; BRAUNSTEIN, YALE M.; FISHER, DIETRICH M.; ORDOVER, JANUSZ A. 1981. Manual of Pricing and Cost Determination for Organizations Engaged in Dissemination of Knowledge. New York, NY: New York University; 1981. 112p.

BLAGDEN, JOHN F. 1980a. Do We Really Need Libraries: An Assessment of Approaches to the Evaluation of the Performance of Libraries. London, England: Clive Bingley; 1980. 162p. ISBN: 0-89664-442-1.

BLAGDEN, JOHN F. 1980b. Libraries and Corporate Performance: The Elusive Connection. In: Taylor, Peter J. New Trends in Documentation and Information: Proceedings of the Federation Internationale de Documentation (FID) 39th Congress; 1978 September 25-28; Edinburgh, Scotland. London, England: Aslib; 1980. 379-382. ISBN: 0-85142-128-8.

BLICK, A. R. 1977a. Evaluating an In-House or Bought-In Service. Aslib Proceedings. 1977; 29(9): 310-319. ISSN: 0001-253X.

BLICK, A. R. 1977b. The Value of Measurement in Decision-Making in an Information Unit– A Cost-Benefit Analysis. Aslib Proceedings. 1977 May; 29(5): 189-196. ISSN: 0001-253X.

BORKO, H. 1983. Information and Knowledge Worker Productivity. Information Processing & Management. 1983; 19(4): 203-212. ISSN: 0306-4573.

BREMBER, V. L.; LEGGATE, P. 1985. Linking a Medical User Survey to Management for Library Effectiveness: 1, The User Survey. Journal of Documentation. 1985 March; 41(1): 1-14. ISSN: 0022-0418.

BRITTAIN, M. 1984. Consensus in the Medical Sciences and the Implications for Information Systems. In: Dietschmann, H. J., ed. Representation and Exchange of Knowledge as a Basis of Information Processes: Proceedings of the 5th International Research Forum in Information Science; 1983 September 5-7; Heidelberg, F. R. G. Amsterdam, The Netherlands: North Holland, Elsevier Science Publishers B.V.; 1984. 165-174. ISBN: 0-444-87563-8.

CAPLAN, NATHAN. 1984. Research on Knowledge Utilization. Lessons and Observations. In: Proceedings of the American Society for Information Science (ASIS) 47th Annual Meeting; 1984 October 21–25; Philadelphia, PA. White Plains, NY: Knowledge Industry Publications, Inc. for the ASIS; 1984. 239–242. ISBN: 0-86729-115-X.

CASPER, CHERYL A. 1983. Economics and Information Science. In: Debons, A., ed. Information Science in Action: System Design. London, England: Martinus Nijhoff Publishers; 1983. 565–572. ISBN: 90-247-2807-X.

CHRISTIE, BRUCE. 1981. Face to File Communication: A Psychological Approach to Information Systems. New York, NY: John Wiley & Sons; 1981. 306p. ISBN: 0-471-27939-0.

CLEVELAND, HARLAN. 1982. Information as a Resource. Futurics. 1982; 6(3–4): 1–5. ISSN: 0164-1770.

COLLETTE, A. D.; PRICE, J. A. 1977. A Cost/Benefit Evaluation of Online Interactive Bibliographic Searching in a Research and Engineering Organization. The Value of Information: Collection of Papers Presented at: American Society for Information Science (ASIS) 6th Mid-Year Meeting; 1977 May 19–21; Syracuse University, Syracuse, NY. Washington, DC American Society for Information Science; 1977. 24–34.

COOPER, M. D. 1983. The Structure and Future of the Information Economy. Information Processing & Management. 1983; 19(1): 9–26. ISSN: 0306-4573.

CRONIN, BLAISE. 1982. Taking the Measure of Service. Aslib Proceedings. 1982; 34(6–7): 272–294. ISSN: 0001-253X.

CRONIN, BLAISE. 1984. Information Accounting. In: van der Laan, A.; Winters, A. A., eds. The Use of Information in a Changing World: Proceedings of the FID 42nd Congress; 1984 September 24–27; The Hague, The Netherlands. Amsterdam, The Netherlands: North-Holland, Elsevier Science Publishers B. V.; 1984. 409–416. ISBN: 0-444-87554-9.

CRONIN, BLAISE. 1985. The Economics of Information. Paper presented at: 3rd Victorian Association for Library Automation (VALA) National Conference on Library Automation; 1985 November 28–December 1; Melbourne, Australia. Melbourne, Australia: VALA; 1985. 10p. ISBN: 0-908-47805-4.

CRONIN, BLAISE. 1986. Towards Information-Based Economies. Journal of Information Science. 1986; 12(3): 129–137. ISSN: 0165-5515.

CRONIN, BLAISE; GUDIM, M. 1986. Information and Productivity: A Review of Research. International Journal of Information Management. 1986; 6(2): 85–101. ISSN: 0268-4012.

DAMMERS, H. F. 1974. Economic Evaluation of Current Awareness Systems. In: Batten, E., ed. EURIM: Proceedings of European Conference on Research into the Management of Information Services and Libraries; 1973 November 20–22; Paris, France. London, England: Aslib Publications; 1974. 107–124. Available from: Aslib, 3 Belgrave Square. London SW1A 8PG, England. ISBN: 0-85142-059-1.

DERR, RICHARD L. 1983. A Conceptual Analysis of Information Need. Information Processing & Management. 1983; 19(5): 273–278. ISSN: 0306-4573.

DRAPER, J. M. 1981. Costs and Benefits of Database Management: Federal Exprience. Washington, DC: U.S. Department of Commerce, Computer Science and Technology; 1981. 101p. (NBS Publication; 500-84).

ELCHESEN, DENNIS R. 1978. Cost-Effectiveness Comparison of Manual and On-line Retrospective Bibliographic Searching. Journal of the American Society for Information Science. 1978 March; 29(2): 56–66. ISSN: 0002-8231.

EUROPEAN INFORMATION PROVIDERS ASSOCIATION. 1981. The Electronic Information Marketplace: Proceedings of the 1981 European Information Providers Association (EURIPA) Symposium; 1981 January; Luxembourg. London, England: EURIPA; 1981. 74p. ISBN: 0-907935-00-1.

EUROPEAN INFORMATION PROVIDERS ASSOCIATION. 1983. Making Money Out of Information: Proceedings of the European Information Providers Association (EURIPA) Symposium; 1983 March 16–17; Luxembourg. London, England: EURIPA; 1983. 92p. ISBN: 0-907935-04-4.

FLOWERDEW, A. D. J.; WHITEHEAD, C. M. E. 1974. Cost-Effectiveness and Cost/Benefit Analysis in Information Science. London, England: London School of Economics and Political Science; 1974. 71p. (OSTI Report no. 5206). Available from: British Library Research and Development Division, Sheraton House, Great Chapel St., London W1V 4BH, England.

GOLDBERG, R.; LORIN, H. 1982a. The Economics of Information Processing. Volume 1. New York, NY: John Wiley & Sons; 1982. 238p. ISBN: 0-471-09206-1.

GOLDBERG, R.; LORIN, H. 1982b. The Economics of Information Processing. Volume 2. Operations, Programming and Software Models. New York, NY: John Wiley & Sons; 1982. 185p. ISBN: 0-471-09206-1.

GRIFFITHS, JOSÉ-MARIE. 1982. The Value of Information and Related Systems, Products and Services. In: Williams, Martha E., ed. Annual Review of Information Science and Technology: Volume 17. White Plains, NY: Knowledge Industry Publications for the American Society for Information Science; 1982. 269–283. ISSN: 0066-4200; ISBN: 0-86729-032-3.

HALL, JAMES L. 1977. On-line Information Retrieval Sourcebook. London, England: Aslib; 1977. 267p. ISBN: 0-85142-094-X.

HANNABUSS, STUART. 1983. Measuring the Value and Marketing the Service: An Approach to Library Benefit. Aslib Proceedings. 1983 October; 35(10): 418–427. ISSN: 0001-253X.

HARMON, G. 1984. The Measurement of Information. Information Processing & Management. 1984; 20(1-2): 193–198. ISSN: 0306-4573.

HARMON, G. 1985. Information Measurement in Natural and Artificial Systems. In: Agrawal, J. C.; Zunde, P., eds. Empirical Foundations of Information and Software Science. New York, NY: Plenum Press; 1985. 303–309. ISBN: 0-306-42091-0.

HAWGOOD, J.; MORLEY, R. 1969. Project for Evaluating the Benefits from University Libraries. Durham, England: University of Durham; 1969. (OSTI Report 5584).

HAYES, ROBERT M.; BORKO, H. 1983. Mathematical Models of Information System Use. Information Processing & Management. 1983; 19(3): 173–185. ISSN: 0306-4573.

HAYES, ROBERT M.; ERICKSON, TIMOTHY. 1982. Added Value as a Function of Purchases of Information Services. The Information Society. 1982; 1(4): 307–339. ISSN: 0197-2243.

HENRY, W. M.; LEIGH, J. A.; TEDD, L. A.; WILLIAMS, P. W. 1980. On-line Searching: An Introduction. London, England: Butterworth & Co. Ltd.; 1980. 203p. ISBN: 0-408-10696-4.
HINDLE, A.; RAPER, D. 1976. The Economics of Information. In: Williams, Martha E., ed. Annual Review of Information Science and Technology: Volume 11. Washington, DC: American Society for Information Science; 1976. 27-54. ISSN: 0066-4200; ISBN: 0-87715-212-8.
HORTON, FOREST W., JR. 1979. Information Resources Management: Concept and Cases. Cleveland, OH: Association for Systems Management; 1979. 343p. ISBN: 0-934356-01-7.
HORTON, FOREST W., JR. 1982. IRM: The Invisible Revolution. In Depth. Computerworld. 1982; (16): 7. ISSN: 0010-4841.
HUNTER, J. A. 1984. What Price Information. Information Services & Use. 1984; 4(4): 217-223. ISSN: 0167-5265.
JÄRVELIN, KALERVO; REPO, AATTO J. 1984. On the Impacts of Modern Information Technology on Information Needs and Seeking: A Framework. In: Dietschmann, H. J., ed. Representation and Exchange of Knowledge as a Basis of Information Processes. Amsterdam, The Netherlands: Elsevier Science Publishers; 1984. 207-230. ISBN: 0-444-87563-8.
KABI, A. 1972. Use, Efficiency and Cost of External Information Services. Aslib Proceedings. 1972; 24(6): 356-363. ISSN: 0001-253X.
KING, DONALD W.; GRIFFITHS, JOSÉ-MARIE; SWEET, ELLEN A.; WIEDERKEHR, ROBERT R. V. 1984. A Study of the Value of Information and the Effect on Value of Intermediary Organiztions, Timeliness of Services & Products, and Comprehensiveness of the EDB. Volume 1: The Value of Libraries as an Intermediary Information Service. Oak Ridge, TN: Office of Scientific and Technical Information; 1984. 122p. DOE/NBM-1078 (DE 85003670).
KING, DONALD; GRIFFITHS, JOSÉ-MARIE; RODERER, NANCY K.; WIEDERKEHR, ROBERT R. V. 1982. Value of the Energy Data Base. Oak Ridge, TN: Technical Information Center, United States Department of Energy; 1982. 81p. NTIS: DE 82 014250.
KING, DONALD W.; RODERER, NANCY K.; OLSEN, HAROLD A., eds. 1983. Key Papers in the Economics of Information. White Plains, NY: Knowledge Industry Publications; 1983. 372p. ISBN: 0-86729-040-4.
KING, WILLIAM R.; EPSTEIN, BARRY JAY. 1976. Assessing the Value of Information. Management Datamatics. 1976; 5(4): 171-180. ISSN: 0377-9149.
KLEIJNEN, JACK P. C. 1980. Economic Frame-Work for Information Systems. Tilburg, The Netherlands: Tilburg University, Department of Business and Economics; 1980. 14p. (no. 80.99).
KRAMER, J. 1971. How to Survive in Industry. Special Libraries. 1971 November; 62(11): 487--489. ISSN: 0038-6723.
LAMBERTON, D. M. 1984. The Economics of Information and Organi-zation. In: Williams, Martha E., ed. Annual Review of Information Science and Technology: Volume 19. White Plains, NY: Knowledge Industry Publications, Inc. for the American Society for Information Science, 1984. 3-30. ISSN: 0066-4200; ISBN: 0-86729-093-5.
LANCASTER, F. W. 1971. The Cost-Effectiveness Analysis of Information Retrieval and Dissemination Systems. Journal of the American Society for Information Science. 1971 January/February; 22(1): 12-27. ISSN: 0002-8231.

LANCASTER, F. W. 1977. The Measurement and Evaluation of Library Services. Arlington, VA: Information Resources Press; 1977. 395p. ISBN: 0-87815-017-X.

LANDAU, HERBERT B. 1969. The Cost Analysis of Document Surrogation: A Literature Review. American Documentation. 1969 October; 20(4): 302–310. ISSN: 0002-8231.

LANDAU, HERBERT B. 1985. Information Processing (Data-Base Production) as a Business. Information Services & Use. 1985; 4(6): 389–396. ISSN: 0167-5265.

LJUNGBERG, SIXTEN. 1978. Monetary Value of Information. Paper presented at: 1977 Annual Meeting, FID/II; Lisbon, Portugal. Tidskrift för Dokumentation. 1978; 34(3): 43–44. ISSN: 0040-6872.

LJUNGBERG, SIXTEN; TULLGREN, ÅKE. 1977. Investigation on the Use of Information in R&D in a Research Intensive Company. In: Fry, B. M.; Shepherd, C.A., comps. Information Management in the 1980s: Proceedings of the American Society for Information Science (ASIS) 40th Annual Meeting: Volume 14, Part 1; 1977 September 26–October 1; Chicago, IL. White Plains, NY: Knowledge Industry Publications for ASIS; 1977. 12. ISBN: 0-914236-12-1.

MACHLUP, FRITZ. 1962. The Production and Distribution of Knowledge in the United States. Princeton, NJ: Princeton University Press; 1962. 416p. LC: 63-7072.

MACHLUP, FRITZ. 1980. Knowledge: Its Creation, Distribution and Economic Significance: Vol. 1, Knowledge and Knowledge Production. Princeton, NJ: Princeton University Press; 1980. 272p. ISBN: 0-691-04226-8.

MARKEE, KATHERINE M. 1981. Economies of Online Retrieval. Online Review. 1981; 5(6): 439–444. ISSN: 0309-314X.

MARTYN, JOHN. 1986. Draft: Literature Searching Habits and Attitudes of Research Scientists. London, England: The Research Group; 1986. 25p.

MARTYN, JOHN; FLOWDERDEW, A. D. J. 1983. The Economics of Information. London, England: The British Library Board; 1983. 39p. ISSN: 0236-1709.

MASON, D. 1973. Programmed Budgeting and Cost Effectiveness. Aslib Proceedings. 1973; 25(3): 100–110. ISSN: 0001-253X.

MASON, ROBERT M.; SASSONE, PETER G. 1978. A Lower Bound Cost Benefit Model for Information Services. Information Processing & Management. 1978; 14(2): 71–83. ISSN: 0306-4573.

MICK, C. K. 1979. Cost Analysis of Information Systems and Services. In: Williams, Martha E., ed. Annual Review of Information Science and Technology: Volume 14. White Plains, NY: Knowledge Industry Publications Inc., for the American Society for Information Science; 1979. 37–64. ISSN: 0066-4200; ISBN: 0-914236-44-X.

MITCHELL, J. H. 1980. Justifying the Electronic Office. The Need for an "Added Value" Approach. In: Proceedings of the Electronic Office; 1980 April 22–25; London, England. London, England: Institution of Electronic and Radio Engineers; 1980. 249–259. ISBN: 0-903-74840-1.

MOISSE, E. 1976. Costing Information in an Independent Research Organization. The Information Scientist. 1976 June; 10(2): 57–68. ISSN: 0020-0263.

MOREHEAD, DAVID R.; PEJTERSEN, ANNELISE M.; ROUSE, WILLIAM B. 1984. The Value of Information and Computer-Aided Information

Seeking: Problem Formulation and Application to Fiction Retrieval. Information Processing & Management. 1984; 20(5-6): 583-601. ISSN: 0306-4573.

OLDMAN, CHRISTINE M. 1978. The Value of Academic Libraries: A Methodological Investigation. Cranfield, England: Institute of Technology, School of Management; 1978. 184p. (Ph.D. dissertation). Available from: British Library Document Supply Centre.

REPO, AATTO J. 1986a. Analysis of the Value of Information: A Study of Some Approaches Taken in the Literature of Economics, Accounting and Management Science. Sheffield, England: The University of Sheffield; 1986. 43p. (CRUS Working Paper; 7). ISBN: 0-906088-29-1.

REPO, AATTO J. 1986b. The Dual Approach to the Value of Information. An Appraisal of Use and Exchange Values. Information Processing & Management. 1986; 22(5): 373-383. ISSN: 0306-4573.

REPO, AATTO J. 1987. Pilot Study of the Value of Secondary Information: Discussions from the Viewpoints of Information Providers and Users. Aslib Proceedings. 1987 April; 39(4): 135-147. ISSN: 0001-253X.

ROBERTS, STEPHEN A. 1985. Cost Management for Library and Information Services. Cambridge, England: Butterworth & Co.; 1985. 181p. ISBN: 0-408-01376-1.

RODERER, NANCY K.; KING, DONALD W.; BROUARD, SANDRA E. 1983. The Use and Value of Defense Technical Information Center Products and Services. Alexandria, VA: Defense Technical Information Center; 1983. 59p.

ROTHENBERG, JEROME W. 1975. Cost-Benefit Analysis: A Methodological Exposition. In: Struening, Elmer L.; Guttentag, Marcia. Handbook of Evaluation Research: Volume 2. Beverly Hills, CA: Sage Publications; 1975. 55-88. ISBN: 0-803-90428-2.

SHANNON, C. E.; WEAVER, W. 1949. Mathematical Theory of Communication. Urbana, IL: University of Illinois Press; 1949. 117p.

STRASSMANN, PAUL A. 1985. Information Payoff: The Transformation of Work in the Electronic Age. New York, NY: Free Press; 1985. 289p. ISBN: 0-02-931720-7.

TAYLOR, ROBERT S. 1982a. Value-Added Processes in the Information Life Cycle. Journal of the American Society for Information Science. 1982 September; 33(5): 341-346. ISSN: 0002-8131.

TAYLOR, ROBERT S. 1982b. Information and Productivity. On Defining Information Output. Social Science Information Studies. 1982; 1(2): 131-138. ISSN: 0143-6236.

TAYLOR, ROBERT S. 1984a. Information and Productivity: On Defining Information Output (II). Social Science Information Studies. 1984; 4(1): 31-41. ISSN: 0143-6236.

TAYLOR, ROBERT S. 1984b. Value-Added Processes in Document-Based Systems: Abstracting and Indexing Services. Information Services & Use. 1984; 4(3): 127-146. ISSN: 0167-5265.

TAYLOR, ROBERT S. 1986. Value-Added Processes in Information Systems. Norwood, NJ: Ablex Publishing Corporation; 1986. 288p. ISBN: 0-89391-273-5.

VAN HOUSE, NANCY A. 1984. Research on the Economics of Libraries. Library Trends. 1984; 32(4): 407-423. ISSN: 0024-2594.

VARLEJS, JANA, ed. 1982. The Economics of Information. London, England: McFarland & Co., Inc., Publishers; 1982. 92p. ISBN: 0-89950-059-5.

VICKERS, PETER. 1976. Ground Rules for Cost-Effectiveness. Aslib Proceedings. 1976; 28(6-7): 224-229. ISSN: 0001-253X.

WHITE, HERBERT S. 1985. Cost Benefit Analysis & Other Fun & Games. Library Journal. 1985 February 15; 110(3): 118-121. ISSN: 0363-0277.

WILKINSON, J. B. 1980. Economics of Information: Criteria for Counting the Cost and Benefit. Aslib Proceedings. 1980 January; 32(1): 1-9. ISSN: 0001-253X.

WILLIAMS, P. W. 1977. The Role and Cost Effectiveness of the Intermediary. In: Proceedings of 1st International Online Meeting; 1977 December 13-15; London, England. Oxford, England: Learned Information Ltd.; 1977. 53-63. ISBN: 0-904933-10-5.

WILLS, GORDON; CHRISTOPHER, MARTIN. 1970. Cost/Benefit Analysis of Company Information Needs. Unesco Bulletin for Libraries. 1970; 24(1): 9-22. ISSN: 0041-5243.

WILLS, GORDON; OLDMAN, CHRISTINE M. 1977. The Beneficial Library. In: Oldman, Christine M. The Value of Academic Libraries: A Methodological Investigation. Cranfield, England: Institute of Technology, School of Management; 1977. 163p. (Report to British Library Research and Development Department on Project SI/G/016).

WOLFE, J. N.; BRYDON, D. H.; SCOTT, A.; YOUNG, R. 1971. Economics of Technical Information Systems. A Study in Cost-Effectiveness. Edinburgh, Scotland: The University of Edinburgh; 1971. 443p. (OSTI Report no. 5103).

YOVITS, M. C.; FOULK, C. R. 1985. Experiments and Analysis of Information Use and Value in a Decision-Making Context. Journal of the American Society for Information Science. 1985; 36(2): 63-81. ISSN: 0002-8231.

II

Basic Techniques
and Technologies

Section II includes five chapters: "Artificial Intelligence and Information Retrieval," "Natural Language Processing," "Retrieval Techniques," "Statistical Methods in Information Science Research," and "Electronic Image Information." In the chapter on "Artificial Intelligence and Information Retrieval" Linda C. Smith of the University of Illinois at Urbana-Champaign reviews progress made over the past seven years, updating the reader from the initial *ARIST* chapter on artificial intelligence (AI), which she authored in 1980. The chapter concentrates on AI as it is applied to the problems of libraries and information systems. Smith opens with background information and discussions of the AI literature, AI in the marketplace, and relevant definitions. She identifies points of intersection between AI and information retrieval, and discusses AI tools, system components, knowledge representations/knowledge bases, learning and user modeling, and cognitive science. A portion of the chapter is devoted to expert systems, which is the major category of applications to date. Descriptions of characteristics of expert systems are provided together with examples of their use in reference, cataloging, and as expert online search intermediaries. Smith closes the chapter by looking toward the future of AI in terms of the limits and impacts of AI, roles for the information professional, and probable developments in system design.

The chapter on "Natural Language Processing" by Amy J. Warner of the University of Wisconsin in Madison covers developments since the prior *ARIST* chapter on the topic by David Becker in 1981. Warner's chapter surveys recent developments in computerized parsing of English-language text. Natural Language (NL) parsing means translating input (English) words/sentences into an internal knowledge representation suitable for subsequent processing. The subsequent processing often involves answering questions, doing logical inference, doing retrieval, and otherwise responding to commands.

Some of the major questions pertinent to NL parsing that are surveyed here are: what is the performance of current NL parsers, what representation languages are being used, to what extent is it desirable or necessary for the computer representation languages to simulate human cognitive structures, and to what degree are NL parsers being used commercially? In general, the answers to these questions are that: current NL parsers are useful only in very limited domains, the only commercial applications of NL parsers are as database front ends, many representation languages are being used, and it is an open and fascinating question as to whether computer representation languages must simulate human cognitive structures.

The chapter on "Retrieval Techniques" by Nicholas J. Belkin of Rutgers University and W. Bruce Croft of the University of Massachusetts reviews the literature on research in retrieval techniques for information retrieval systems, from about 1980 to date, within the framework of an original classification of retrieval techniques. By retrieval techniques Belkin and Croft mean the comparison of representations of queries or problems with representations of texts, for the purpose of identifying, retrieving, and/or ranking texts that might be relevant to a query or problem. The review stresses the contradictions between operational and experimental techniques, considers the issues of tailoring techniques to queries and using multiple techniques for single queries, and discusses the place of retrieval techniques in expert information systems and integrated information systems. The authors conclude with suggestions for research in retrieval techniques.

The first *ARIST* chapter on "Statistical Methods in Information Science Research" is by Mark T. Kinnucan, Michael J. Nelson, and Bryce L. Allen, all of the University of Western Ontario. Statistical methods are used in a wide array of information science research. Kinnucan, Nelson, and Allen survey the types and categories of such methods and then analyze the use of statistical methods in information science. Inferential statistics are emphasized. The distinction between parametric and nonparametric methods is explored, especially in the light of current thought about the difference between Gaussian and Zipfian distributions in bibliometrics. Recent research is reviewed to provide an overview of current statistical practice in the field. The main families of inferential statistics covered are Analysis of Variance (ANOVA) and related techniques, regression and correlation, and contingency tables. Goodness-of-fit tests, time series analysis, and dimensionality-reduction techniques (such as cluster analysis and multidimensional scaling) are also surveyed. In each case, the review identifies the kind of research that typically makes use of the particular statistical method as well as problems that are sometimes encountered.

Another first for *ARIST* is the chapter on "Electronic Image Information" by Lois F. Lunin of Herner and Company and Cornell University Medical College. Lunin notes that digital image processing occurs in many fields, among them industry, business, science, medicine, library, and museum. The recent surge of activity in image processing is due to many factors, such as the need for access to growing collections of digitized images, advances in microelectronic technology, and the development of sophisticated algorithms. One of the main problems of image handling is how to represent structural information in a database mode and allow users to retrieve by the language within the image itself. This chapter describes: elements of electronic images; image processing (input, storage, processing, output, communications); hardware; software; image database criteria, concepts, and challenges; applications; systems design; standards; and trends, markets, and predictions. It includes a glossary of electronic image terminology and a table comparing conventional and image databases.

2 Artificial Intelligence and Information Retrieval

LINDA C. SMITH
University of Illinois, Urbana-Champaign

INTRODUCTION

Information retrieval (IR) has not generally attracted the attention of workers in AI [artificial intelligence]. But AI techniques
should contribute to IR, if only in the long run, and IR offers
very worthwhile problems as well as some techniques to AI.

(SPARCK JONES, 1987, p. 419)

Although researchers and system developers in artificial intelligence (AI)
do not yet consider information retrieval (IR) to be a major application area,
IR does offer a potentially fruitful application domain for AI concepts and
techniques. In addition, there is some recognition that IR can provide techniques useful in certain other AI application areas. This chapter reviews progress over the past seven years in exploring AI applications in libraries and
information systems and concludes with an indication of the direction that
work in this area is likely to take in the near future.

Scope

The first *ARIST* chapter devoted to AI applications appeared in 1980
(SMITH, 1980). In the interim, related chapters have dealt with programming
languages for text and knowledge processing (LESK) and expert systems
(SOWIZRAL). The chapter on the software interface (VIGIL) included discussion of intelligent interfaces. Related chapters in this current volume of
ARIST include that on retrieval techniques (BELKIN & CROFT), especially
the section on expert information intermediaries and systems, and that on

Annual Review of Information Science and Technology (ARIST), Volume 22, 1987
Martha E. Williams, Editor
Published for the American Society for Information Science (ASIS)
by Elsevier Science Publishers B.V.

natural language processing (WARNER). This chapter does not explore
natural language processing except as it relates to the two roles identified by
GRISHMAN: providing a friendly, easily learned interface to IR systems and
automatically structuring texts so that their information can be more easily
processed and retrieved. DOSZKOCS offers a recent review of natural language processing in the context of large operational IR systems and services.

The 1980 review considered three AI application areas: reference (document) retrieval, data retrieval, and computer-assisted instruction (CAI). The
latter two areas lie outside the scope of the present review because each has
developed a large literature of its own. As starting points for investigating
those application areas, DEDE reviews recent research in intelligent computer-assisted instruction (ICAI), and KERSCHBERG and BRODIE &
MYLOPOULOS deal with the integration of AI and database technologies to
create systems known variously as "expert database systems" and "knowledge
base management systems."

CERCONE & MCCALLA subdivide AI into the following major areas of
investigation: 1) natural language understanding, 2) computer vision, 3)
expert systems, 4) search, problem solving, planning, 5) theorem proving,
logic programming, 6) knowledge representation, and 7) learning. An introduction to computer vision and related technologies (robotics, speech synthesis, speech recognition, tactile sensing) can be found in STAUGAARD.
These technologies are excluded from this review because they have less
immediate application in library and information science although VEANER
has suggested the possibility of robots for stack maintenance and of speech
recognition for query entry in retrieval systems. The other areas identified by
Cercone and McCalla are discussed, with particular emphasis on expert
systems, knowledge representation, and learning in the context of libraries
and information systems. Sources covered are predominantly from the United
States and Great Britain and are largely journal articles, conference papers,
and books published between 1980 and mid-1987. A few technical reports
have also been cited when the contents have not subsequently appeared in
another form. That literature characterized by hype, exaggeration, or unrealistic expectations has been excluded.

Artificial Intelligence Literature

As MILLS has recently documented, the literature of AI is scattered over
many books, conference proceedings, and journals, and its coverage is
scattered over many databases. Much relating to AI appears in publications
dealing with the area of application rather than being confined to the computer science literature. This chapter cites only a few items from the AI literature, emphasizing instead discussions of applications drawn from the library
and information science literature. Several recently published guides to the AI
literature can be consulted for more thorough coverage (HILKER; RODGERS;
SMITH, 1985). AMSLER provides a guide arranged by decades and notes that
only in the 1980s has there been an effort to package the literature of AI for
easier access. This includes the publication of reference books on AI, such as
the two-volume encyclopedia edited by SHAPIRO. The need for better

bibliographic control of the AI literature has led at least one AI research organization, the Turing Institute in the United Kingdom, to create a database devoted to the literature (WILKINSON).

Artificial Intelligence in the Marketplace

As article titles in the popular press suggest ("AI Enters the Mainstream," "AI is Here," "AI Comes of Age," and "AI Goes to Work"), products based on AI research and development are now moving into the marketplace. An additional factor generating interest in AI is the Japanese research program on fifth-generation computers with advanced AI capabilities. The Japanese plans have spurred responses by government and industry in both the United States and Europe. BISHOP provides a tutorial on the basic concepts associated with fifth-generation computers (including hardware, software engineering, programming languages, intelligent knowledge-based systems, and intelligent user interfaces), and BRAMER & BRAMER have compiled a bibliography of related literature. As SPARCK JONES (1983b, p. 87) observes, "the pressure and enthusiasm for R & D in these areas is coming from outside the documentation community," but the results are likely to be of considerable interest to librarians and information scientists.

Areas of commercial application of AI have been described by HARRIS & DAVIS and WINSTON & PRENDERGAST. AI products include packaged expert systems, expert system toolkits, natural language processors, AI languages, and symbolic processors (D. B. DAVIS). BUNDY has compiled what he terms a "catalogue of artificial intelligence tools," which includes descriptions of AI techniques and portable AI software. The Information Industry Association has tried to keep its members informed of these developments and to help them assess the implications for information products and services (CHASE & LANDERS). The list of AI applications in the service sector prepared by KAPLAN clearly has relevance to information services; such applications can include automation of routine advice, standardization of service, quality control, worker productivity, advice on accessing and using systems, and training. The various applications described later in this chapter demonstrate that many of these possibilities are already being realized by libraries and other information services. One negative outcome of the commercial interest in AI is the growing tendency to label hardware and software as "intelligent," whether or not such a label is justified. In the past when AI was only an area of research, it suffered "from both too-vigorous promotion and uninformed criticism" (CAMPBELL, p. 270), and there is a danger that this pattern will repeat itself with regard to AI applications.

Definitions

In introducing the *Encyclopedia of Artificial Intelligence*, SHAPIRO states that "the object of research in AI is to discover how to program a computer to perform the remarkable functions that make up human intelligence" (p. xiii). Its goals are twofold: 1) to make computers more useful, and 2) to understand the principles that make intelligence possible. Interesting recent discussions of the scope of AI are the enumeration of intellectual issues in the history of AI by NEWELL and the comments on the past, present, and future of AI, gathered from researchers and knowledgeable observers and edited by BOBROW & HAYES. Both demonstrate the range of viewpoints within the field.

There is some dissatisfaction with the term "artificial intelligence" because of the connotations of "artificial." HAUGELAND suggests that AI should be called "synthetic intelligence," noting by analogy that artificial diamonds are fake imitations whereas synthetic diamonds are genuine diamonds but are manufactured instead of mined. BOLTER advocates the use of the term "synthetic intelligence" in a different sense: a synthesis of man and computer rather than the replacement of man by machine. This contrast is voiced by others as well, though with varying terminology: computer-enhanced vs. computer-replaced (PENZIAS), cooperative vs. autonomous, exhibiting machine-aided intelligence vs. machine intelligence (SMITH, 1986b). In domains where it is not yet known how to construct programs that perform well on their own, it may be possible to build programs that significantly assist people in their performance of a task. As more is learned about the task, responsibility for problem solving may gradually shift from human-controlled to machine-controlled systems. RICH (1984) has termed this development "the gradual expansion of artificial intelligence," noting that AI can play a useful role even before it is possible to build programs that solve hard problems on their own. One can interpret the difference between human-controlled and machine-controlled systems as the distinction between augmentation and delegation. In augmentation, the computer assists the user in the substance of his task; in delegation, decisions are made by the system itself using programmed criteria.

One of the difficulties in defining the scope of AI is that the standards as to what constitutes AI are constantly changing (KURZWEIL). As a process begins to be understood well enough, one begins to consider it to be just a rote technique and not an example of intelligence. Another difficulty is the use of alternative terms. ANDRIOLE recognizes that in the past, approaches that would now be called "intelligent" IR were then called by such terms as "adaptive," "automated," or "responsive." He suggests that the depth of applied intelligence will vary across applications, with some being very unsophisticated and others requiring representation of great amounts of knowledge and powerful inferential capabilities.

Because there is as yet no consensus as to what "intelligent IR" denotes, this chapter adopts a broad definition in identifying applications for inclusion. Some applications actively borrow AI concepts and techniques while others program computers to perform intelligent functions through other means. Other applications included have the objective of augmentation (machine-aided intelligence) rather than delegation (AI). Although AI is a part of computer science, it also contributes to the newly emerging discipline of cognitive science, the search for understanding of cognition. Thus, discussions of cognitive science in relation to library and information science are also within the scope of this chapter.

Organization

Rather than reviewing AI concepts and techniques in isolation, this chapter focuses primarily on applications. The next section identifies points of intersection between AI and IR, with discussions of AI tools, system components, knowledge representations/knowledge bases, learning and user modeling, and cognitive science. This is followed by a section devoted to expert systems- the major category of applications to date- with a description of their characteristics and examples of their use in reference, cataloging, and as expert search intermediaries. The section on future developments explores the limits and impacts of AI, roles for the information professional, and probable developments in system design.

POINTS OF INTERSECTION

The earlier review on this topic (SMITH, 1980) concluded with the observation that: "An awareness of and support for possible AI applications in information systems are thus not widespread in the information science community" (p. 94). As the discussion of specific applications will demonstrate, this situation has changed considerably in the intervening seven years. For example, FINN asserts that intelligent information systems are a practical possibility, created by advances in computer science and AI research. Before turning to a discussion of specific applications, more general surveys identifying points of intersection between AI and IR are considered.

At the most general level, AI is juxtaposed with library and information science—e.g., in their inclusion as major fields making up studies of information in the compilation by MACHLUP & MANSFIELD. A recent collection of papers concerned with the foundations of information science includes a series of papers on the place of AI in information science (HEILPRIN). HJERPPE identifies two application areas of AI in libraries: organization of the knowledge resource and application of methods for knowledge representation to the problems of bibliographic description and control and to the design of classification and indexing systems. The possibility for greater exchange of ideas between AI and library and information science is hindered by the lack of shared terminology (HUMPHREY).

Some authors have explored the more specific theme of the relationship between AI and IR. HICE & ANDRIOLE suggest ways in which AI and videotex could be merged, including use of natural language processing and the provision of problem-solving aids. DEJONG proposes that AI applications to IR fall into four broad categories: 1) human–database interfaces; 2) conceptual indexing (organization of a database in terms of the meaning of its entries); 3) automated data entry; and 4) active memory techniques, in which the memory itself plays an active role in the update and retrieval functions. As COOPER observes, document retrieval systems constitute only one class of information retrieval systems; database management systems and expert systems are others. If IR systems of truly advanced design, capable of logical deduction of information from a knowledge base supplied as free natural language prose, are to be built, Cooper believes that an interdisciplinary theory of language and logic must first be developed. Such a theory would draw on logic, linguistics, AI, and information science.

Important in the emergence of the concept of intelligent information retrieval are the Aslib Information Group and its Informatics conference series, notably those on "Intelligent Information Retrieval" in 1983 (JONES, 1983) and "Advances in Intelligent Retrieval" in 1985 (ASLIB). JONES (1985) reviews the topics covered at the first seven Informatics conferences, including those papers dealing with AI. In her paper presented at Informatics 7, SPARCK JONES (1983a) identifies expert interface systems as a research area worth pursuing. In 1986 the first collection of papers on intelligent information systems appeared, with sections on database creation and cataloging, information retrieval, referral and user modeling, and cognitive science and information science (DAVIES, 1986b). As IR systems develop new capabilities, it will be necessary to develop terminology for classifying them. New terms are already being proposed, such as "active" (MONTGOMERY), "proactive" (HJERPPE), and "intra-active" (CAVANAGH, 1986), but there is not yet a well-developed taxonomy associating forms of "intelligence" with appropriate labels.

Not all researchers are convinced that AI has relevance to the design of document retrieval systems. SALTON (1986b) examines the main components of advanced AI systems and concludes that AI methods are likely to be even more difficult to apply to normal document retrieval environments than the automated text analysis methods developed in the past by IR researchers. Specifically, he comments on the difficulties in generating meaning representations from texts, the lack of consensus on methods for supplying domain and world knowledge, the problems in developing user models applicable to large populations of diverse users, and the inadequacy of expert system approaches in unstructured environments covering heterogeneous tasks. These are indeed difficult problems and suggest that AI techniques will have more impact in the short term in IR systems with more limited objectives than general-purpose document retrieval systems. The sections that follow identify numerous studies already under way to identify the limits and possibilities of AI applications in various types of IR systems and as part of various system components.

Artificial Intelligence Tools

The discussion of specific applications will demonstrate that a number of AI concepts, techniques, and tools have been used in building more powerful retrieval systems. A few authors have emphasized the usefulness of AI programming languages and techniques, particularly PROLOG and logic programming, in information retrieval. HERTHER provides a brief tutorial on PROLOG and presents information on versions available for personal computers. EASTMAN advocates the use of languages such as PROLOG in IR experimentation because of their possibility for enhancing software development productivity. WATTERS ET AL. present a prototype document-retrieval system programmed in PROLOG with enhanced retrieval capabilities through the application of deductive reasoning. Their system can be scaled up only if they exploit special-purpose hardware or extended software capabilities, both of which they are investigating. SMALLTALK-80 and related languages have attracted the attention of developers of thesauri (KLEINBART) and terminology data banks (NEDOBITY), who believe that features of these languages would allow improved capabilities for building, accessing, and browsing files of interrelated terms.

System Components

One approach to making systems more "intelligent" is to enhance the capabilities of particular components. For example, YANNAKOUDAKIS & FAWTHROP describe an intelligent spelling corrector that could be applied in editing records and queries. COYLE & GALLAHER-BROWN describe the procedure used to improve the performance of an algorithm for identifying duplicate records in the process of building an online union catalog for the University of California. The algorithm was refined until it made approximately the same decisions as human experts, with the machine's judgment of a match based on a weighted evaluation of data elements. FRAKES has demonstrated through experiments that automatic stemming can lead to retrieval performance comparable with that achieved by expert searchers using truncation of terms.

Approaches to automating the indexing process continue to receive some attention. As SALTON (1986a) describes, there is a long history of IR research on automatic indexing, and a "blueprint" for automatic indexing can be prescribed based on the results of numerous retrieval experiments. In operational systems using controlled vocabularies for indexing, there is interest in new approaches to machine-aided indexing. BRENNER ET AL. describe the American Petroleum Institute's approach, using the existing thesaurus as a basis for the knowledge base but refining it with the development of additional rules to improve the accuracy of index-term assignment. HUMPHREY & MILLER describe the ambitious Indexing Aid Project at the National Library of Medicine, which is developing an interactive knowledge-based system to assist indexers in indexing the periodical literature of medicine.

Knowledge Representations/Knowledge Bases

The term "knowledge representation," used in AI to designate the formalism for knowledge possessed by a computer-based system, has not yet gained currency in library and information science despite the need for a more comprehensive term to cover the products of abstracting, indexing, classifying, cataloging, annotating, and content analyzing (CREMMINS). SMITH & WARNER describe the major categories of representation in document-retrieval system design (objects—documents and queries; relationships—document/query, term/term, document/document) and identify the need for comparative evaluations of alternative representations. FRAKES ET AL. describe the results of one such study comparing alternative document representations (e.g., titles, abstracts, descriptors), which demonstrated that different representations led to retrieval of different document sets in response to a given query. B. C. VICKERY believes that information scientists can learn from knowledge representations developed in other disciplines (AI, psychology, linguistics), and his review highlights major features of these representations.

Although much effort to date in IR has concentrated on retrieval systems with documents represented by citations, abstracts, and descriptors, there is a growing interest in the creation of knowledge bases that could be used as a basis for question answering and fact retrieval. Such knowledge bases could be compiled by human effort or derived automatically. A good example of a knowledge base requiring construction and maintenance by humans using machine aids is the Hepatitis Knowledge Base sponsored by the National Library of Medicine, which synthesized the information contained in many documents and was made available for consultation by physicians (SIEGEL). Examples of experimental systems building knowledge bases automatically are CyFr (CYRUS + FRUMP), a system for reading and summarizing news stories from the UPI wire service, producing conceptual representations of events in stories, organizing these in memory, and answering questions (KOLODNER; SCHANK ET AL.); TOPIC (Text-Oriented Procedures for Information management and Condensation of expository texts), a system creating text condensations from German texts (HAHN & REIMER); and RESEARCHER, a system for reading patent abstracts, adding information to long-term memory, learning through generalization, and answering questions (LEBOWITZ, 1983; 1986). Other retrieval systems with knowledge representations influenced by AI approaches are RUBRIC for news stories (TONG ET AL.) and RESEDA for biographical data (ZARRI). SAGER & KOSAKA describe linguistically-based procedures for analyzing the texts of medical articles to build a structured database to support question answering.

Some of the knowledge base projects have included research on how best to give users access to the contents of these knowledge bases. BERNSTEIN & WILLIAMSON describe ANNOD (A Navigator of Natural language Organized Data), which could be used to identify passages in the Hepatitis Knowledge Base most likely to be relevant to answering natural language queries posed by users. THIEL & HAMMWÖHNER describe what they term "informational zooming" in TOPIC. Using a graphical retrieval language, a searcher is given access to several layers of specificity, from topic descriptors to full text. Other sources of knowledge bases are expert systems. FIRST ET AL. have successfully used a knowledge base for a medical expert system (Internist-1/ Caduceus) as an electronic textbook of medicine by creating QUICK (QUick Index to Caduceus Knowledge), a user-friendly access system. Research reported by KINNELL & CHIGNELL suggests that several perspectives on a topic may need to be considered when building a conceptual representation for an intelligent interface to serve as a map and guide to a database. In their application of building an interface to assist students in understanding and discussing constitutional issues, they found that they had to take into account the perspectives of the student, the domain expert, the document indexer, the database developer, and the curriculum instructor. Interest in building knowledge bases and providing new means of access to them is likely to grow as increasing amounts of raw material in the form of machine-readable texts become available.

Learning and User Modeling

Machine learning is an important area of activity within AI. Systems with learning capabilities are necessary if a system is to improve its performance over time without always relying on explicit changes made by human programmers. There are many different approaches to machine learning, which can be organized with the aid of a taxonomy proposed by CARBONELL ET AL. These authors suggest that three dimensions can be used to classify learning systems; they can be classified on the basis of the underlying learning strategies used, the representation of knowledge, or the application domain. Underlying learning strategies can include rote learning, learning from instruction, learning by analogy, learning from examples, and learning from observation and discovery. Representation of knowledge can take the form of parameters in algebraic expressions, decision trees, formal grammars, production rules, formal logic-based expressions, graphs and networks, frames and schemas, computer programs, taxonomies, or some combination of these.

L. C. SMITH (1980) suggests that learning in retrieval systems can have short-term or long-term effects. Short-term learning is the modification of system response during the processing of a particular query to better meet the needs of the user. Long-term learning could involve modifying and/or extending document representations to improve system response over time. Both approaches continue to be explored. For example, PIETILÄINEN describes an approach to automated query expansion, and GORDON explains the use of an AI learning algorithm to modify document descriptions over time. RADA offers a useful distinction between knowledge-sparse and knowledge-rich learning. He considers learning to be knowledge-sparse when the feedback used is very simple, such as relevance judgments by searchers or terms found to co-occur in documents with terms used in the query. Knowledge-rich learning would use more complex sources of feedback. FORSYTH & RADA have published a book on machine learning, including a tutorial on various learning strategies and detailed case studies on learning experiments in IR systems.

The topic of user modeling is closely related to learning because there is an interest in finding ways in which systems can learn about users and build models to tailor the interaction and responses according to the characteristics of a particular user or user group. RICH (1986) proposes a taxonomy of user models with three dimensions: 1) one model of a single canonical user vs. a collection of models of individual users; 2) models specified either by the system designer or by the users themselves vs. models inferred by the system on the basis of user behavior; 3) models of fairly long-term user characteristics, such as areas of interest or expertise, vs. models of relatively short-term user characteristics, such as the problem the user is currently trying to solve. SPARCK JONES (1985) describes the possible sources of modeling information and cautions that it may be difficult in practice to get much modeling information from the user. The resulting models may be very simple, with limited predictive value. Some experimental retrieval systems have been designed to exploit user models in responding to user queries. The I^3R system (Intelligent Intermediary for Information Retrieval) has been designed to act as an expert intermediary (CROFT & THOMPSON; THOMPSON & CROFT). The system improves the effectiveness of retrieval by acquiring detailed models of users and their information needs and selecting search strategies based on these models. KORFHAGE terms his user models "user profiles" with three major components: 1) keywords from the query, 2) user information habits and experience (languages read, sources scanned regularly, preference for high recall or high precision, degree of familiarity with the system), and 3) a trace of search history. These profiles are used in tailoring system output to better match the needs of particular users. DANIELS should be consulted for a much more complete review of work on user models and their potential applicability in IR systems.

Cognitive Science

Not surprisingly, increasing interest in AI among librarians and information scientists relates to increasing interest in cognitive science of which AI is a contributing part. As JOHNSON, a science journalist and author of a book on AI, observes: "One of the most engaging things I found about artificial intelligence is that it drives you to introspect, to turn your powers of observation inward and wonder how you do what you do-- and how you would describe it precisely enough to get a computer to do it" (p. x). GARDNER has written a readable history of cognitive science, including discussions of developments in AI, psychology, philosophy, linguistics, anthropology, and neuroscience.

Reviews providing good coverage of a cognitive science perspective include that by INGWERSEN on the contribution cognitive science can make toward understanding IR processes and that by BELKIN & VICKERY on interaction in information systems, including cognitive aspects of various stages of the search process. GRIFFITH asserts that information science must use and extend knowledge of human information processing to improve user-machine interaction. Other areas that have received the attention of information scientists are cataloging, indexing, and abstracting. ERCEGOVAC (1986) proposes that one might be able to suggest different ways to organize and display cataloging rules based on studies of catalogers' cognitive behavior. Drawing on hypotheses and suggestions of cognitive science, ANDERSON discusses possible improvements in the design of indexing systems for information retrieval based on attributes of the mind's indexing and retrieval system. KUHLEN has analyzed the sets of abstracting rules used by human abstractors as a basis for assessing the feasibility and appropriateness of using them as a model for machine processes.

One research team has developed cognitive models of scientific work and identified their implications for design of computer-based, interactive, knowledge-delivery systems (MAVOR ET AL.; VAUGHAN & MAVOR). They model the process of doing research as a procedural script and the researcher's conceptualization of research as a set of interrelated schemata or frames. In building and testing these models, they used verbal data, which they termed "think-aloud-about-your-research" sessions. The use of verbal reports as data is well established as the method of protocol analysis in cognitive science. ERICSSON & SIMON discuss the assumptions, techniques, and limitations of this method, which could prove useful to other researchers seeking to explore cognitive science questions in library and information science.

EXPERT SYSTEMS

Characteristics of Expert Systems

Although "expert system" is the term that has gained currency, HAYES-ROTH suggests that "knowledge system" is a better label for the large class of systems that apply large knowledge bases to solve practical problems. Knowledge systems in general do not necessarily mimic human experts but do provide the electronic means to collect, store, distribute, reason about, and apply knowledge. Expert systems are a special case, incorporating know-how gathered from experts and designed to perform as human experts do. Roles for expert systems include: preserve otherwise perishable human expertise; distribute otherwise scarce expertise; reduce costs of mediocre or poor human performance; and provide help to humans trying to access information and use computers. Appropriate problem domains for expert systems are valued, bounded, routine, and knowledge-intensive (BOBROW ET AL.). The work involved in designing and building expert systems is often termed "knowledge engineering" and includes knowledge acquisition, knowledge system design, knowledge programming (implementation of a knowledge base and inference engine), and knowledge refinement (HAYES-ROTH). Expert systems are generally thought of as advice-givers and usually have an explanatory capability so that users can understand the rationale on which the advice is based.

While work on what have come to be called expert systems dates back to the 1960s, commercial applications on a wide scale have emerged only recently. In addition to marketing advisory systems for specific applications, a growing number of firms have developed expert system "shells," software frameworks that can be used for in-house development of expert systems. WATERMAN provides an extensive catalog and bibliography of expert systems and tools developed through 1986. In spite of the commercial successes, there are a number of open problems in expert system design, such as the need for improved methods of knowledge acquisition, maintenance and organization of large knowledge bases, learning, and validation.

Although the advisory role of expert systems is the one most often emphasized, MICHIE & JOHNSTON argue that this view is too narrow. The knowledge base itself is a valuable product. Expert systems can "help to codify and improve expert human knowledge, taking what was fragmentary, inconsistent and error-infested and turning it into knowledge that is more precise, reliable and comprehensive" (p. 129). While in principle such knowledge bases should prove useful for teaching, in practice this may not be the case. As R. DAVIS comments, the extension to instruction is not necessarily easy because teaching sometimes requires deeper understanding than doing and also requires pedagogical skills.

Expert systems have captured the interest of librarians and information scientists. ADDIS and THORNBURG review features of expert systems and issues in their design and use. BROOKS cautions that the process of applying expert system techniques to IR systems is at a very early stage. NOWAK & SZABLOWSKI suggest that expert systems can be thought of as a new generation of scientific and technical information systems. Several authors have identified areas of activity in which expert systems may have an impact, such as cataloging, classification, and assistance in online searching (CLARKE & CRONIN; ERCEGOVAC (1984); JONES (1984); YAGHMAI & MAXIN). The proceedings of a conference on expert systems in libraries, held in 1985, have now appeared (GIBB). The volume includes a lengthy bibliography and papers covering basic concepts of expert systems, expert system development tools, and applications in reference, classification, and cataloging. The sections that follow illustrate work already completed in exploring expert systems in relation to cataloging, reference, and online searching.

Expert Systems and Cataloging

According to DAVIES (1986a), cataloging is a possible domain of application for expert systems because it has certain characteristics: there are recognized experts, the experts are demonstrably better than amateurs, the task takes an expert a few minutes to a few hours, the task is primarily cognitive, and the skill is routinely taught to neophytes. Because cataloging costs constitute a significant proportion of the budgets of many libraries, it would appear to be a worthy domain to explore. Projects reported to date are simply at the prototype stage. Although some may ultimately lead to useful tools for catalogers, in the interim they are proving to be a vehicle for better understanding the expertise that underlies cataloging and the extent to which it can be made explicit. The existence of cataloging codes, such as *AACR2* (*Anglo-American Cataloguing Rules*, 2nd edition) suggests that cataloging can be thought of as a rule-based domain. However, the automation of the rules would not make cataloging a largely automatic process because at present the cognitive processes involved in the interpretation of titles pages are not fully understood. DAVIES (1987) suggests some heuristics for interpreting title pages (e.g., the title is usually the feature nearest the top of the title page), but they require further study.

An early project by DAVIES & JAMES written in PROLOG demonstrated that *AACR2* rules for determining access points could be restated as production rules and explanation facilities could be incorporated. More recent projects have explored the feasibility of using expert system building tools.

BLACK ET AL. have used ES/P ADVISER and SAGE to build cataloging advisory systems, providing advice on completing catalog data entry forms. BORKO (1987) is using the EXSYS expert system development package to build a prototype expert system in the domain of map cataloging, capable of advising on the completion of the MARC (machine-readable cataloging) map format by selection of appropriate *AACR2* rules. The domain of map cataloging was selected because it is bounded and is a complex subject with few experts. Project ESSCAPE (Expert Systems for Simple Choice of Access Points for Entries), described by HJERPPE ET AL., has used EMYCIN and EXPERT-EASE to build systems that either construct simple catalog records or provide pointers to relevant rules in *AACR2*. They have chosen the rules in Chapter 21 of *AACR2*–"Choice of Access Points"–as their domain and view this approach as a means for elucidating sets of rules and their application. They also emphasize the need to understand the ways in which interpretation contributes to the cataloging process.

Authority control, methods by which authoritative forms of names, subjects, and other headings in a catalog are consistently applied and maintained, is an important responsibility of catalogers. BURGER discusses present implications of AI for authority work and speculates on future applications of AI for authority control The products of the cataloger's efforts make up a library's catalog. To date there has been little investigation of the possibilities of building intelligent interfaces to online catalogs, but work by MITEV & WALKER suggests one approach worthy of further study. They are investigating techniques for automated intelligent search sequencing, taking terms entered by the user and manipulating them in various ways to identify likely records to retrieve.

Expert Systems and Reference

Reference work in libraries is made up of various tasks, some of which are more amenable to the application of expert systems than others. PARROTT suggests three criteria to be used in identifying candidate tasks for expert systems in reference. Such tasks should: 1) qualify reference librarians as experts; 2) require so much time that the expense of constructing expert systems can be justified to reduce the load; and 3) be sufficiently low level that expert systems can be constructed in a reasonable amount of time. He enumerates the tasks involved in responding to requests for factual information and literature searches and requests for help in interpreting bibliographic references and in obtaining the corresponding physical items. Expert systems designed to carry out some of these tasks have been developed at the University of Waterloo Library.

A number of investigators have developed expert systems designed to direct users to sources likely to answer specific categories of reference questions. K. F. SMITH describes POINTER, an expert system designed to provide reference assistance for federal documents in a separate government documents department in an academic library. In a 1986 meeting of the AI/Expert Systems Group of the Library and Information Technology Association, Howard D. White of Drexel University described an expert system that would present the user with a ranked list of plausible reference sources to consider, based on a match of reference book attributes and query attributes (ALURI). Researchers at the National Agricultural Library have created Answerman, which supplements pointers to reference books likely to contain the answer to specific kinds of questions with links to external programs providing online access to databases of bibliographic citations and to full-text files that could provide the answers to some questions (WATERS).

Although much of the work in reference expert systems to date has involved systems with relatively shallow models of reasoning, there are a few examples of efforts to build more detailed models. SLOCUM ET AL. demonstrate the use of inference nets for modeling the heuristic reasoning process followed by a geoscience reference specialist in working from a question to recommendations on sources likely to answer that question. DESALVO & LIEBOWITZ are trying to capture the knowledge of experts in the National Archives and Records Administration regarding strategies for "navigating" through the archives' holdings to find record sets likely to contain the information sought. Their approach may be applicable to the development of expert systems that could advise users on how to exploit other specialized collections.

Other aspects of reference work have been addressed in proposals or prototype systems. MICCO & SMITH describe the design for a system that could help the user refine a search strategy by identifying resources to be searched and developing subject terms. CAVANAGH (1987) suggests that an expert system could function as an automated readers' advisor, helping patrons select materials responsive to their interests and intellectually accessible. OBERMEIER & COOPER propose a system that could satisfy the needs of the business community by developing plans of action, identifying the resources to consult in response to a request and in what order. BIANCHI & GIORGI are creating a system that will serve as a consultant for anyone wishing to set up an online search service. According to the needs of the requester, this expert system will provide all the relevant information in terms of databases, host computer, costs, document suppliers, telecommunications networks, modems, and equipment required.

The referral function in reference work involves connecting a requester with an appropriate personal or organizational source of specialized information. PLEXUS, an expert system for referral in the area of gardening, was created over a 20-month period and will be tested in a public library setting (VICKERY & BROOKS; VICKERY ET AL.). In studying the referral process the investigators identified seven kinds of knowledge that an expert system for referral must embody: librarianship and library knowledge, knowledge of information-retrieval techniques, knowledge of the subject area, knowledge of referral resources, knowledge of the structure of world knowledge, knowledge of the library users, and knowledge of problem-statement development. Knowledge is represented in PLEXUS in four ways: 1) the Broad System of Ordering (BSO), which provides the world knowledge framework, semantic categories, or facets; 2) production rule sets, which direct problem-statement development, search strategy modification, and evaluation of output; 3) the semantic context of each dictionary term; and 4) the content of the records of referral sources contained within the database. The investigators provide detailed examples from a consultation session. Although developed for a particular subject area, the approach should be applicable to the development of referral systems in other domains.

Expert Search Intermediaries

The area of most activity, having the longest history and the largest number of R&D activities, is work on expert search intermediaries. This has been a focus of activity in recognition of the complexity of the online environment with its diverse systems and databases to which users wish to gain access. The intent of much of the work is to make online systems directly accessible to end users without the need to rely on human intermediaries. Some of the projects can also provide more powerful tools for human intermediaries.

Although such efforts are often termed "expert" search intermediaries, they differ from expert systems that aim to capture the expertise of one or several experts in a restricted domain. As PAICE observes, the expertise of human intermediaries is centered not on the subject of queries but on the tools and techniques for finding that information. The domain, the subjects of interest, is usually much wider than that for a typical expert system. In designing an expert search intermediary, or intelligent interface, for an IR system, the ultimate objective is to do the same functions that a good human intermediary would in the same situation (BROOKS ET AL., 1985). Although some research encompasses this objective, many research projects and implemented interface systems have a much narrower objective, so as to make the problem manageable.

This chapter does not attempt to identify all intermediary systems but instead focuses on those projects that have tried to go beyond the mechanical aspects of searching to address the intellectual aspects. DAYTON ET AL. describe approaches for online search guidance explored up to 1980, and KEHOE presents a historical review. HUSHON & CONRY provide a comparative analysis of the features of five available commercial products, and JACOBSON & WITGES have compiled the proceedings of a recent conference exploring many aspects of the design and implementation of computer intermediaries for information retrieval.

As M. E. WILLIAMS explains, there is a terminology problem because research efforts in this area have been designated by many different terms: front end, interface, intermediary system, post processor, gateway, and transparent system. She provides a detailed analysis of retrieval-related functions that could be automated. TOLIVER explains that intermediary software can be inserted at four places in the retrieval process (the host mainframe, a mainframe remote from both the host and the user, a mainframe to which the user has direct access, or a personal computer) and summarizes the advantages and disadvantages of each approach.

Often in expert system development, the knowledge engineer responsible for designing the system relies heavily on human experts in building the knowledge base and deriving the inference rules. A series of research projects completed by BROOKS ET AL. (1986) have derived specifications for an intelligent interface from a study of real life human user–human intermediary information interactions. They have used a variety of methods, including functional discourse analysis. The results of these investigations have led to the specification for a distributed expert system architecture, identification and specification of the necessary functions, and proposals for a problem-structure driven human–computer dialog. BELKIN argues that search terms and databases cannot be selected without an understanding of the user's problem, which is to be developed through a highly interactive, cooperative dialog with the user. In contrast to the research strategy followed by Brooks et al., HARTER & PETERS have used the IR and online searching literature as a basis for identifying heuristics, general rules of thought or action that tend to produce useful results in online searching. They present a typology of heuristics in six categories: 1) those relating to overall philosophy and approach, 2) those relating to the language of problem description, 3) those based on file structures, 4) those for concept formulation and reformulation, 5) those for increasing or decreasing recall and precision, and 6) those for cost effectiveness. This enumeration of heuristics is suggestive of the knowledge and inference rules that an expert search intermediary would have to incorporate if it were to match completely the range of expertise possessed by a human intermediary. Other lists of desiderata for intelligent interfaces have been compiled by A. VICKERY, OPPENHEIM, and WALKER & JANES.

Although the long-term goal may be to match fully the capabilities of a human intermediary, in the short term researchers and system developers have identified various ways to constrain the problem and still provide useful tools for end users. An important component of these efforts is evaluation to demonstrate that the intermediary system does indeed perform at a level helpful to an end user.

One approach to constraining the problem is to create an expert intermediary for handling certain classes of questions on particular databases. This approach has been followed by POLLITT (1986; 1987) in designing CANSEARCH, an intermediary system for handling cancer therapy questions in searches of MEDLINE; by SMITH & CHIGNELL in designing the Natural Products Expert (NP-X) for searches of CA Search; and by KRAWCZAK ET AL. in designing the Environmental Pollution Expert (EP-X). These authors explain how such intermediary systems can assist users in formulating queries in specialized subject domains.

Another approach to constraining the problem is to focus on one type of decision in the search strategy formulation process, such as the selection of databases or search terms. THORNBURG is building a prototype expert system to advise human intermediaries on choice of databases, constructing rules based on interviews with expert searchers in the life sciences. FIDEL has used the behavior of experienced search intermediaries as a basis for building a decision tree of rules for selecting search keys, deciding whether to use controlled vocabulary, free text, or a combination of both. SHOVAL (1983; 1985; 1986) developed a prototype system to assist users in selecting the right vocabulary terms for a database search. The knowledge base used originated mainly from knowledge provided in a thesaurus and represented as a semantic network, where nodes are terms or concepts and links are various types of relations between them. During the search stage, relevant knowledge in the semantic network is activated, and search and evaluation rules are applied to find appropriate vocabulary terms. During the suggest stage, those terms are further evaluated and suggested to the user. Shoval compared three different decision-support strategies for selecting vocabulary terms: 1) the conventional strategy in which the system provides information only at the user's request, 2) a participative strategy in which there is interaction between the system and the user with the user evaluating intermediate findings, and 3) an independent strategy in which the user is presented with the findings at the end. The participative and independent strategies are examples of augmentation vs. delegation as explained above in the section on definitions. Shoval found that the participative and independent strategies were similar in performance and significantly better than the conventional strategy.

Yet another approach to constraining the problem is to keep the user in control of the search process but allow the system to offer advice. One experimental system of this type is IIDA (Individualized Instruction for Data Access), which was designed to provide diagnostic analysis of the users' performance and intervene by pointing out errors, offering advice, or giving instruction (MEADOW ET AL., 1982a; 1982b). Evaluations of IIDA demonstrated that end users could successfully carry out searches with the system's assistance.

As is true in other expert system application areas, designers must select an appropriate knowledge formalism for the system. Various formalisms have been exploited, including frames (JAKOBSON) and production rules (P.W. WILLIAMS). Designers of many expert intermediary systems have borrowed and applied formalisms already well developed in other AI applications. Approaches to natural language processing have also been used in designing interfaces to IR systems (GUIDA & TASSO).

The CONIT (COnnected Network for Information Transfer) system, the subject of development and experimentation by MARCUS (1983; 1986) for many years at MIT, demonstrates the evolution of an intermediary system with increasingly expert capabilities. Based on his experience with CONIT, Marcus concludes that to provide a truly comprehensive expert retrieval assistant a very extensive knowledge base development is required. In addition, "although expert retrieval assistance development is difficult, it shows promise for deepening our understanding of the retrieval process from a basic scientific viewpoint as well as for improving search techniques themselves" (p. 185).

Given the complexity of developing a truly expert search intermediary, it is appropriate to be somewhat skeptical of commercial products that claim to be expert. In assessing an intermediary system, the following questions should be posed: 1) for which decisions in formulating the search strategy does the interface offer assistance? 2) what form does this assistance take—i.e., is the decision made automatically by the computer or interactively with the computer offering advice to the user? 3) what expertise underlies the assistance offered by the interface? 4) how well does the system model the expertise of a skilled human intermediary, and in what ways is it deficient?

FUTURE DEVELOPMENTS

Limits and Impacts of Artificial Intelligence

A thorough review of a new technology such as AI and its applications in IR should also assess the limits and impacts of this technology. Although some of these limits and impacts are specific to the particular area of application, more general discussions drawn from the AI literature can also provide useful insights. Recent analyses addressing various aspects of the limits of AI have been prepared by SLOMAN, SCHWARTZ, E. DAVIS, and WINOGRAD & FLORES. Sloman provides an overview of some unsolved problems in AI. Following a discussion of concepts from philosophy, biology, and linguistics, Winograd and Flores argue that one cannot program computers to be intelligent and therefore that different directions must be pursued for the design of powerful computer technology. Schwartz distinguishes various types of limits: fundamental limits, limits due to computational complexity, limitations of the present state of knowledge in AI, and moral limits. The first three types are concerned with an assessment of what AI *can* do, while the fourth addresses the question of what AI *should* do. He cautions that the methodological level of research in AI is often low, with publications simply describing the structure of some program believed by its authors to embody some function mimicking an aspect of intelligence or tracing some program's internal activity. Published research on AI applications in IR is not immune from this criticism. Davis contrasts AI programs—"slow, fragile, terribly limited solutions to particular problems" (p. 105)—with the capabilities of the human mind. SHORE, in a passage entitled "Crunching Numbers and Bashing Symbols Isn't Enough," cites an observation by Alan Perlis that "good work in AI concerns the automation of things we know how to do, not the automation of things we would like to know how to do" (p. 245). As DRAY comments, to date the serious literature and thinking on AI's social impacts is skimpy. He discusses the possible social impacts of AI using the following categories of issues: economic, political and control, psychological and sociological, and technical spinoff.

The limitations of expert systems in particular have come under increasing scrutiny. As SCHANK & CHILDERS observe: "As a descriptive term used by AI researchers, an expert system is a fairly clearly defined type of program. But, like most AI terms, the words 'expert system' are loaded with a great deal more implied intelligence than is warranted by their actual level of sophistication" (p. 33). DREYFUS ET AL. have developed a five-stage model of skill acquisition (novice, advanced beginner, competent, proficient, expert) to demonstrate that there is more to expert performance than calculative rationality. They suggest that computer-based "expert systems" can only reach the competent stage and should be used as advisors to, rather than replacements for, experts. To use an expert system effectively, a user must understand its scope (the broad class of problems that the system is designed to solve) and its limitations (the human capabilities that are successfully modeled). Straying outside these boundaries may lead to results that will be difficult for the user to interpret (HARRIS & HELANDER). Unfortunately, it is difficult to make the scope and limitations explicit.

In identifying the limitations of AI in the context of IR, a framework developed by MIKSA based on Machlup's categories of knowledge is helpful. Miksa distinguishes among three categories of knowledge: 1) practical or instrumental knowledge, which is useful for work, decisions, and actions; 2) intellectual knowledge, which satisfies intellectual curiosity; and 3) small-talk and pastime knowledge, which satisfies a desire for light entertainment and emotional stimulation. Miksa cautions that the discovery of the capacity of libraries and information services to supply practical knowledge does not mean that this is the only knowledge-retrieval function that is valuable or necessary or in need of being studied and understood. Clearly expert systems emphasize the provision of practical knowledge. Miksa suggests that the challenge remains to study and learn how to make the best possible systems for providing pastime or intellectual knowledge as well. It remains to be seen whether AI can contribute to this effort. SMITH (1986a) expresses concern that an emphasis on autonomous systems that simply provide answers ignores the need for comprehensibility and identification of the sources of information. When information is sought from printed sources or from other people, the user is aware of the process used and has some basis for judging the authoritativeness of the material or the response. If the system is a black box, the user has no clues as to the adequacy of the search or the reliability of the answer. Careful thought should be given to these concerns in designing intelligent information systems, as the incorporation of an explanatory facility in expert systems suggests.

Roles for the Information Professional

It is evident from the applications identified in this chapter that librarians and information scientists at some institutions are already taking an active role in designing and developing expert systems and other AI applications. They have gone beyond the recommendation of DAVIES (1983) that libraries include provisions for accessing expert systems to the creation of such systems in-house for particular purposes. MOLHOLT draws an analogy between the reference interview and the interview between a knowledge engineer and the expert in the process of knowledge acquisition in building expert systems. She asserts that knowledge engineering is a possible new role for librarians and recommends that schools of library and information science should consider developing curricula that would allow graduates to fill this role. Library and information science educators in both the United States (BORKO, 1985) and the United Kingdom (BRITTAIN) also argue that it is the responsibility of schools to carry out teaching and research in AI and expert systems. In this regard, it is appropriate to recall the observation of SIMON: "The professional schools will reassume their professional responsibilities just to the degree that they can discover a science of design, a body of intellectually tough, analytic, partly formalizable, partly empirical, teachable doctrine about the design process" (p. 132). Schools of library and information science are only beginning to address the question of how developments in AI should be incorporated in their curricula.

Future System Design

In identifying likely characteristics of future information systems, SALTON (1985) predicts a unified approach to several different information-processing tasks (database management, bibliographic reference retrieval, question answering) and more user-friendly processing environments. Proposals for system architectures that would allow the integration of data, knowledge, and information bases have already appeared (BELL). Developments in various areas of AI can also contribute to improved human–computer interaction, as described by JACOB. Considerable R&D is still required to explore various possibilities. FOX (1986a; 1986b) describes the CODER (Composite Document Expert/extended/effective Retrieval) project developed as a testbed for prototyping and validating AI approaches for creating knowledge representations from document collections and for interacting closely with users. DEFUDE (1984; 1985) is also experimenting with a variety of AI techniques in building an expert retrieval system.

Visions of future systems encompass far more than enhanced document-retrieval systems or expert advisory systems. Authors recall H. G. Wells's World Brain (MICHIE) and Vannevar Bush's Memex (BRUNELLE & MCCLELLAND) as earlier visions of sophisticated information systems but suggest the need to develop new views of systems made possible by developments in AI and other information technologies. A number of authors have explored the range of possibilities for supporting knowledge work, such as the taxonomy of knowledge work-support tools constructed by JÄRVELIN & REPO. WALKER (1981; 1986) identifies the need to develop systems for experts that will support specialists in the search for knowledge. Work by him and others at Bell Communications Research is devoted to developing knowledge resource tools for use in a workstation environment by experts who organize and use information from documents in the course of their work. STEFIK suggests that such an approach is consistent with a shift in AI from a focus on mechanisms of intelligence to the role of knowledge in intelligence to the augmentation of knowledge processes in a medium. Development of a knowledge medium will rely on work in computer science, such as databases and network technology, as well as core work in AI on language processing, knowledge representation, and problem solving.

As suggested in the introduction to this chapter, the technology transfer between AI and IR need not be confined to a one-way borrowing of AI concepts and techniques by IR. Nevertheless, the literature reviewed for this chapter included very few articles explicitly identifying IR techniques that could be useful in other AI applications. KANTOR, WINETT & FOX, and GEBHARDT discuss possibilities of using certain IR techniques in the design of expert systems. As researchers in IR become more familiar with work in AI, there may be more such contributions from IR to AI in the future.

CONCLUSION

The earlier review on this topic (SMITH, 1980) concluded by identifying five trends suggested by the literature. It is instructive to enumerate and comment on these trends so as to assess the progress made in the past seven years:

- As retrieval systems move from storage of document representations to storage of full text, it is likely that some of the distinctions between reference and data retrieval systems will begin to blur;
- The emphasis in interactive systems is likely to continue to be on machine-aided intelligence rather than machine intelligence alone;
- Developments in hardware and software are likely to make AI applications more feasible;
- As computer systems grow in power, sophistication, and complexity, it becomes more difficult to become (or even remain) expert in their use. Therefore, "intelligent" online assistants and tutors will be needed; and
- The emerging disciplines of cognitive science and knowledge engineering suggest challenging new roles for information scientists in the investigation of human information processing and in the construction of knowledge-based systems.

These five trends noted in 1980 are still valid observations in trying to assess the impact of AI on IR. There continues to be interest in the goal of providing coordinated access to multiple resources (data, documents, knowledge bases), and some progress has been made in the form of expert search intermediaries. This area requires continued research and development to lead to an understanding of how to relate each question to the resource(s) most appropriate to responding to that question. Although the available expert search intermediaries represent various divisions of labor between the human searcher and the computer, there is an emphasis on developing tools for the searcher (machine-aided intelligence) rather than completely replacing the searcher. The growing complexity of the online environment has made the development of "intelligent" online assistants a major R&D area. Various applications have used the AI hardware and software tools that have become available over the past seven years, such as programming languages and expert system shells. There is increasing interest in cognitive science and knowledge engineering and recognition of the need for educational programs to support new roles.

From the vantage point of 1987, a few additional observations can be made. For the moment designers are concentrating on trying to build more intelligent interfaces to existing systems, which require considerable expertise to use unaided. In the future it is likely that new system designs will reflect the increasing familiarity with AI concepts and techniques. There is a growing body of researchers and system developers actively exploring topics in intelligent IR. This chapter has cited representative published reports from the work of several individuals and research groups; work in progress by others should lead to publications that can be the subject of the next review on this topic. One of the noteworthy developments in the past seven years is the emergence of work in libraries and the growing interest in the topic by librarians and information scientists, as evidenced, for example, by the formation of an AI/Expert Systems Group within the Library and Information Technology Association of the American Library Association.

As the discussion of limits of AI made clear, it is necessary to avoid false optimism and unrealistic expectations or risk disenchantment. There is a need for explicit discussion of the forms "intelligent information retrieval" can take, so that new developments in experimental and operational systems can be put within some framework. The definition section and the variety of applications demonstrate the range of possibilities: automated vs. machine-aided, borrowing from AI vs. developing techniques unique to IR, focusing on document retrieval vs. encompassing data and knowledge bases, and systems for experts vs. expert systems. Greater precision in terminology should be coupled with an emphasis on evaluation of performance. The latter is an important part of the tradition in experimental IR and should be carried forward into the era of "intelligent IR," no matter how our understanding of that label evolves:

> Though simply implementing AI techniques in IR [information retrieval] is hard enough, there is the further problem of establishing whether the results are of use. IR system performance evaluation calls for very demanding experiments, presenting many problems of sampling, variable control and interpretation and implying testing on a scale and with a rigor far beyond that ordinarily considered, let alone practiced, in AI.
>
> (SPARCK JONES, 1987, p. 421)

BIBLIOGRAPHY

ADDIS, T. R. 1982. Expert Systems: An Evolution in Information Retrieval. Information Technology: Research and Development. 1982; 1(4): 301–324. ISSN: 0144-817X.

ALURI, RAO. 1987. Artificial Intelligence and Expert Systems. LITA Newsletter. 1987 Winter; 27: 11–12. ISSN: 0196-1799.

AMSLER, R. A. 1987. Literature, AI. In: Shapiro, Stuart C., ed. Encyclopedia of Artificial Intelligence: Volume 1. New York, NY: John Wiley & Sons, 1987. 530–536. ISBN: 0-471-80748-6.

ANDERSON, JAMES D. 1985. Indexing Systems: Extensions of the Mind's Organizing Power. In: Ruben, Brent D., ed. Information and Behavior: Volume 1. New Brunswick, NJ: Transaction Books; 1985. 287–323. ISSN: 0740-5502.

ANDRIOLE, STEPHEN J. 1985. Applied Artificial Intelligence in Perspective. In: Andriole, Stephen J., ed. Applications in Artificial Intelligence. Princeton, NJ: Petrocelli Books; 1985. 5–13. ISBN: 0-89433-219-8.

ASLIB. 1985. Informatics 8: Advances in Intelligent Retrieval: Proceedings of a Conference Jointly Sponsored by Aslib, the Aslib Informatics Group, and the Information Retrieval Specialist Group of the British Computer Society; 1985 April 16–17; Oxford, England. London, England: Aslib; 1985. 314p. ISBN: 0-85142-195-4.

BELKIN, NICHOLAS J. 1986. What Does It Mean for an Information System Interface To Be "Intelligent"? See reference: JACOBSON, CAROL E., WITGES, SHIRLEY A., comps. 97–106.

BELKIN, NICHOLAS J.; CROFT, W. BRUCE. 1987. Retrieval Techniques. In: Williams, Martha E., ed. Annual Review of Information Science and Technology: Volume 22. Amsterdam, The Netherlands: Elsevier Science Publishers B.V. for the American Society for Information Science; 1987. ISBN: 0-444-70302-0.

BELKIN, NICHOLAS J.; VICKERY, ALINA. 1985. Interaction in Information Systems: A Review of Research from Document Retrieval to Knowledge-Based Systems. London, England: British Library; 1985. 250p. (Library and Information Research Report 35). ISSN: 0263-1709; ISBN: 0-7123-3050-X.

BELL, D. A. 1985. An Architecture for Integrating Data, Knowledge, and Information Bases. See reference: ASLIB. 240–257.

BERNSTEIN, LIONEL M.; WILLIAMSON, ROBERT E. 1984. Testing of a Natural Language Retrieval System for a Full Text Knowledge Base. Journal of the American Society for Information Science. 1984 July; 35(4): 235–247. ISSN: 0002-8231.

BIANCHI, G.; GIORGI, M. 1986. Towards an Expert System as Intelligent Assistant for the Design of an Online Documentation Service. In: Proceedings of the 10th International Online Information Meeting; 1986 December 2–4; London, England. Oxford, England: Learned Information; 1986. 199–208. ISBN: 0-904933-57-1.

BISHOP, PETER. 1986. Fifth Generation Computers: Concepts, Implementations and Uses. Chichester, England: Ellis Horwood; 1986. 166p. ISBN: 0-85312-923-1.

BLACK, W. J.; HARGREAVES, P.; MAYES, P. B. 1985. HEADS: A Cataloguing Advisory System. See reference: ASLIB. 227–239.

BOBROW, DANIEL G., HAYES, PATRICK J., eds. 1985. Artificial Intelligence: Where Are We? Artificial Intelligence. 1985 March; 25(3): 375–415. ISSN: 0004-3702.

BOBROW, DANIEL G.; MITTAL, SANJAY; STEFIK, MARK J. 1986. Expert Systems: Perils and Promise. Communications of the ACM. 1986 September; 29(9): 880–894. ISSN: 0001-0782.

BOLTER, J. DAVID. 1984. Turing's Man: Western Culture in the Computer Age. Chapel Hill, NC: University of North Carolina Press; 1984. 264p. ISBN: 0-8078-1564-0.

BORKO, HAROLD. 1985. Artificial Intelligence and Expert Systems Research and Their Possible Impact on Information Science Education. Education for Information. 1985; 3: 103–114. ISSN: 0167-8329.

BORKO, HAROLD. 1987. Getting Started in Library Expert Systems Research. Information Processing & Management. 1987; 23(2): 81–87. ISSN: 0306-4573.

BRAMER, MAX; BRAMER, DAWN. 1984. The Fifth Generation: An Annotated Bibliography. Reading, MA: Addison-Wesley; 1984. 119p. ISBN: 0-201-14427-1.

BRENNER, E. H.; LUCEY, J. H.; MARTINEZ, C. L.; MELEKA, ADEL. 1984. American Petroleum Institute's Machine-Aided Indexing and Searching Project. Science & Technology Libraries. 1984 Fall; 5(1): 49–62. ISSN: 0194-262X.

BRITTAIN, MICHAEL. 1987. Implications for LIS Education of Recent Developments in Expert Systems. Information Processing & Management. 1987; 23(2): 139–152. ISSN: 0306-4573.

BRODIE, MICHAEL L.; MYLOPOULOS, JOHN, eds. 1986. On Knowledge Base Management Systems: Integrating Artificial Intelligence and Database Technologies. New York, NY: Springer-Verlag; 1986. 660p. ISBN: 0-387-96382-0.

BROOKS, HELEN M. 1983. Information Retrieval and Expert Systems–Approaches and Methods of Development. In: Jones, Kevin P., ed. Informatics 7: Intelligent Information Retrieval: Proceedings of a Conference Held by the Aslib Informatics Group and the Information Retrieval Group of the British Computer Society; 1983 March 22–23; Cambridge, England. London, England: Aslib; 1983. 65–75. ISBN: 0-85142-187-5.

BROOKS, HELEN M.; DANIELS, P. J.; BELKIN, NICHOLAS J. 1985. Problem Descriptions and User Models: Developing an Intelligent Interface for Document Retrieval Systems. See reference: ASLIB. 191–214.

BROOKS, HELEN M., DANIELS, P. J.; BELKIN, NICHOLAS J. 1986. Research on Information Interaction and Intelligent Information Provision Mechanisms. Journal of Information Science. 1986; 12(1–2): 37–44. ISSN: 0165-5515.

BRUNELLE, BETTE, MCCLELLAND, BRUCE. 1987. As We May Think, Revisited. In: Williams, Martha E.; Hogan, Thomas H., comps. Proceedings of the 8th National Online Meeting; 1987 May 5–7; New York, NY. Medford, NJ: Learned Information, Inc.; 1987. 41–46. ISBN: 0-938734-17-2.

BUNDY, ALAN, ed. 1986. Catalogue of Artificial Intelligence Tools. 2nd revised edition. New York, NY: Springer-Verlag; 1986. 168p. ISBN: 0-387-16893-1.

BURGER, ROBERT H. 1984. Artificial Intelligence and Authority Control. Library Resources & Technical Services. 1984 October/December; 28(4): 337–345. ISSN: 0024-2527.

CAMPBELL, J. A. 1984. Three Uncertainties of AI. In: Yazdani, Masoud; Narayanan, Ajit, eds. Artificial Intelligence: Human Effects. Chichester, England: Ellis Horwood; 1984. 249–273. ISBN: 0-85312-577-5.

CARBONELL, JAIME G.; MICHALSKI, RYSZARD S.; MITCHELL, TOM M. 1983. An Overview of Machine Learning. See reference: MICHALSKI, RYSZARD S.; CARBONELL, JAIME G.; MITCHELL, TOM M., eds. 3–23.

CAVANAGH, JOSEPH M. A. 1986. Intra-Active Retrieval Systems. In: Williams, Martha E.; Hogan, Thomas H., comps. Proceedings of the 7th National Online Meeting; 1986 May 6–8; New York, NY. Medford, NJ: Learned Information, Inc.; 1986. 59–65. ISBN: 0-938734-12-1.

CAVANAGH, JOSEPH M. A. 1987. The Automated Readers' Advisor: Expert Systems Technology for a Reference Function. In: Williams, Martha E.; Hogan, Thomas H., comps. Proceedings of the 8th National Online Meeting; 1987 May 5–7; New York, NY. Medford, NJ: Learned Information, Inc.; 1987. 57–65. ISBN: 0-938734-17-2.

CERCONE, NICK; MCCALLA, GORDON. 1984. Artificial Intelligence: Underlying Assumptions and Basic Objectives. Journal of the American Society for Information Science. 1984 September; 35(5): 280–290. ISSN: 0002-8231.

CHASE, LESLIE R.; LANDERS, ROBERT K., eds. 1984. AI: Reality or Fantasy? Washington, DC: Information Industry Association; 1984 December. 149p. ISBN: 0-942774-19-1.

CLARKE, ANN; CRONIN, BLAISE. 1983. Expert Systems and Library/Information Work. Journal of Librarianship. 1983 October; 15(4): 277–292. ISSN: 0022-2232.

COOPER, W. S. 1984. Bridging the Gap between AI and IR. See reference: VAN RIJSBERGEN, C. J., ed. 259-265.

COYLE, KAREN; GALLAHER-BROWN, LINDA. 1985. Record Matching: An Expert Algorithm. In: Parkhurst, Carol A., ed. ASIS '85: Proceedings of the American Society for Information Science (ASIS) 48th Annual Meeting: Volume 22; 1985 October 20-24; Las Vegas, NV. White Plains, NY: Knowledge Industry Publications, Inc.; 1985. 77-80. ISSN: 0044-7870; ISBN: 0-86729-176-1.

CREMMINS, EDWARD T. 1982. New Terms for Old: Information and Knowledge Representation. Bulletin of the American Society for Information Science. 1982 October; 9(1): 44. ISSN: 0095-4403.

CROFT, W. BRUCE; THOMPSON, ROGER H. 1986. An Overview of the I^3R Document Retrieval System. See reference: JACOBSON, CAROL E.; WITGES, SHIRLEY A., comps. 123-134.

DANIELS, P. J. 1986. Cognitive Models in Information Retrieval: An Evaluative Review. Journal of Documentation. 1986 December; 42(4): 272-304. ISSN: 0022-0418.

DAVIES, ROY. 1983. Documents, Information or Knowledge? Choices for Librarians. Journal of Librarianship. 1983 January; 15(1): 47-65. ISSN: 0022-2232.

DAVIES, ROY. 1986a. Cataloguing as a Domain for an Expert System. In: Davies, Roy, ed. Intelligent Information Systems: Progress and Prospects. Chichester, England: Ellis Horwood; 1986. 54-77. ISBN: 0-85312-896-0.

DAVIES, ROY, ed. 1986b. Intelligent Information Systems: Progress and Prospects. Chichester, England: Ellis Horwood; 1986. 300p. ISBN: 0-85312-896-0.

DAVIES, ROY. 1987. Outlines of the Emerging Paradigm in Cataloguing. Information Processing & Management. 1987; 23(2): 89-98. ISSN: 0306-4573.

DAVIES, ROY; JAMES, BRIAN. 1984. Towards an Expert System for Cataloguing: Some Experiments Based on AACR2. Program. 1984 October; 18(4): 283-297. ISSN: 0033-0337.

DAVIS, DWIGHT B. 1987. Artificial Intelligence Goes to Work. High Technology. 1987 April; 7(4): 16-27. ISSN: 0277-2981.

DAVIS, ERNEST. 1987. Limits and Inadequacies in Artificial Intelligence. In: Davis, Philip J.; Park, David, eds. No Way: The Nature of the Impossible. New York, NY: W. H. Freeman and Company; 1987. 90-109. ISBN: 0-7167-1813-8.

DAVIS, RANDALL. 1986. Knowledge-Based Systems. Science. 1986 February 28; 231(4741): 957-963. ISSN: 0036-8075.

DAYTON, D. L.; LUNDEEN, J. W.; POLLOCK, J. J. 1980. Automated Techniques for Online Search Guidance: A Review. In: Proceedings of the 4th International Online Information Meeting; 1980 December 9-11; London, England. Oxford, England: Learned Information; 1980. 317-333. ISBN: 0-904933-28-8.

DEDE, CHRISTOPHER. 1986. A Review and Synthesis of Recent Research in Intelligent Computer-Assisted Instruction. International Journal of Man-Machine Studies. 1986 April; 24(4): 329-353. ISSN: 0020-7373.

DEFUDE, B. 1984. Knowledge Based Systems versus Thesaurus: An Architecture Problem About Expert Systems Design. See reference: VAN RIJSBERGEN, C. J., ed. 267-280.

DEFUDE, B. 1985. Different Levels of Expertise for an Expert System in Information Retrieval. In: Research and Development in Information Retrieval: [Proceedings of the] Association for Computing Machinery, Special Interest Group on Information Retrieval (ACM SIGIR) 8th Annual International Conference; 1985 June 5-7; Montreal, Canada. New York, NY: ACM, Inc.; 1985. 147-153. ISBN: 0-89791-159-8.

DEJONG, GERALD. 1983. Artificial Intelligence Implications for Information Retrieval. In: Kuehn, Jennifer J., ed. Research and Development in Information Retrieval: [Proceedings of the] Association for Computing Machinery, Special Interest Group on Information Retrieval (ACM SIGIR) 6th Annual International Conference; 1983 June 6-8; Bethesda, MD. New York, NY: ACM, Inc.; 1983. 10-17. ISBN: 0-89791-107-5.

DESALVO, DANIEL A.; LIEBOWITZ, JAY. 1986. The Application of an Expert System for Information Retrieval at the National Archives. Telematics and Informatics. 1986; 3(1): 25-38. ISSN: 0736-5853.

DOSZKOCS, TAMAS E. 1986. Natural Language Processing in Information Retrieval. Journal of the American Society for Information Science. 1986 July; 37(4): 191-196. ISSN: 0002-8231.

DRAY, J. 1987. Social Issues of AI. In: Shapiro, Stuart C., ed. Encyclopedia of Artificial Intelligence: Volume 2. New York, NY: John Wiley & Sons; 1987. 1049-1060. ISBN: 0-471-80748-6.

DREYFUS, HUBERT L.; DREYFUS, STUART E.; ATHANASIOU, TOM. 1986. Mind Over Machine: The Power of Human Intuition and Expertise in the Era of the Computer. New York, NY: Free Press; 1986. 231p. ISBN: 0-02-908060-6.

EASTMAN, C. M. 1983. The Use of Logic Programming in Information Retrieval Experimentation. In: Vondran, Raymond F.; Caputo, Anne; Wasserman, Carol, Diener, Richard A. V., eds. Productivity in the Information Age: Proceedings of the American Society for Information Science (ASIS) 46th Annual Meeting: Volume 20; 1983 October 2-6; Washington, DC. White Plains, NY: Knowledge Industry Publications, Inc; 1983. 58-59. ISSN: 0044-7870; ISBN: 0-86729-072-2.

ERCEGOVAC, ZORANA. 1984. Knowledge-Based Expert Systems: A Profile and Implications. In: Williams, Martha E.; Hogan, Thomas H., comps. Proceedings of the 5th National Online Meeting; 1984 April 10-12; New York, NY. Medford, NJ: Learned Information, Inc.; 1984. 39-46. ISBN: 0-938734-07-5.

ERCEGOVAC, ZORANA. 1986. Artificial Intelligence and Cataloging Reasoning: Promises and Pitfalls. In: Williams, Martha E.; Hogan, Thomas H., comps. Proceedings of the 7th National Online Meeting; 1986 May 6-8; New York, NY. Medford, NJ: Learned Information, Inc.; 1986. 109-117. ISBN: 0-938734-12-1.

ERICSSON, K. ANDERS; SIMON, HERBERT A. 1984. Protocol Analysis: Verbal Reports as Data. Cambridge, MA: MIT Press; 1984. 426p. ISBN: 0-262-05029-3.

FIDEL, RAYA. 1986. Towards Expert Systems for the Selection of Search Keys. Journal of the American Society for Information Science. 1986 January; 37(1): 37-44. ISSN: 0002-8231.

FINN, V. K. 1984. Information Systems and Problems of Upgrading Their Intelligence Level. Automatic Documentation and Mathematical Linguistics. 1984; 18(1): 1-24. ISSN: 0005-1055.

FIRST, MICHAEL B.; SOFFER, LYNN J.; MILLER, RANDOLPH A. 1985. QUICK (QUick Index to Caduceus Knowledge): Using the Internist-1/ Caduceus Knowledge Base as an Electronic Textbook of Medicine. Computers and Biomedical Research. 1985; 18: 137-165. ISSN: 0010-4809.

FORSYTH, RICHARD; RADA, ROY. 1986. Machine Learning: Applications in Expert Systems and Information Retrieval. Chichester, England: Ellis Horwood; 1986. 277p. ISBN: 0-85312-947-9.

FOX, EDWARD A. 1986a. A Design for Intelligent Retrieval: The CODER System. See reference: JACOBSON, CAROL E.; WITGES, SHIRLEY A., comps. 135-153.

FOX, EDWARD A. 1986b. Expert Retrieval for Users of Computer Based Message Systems. In: Hurd, Julie M., ed. ASIS '86: Proceedings of the American Society for Information Science (ASIS) 49th Annual Meeting: Volume 23; 1986 September 28-October 2; Chicago, IL. Medford, NJ: Learned Information, Inc.; 1986. 88-95. ISSN: 0044-7870; ISBN: 0-938734-14-8.

FRAKES, W. B. 1984. Term Conflation for Information Retrieval. See reference: VAN RIJSBERGEN, C. J., ed. 383-389.

FRAKES, W. B.; KATZER, JEFFREY; MCGILL, MICHAEL; TESSIER, JUDITH A.; DASGUPTA, PADMINI. 1981. A Study of the Impact of Representations in Information Retrieval Systems. In: Lunin, Lois F.; Henderson, Madeline; Wooster, Harold, eds. The Information Community: An Alliance for Progress: Proceedings of the American Society for Information Science (ASIS) 44th Annual Meeting: Volume 18; 1981 October 25-30; Washington, DC. White Plains, NY: Knowledge Industry Publications, Inc.; 1981. 301-303. ISSN: 0044-7870; ISBN: 0-914236-85-7.

GARDNER, HOWARD. 1985. The Mind's New Science: A History of the Cognitive Revolution. New York, NY: Basic Books, Inc.; 1985. 423p. ISBN: 0-465-04634-7.

GEBHARDT, FRIEDRICH. 1985. Querverbindungen zwischen Information-Retrieval- und Experten-Systemen [Connections between Information Retrieval and Expert Systems]. Nachrichten für Dokumentation. 1985; 36(6): 255-263. ISSN: 0027-7436.

GIBB, FORBES, ed. 1986. Expert Systems in Libraries: Proceedings of a Conference of the Library Association Information Technology Group and the Library and Information Research Group; 1985 November 8; Birmingham, England. London, England: Taylor Graham; 1986. 97p. ISBN: 0-947568-10-7.

GORDON, MICHAEL D. 1985. A Learning Algorithm Applied to Document Redescription. In: Research and Development in Information Retrieval: [Proceedings of the] Association for Computing Machinery, Special Interest Group on Information Retrieval (ACM SIGIR) 8th Annual International Conference; 1985 June 5-7; Montreal, Canada. New York, NY: ACM, Inc.; 1985. 179-186. ISBN: 0-89791-159-8.

GRIFFITH, BELVER C. 1981. Introduction to Perspectives on Cognition: Human Information Processing. Journal of the American Society for Information Science. 1981 September; 32(5): 344-346. ISSN: 0002-8231.

GRISHMAN, RALPH. 1984. Natural Language Processing. Journal of the
    American Society for Information Science. 1984 September; 35(5):
    291–296. ISSN: 0002-8231.
GUIDA, GIOVANNI; TASSO, CARLO. 1983. An Expert Intermediary
    System for Interactive Document Retrieval. Automatica. 1983; 19(6):
    759–766. ISSN: 0005-1098.
HAHN, UDO; REIMER, ULRICH. 1984. Heuristic Text Parsing in "TOPIC":
    Methodological Issues in a Knowledge-Based Text Condensation System.
    In: Dietschmann, Hans J., ed. Representation and Exchange of Knowl-
    edge as a Basis of Information Processes: Proceedings of the 5th Inter-
    national Research Forum in Information Science (IRFIS 5); 1983
    September 5-7; Heidelberg, Germany. New York, NY: North-Holland;
    1984. 143–163. ISBN: 0-444-87563-8.
HARRIS, LARRY R.; DAVIS, DWIGHT B. 1986. Artificial Intelligence
    Enters the Marketplace. New York, NY: Bantam Books; 1986. 194p.
    ISBN: 0-553-34293-2.
HARRIS, S. D.; HELANDER, M. G. 1984. Machine Intelligence in Real
    Systems: Some Ergonomics Issues. In: Salvendy, G., ed. Human-
    Computer Interaction: Proceedings of the 1st U.S.A.-Japan Conference
    on Human-Computer Interaction; 1984 August 18–20; Honolulu, HI.
    Amsterdam, The Netherlands: Elsevier; 1984. 267–277. ISBN:
    0-444-42395-8.
HARTER, STEPHEN P.; PETERS, ANNE ROGERS. 1985. Heuristics for
    Online Information Retrieval: A Typology and Preliminary Listing.
    Online Review. 1985 October; 9(5): 407–424. ISSN: 0309-314X.
HAUGELAND, JOHN. 1985. Artificial Intelligence: The Very Idea. Cam-
    bridge, MA: MIT Press; 1985. 287p. ISBN: 0-262-08153-9.
HAYES-ROTH, FREDERICK. 1984. The Knowledge-Based Expert System:
    A Tutorial. Computer. 1984 September; 17(9): 11–28. ISSN: 0018-
    9162.
HEILPRIN, LAURENCE B., ed. 1985. Toward Foundations of Information
    Science. White Plains, NY: Knowledge Industry Publications, Inc.; 1985.
    229p. ISBN: 0-86729-149-4.
HERTHER, NANCY K. 1986. PROLOG to the Future: A Glimpse of
    Things to Come in Artificial Intelligence. Microcomputers for Informa-
    tion Management. 1986 March; 3(1): 31–45. ISSN: 0742-2342.
HICE, GERALD F.; ANDRIOLE, STEPHEN J. 1985. Artificially Intelligent
    Videotex. In: Andriole, Stephen J., ed. Applications in Artificial Intelli-
    gence. Princeton, NJ: Petrocelli Books; 1985. 295–309. ISBN:
    0-89433-219-8.
HILKER, EMERSON. 1985. Artificial Intelligence: A Review of Current
    Information Sources. Collection Building. 1985; 7(3): 14–30. ISSN:
    0160-4953.
HJERPPE, ROLAND. 1983. What Artificial Intelligence Can, Could, and
    Can't Do for Libraries and Information Services. In: Proceedings of the
    7th International Online Information Meeting; 1983 December 6-8;
    London, England. Oxford, England: Learned Information; 1983. 7–
    25. ISBN: 0-904933-42-3.
HJERPPE, ROLAND; OLANDER, BIRGITTA; MARKLUND, KARI. 1985.
    Project ESSCAPE—Expert Systems for Simple Choice of Access Points
    for Entries: Applications of Artificial Intelligence in Cataloging.
    Linköping, Sweden: LIBLAB; 1985. 17p. (Paper presented at IFLA
    51st General Conference; 1985 August 18-24; Chicago, IL). ERIC: ED
    262 796.

HUMPHREY, SUSANNE M. 1984. We Need to Bridge the Gap Between Computer and Information Science. Bulletin of the American Society for Information Science. 1984 December; 11(2): 25-27. ISSN: 0095-4403.

HUMPHREY, SUSANNE M.; MILLER, NANCY E. 1987. Knowledge-Based Indexing of the Medical Literature: The Indexing Aid Project. Journal of the American Society for Information Science. 1987 May; 38(3): 184-196. ISSN: 0002-8231.

HUSHON, JUDITH M.; CONRY, THOMAS J. 1986. The Evolution of Gateway Technology. See reference: JACOBSON, CAROL E.; WITGES, SHIRLEY A., comps. 181-201.

INGWERSEN, PETER. 1986. Cognitive Analysis and the Role of the Intermediary in Information Retrieval. In: Davies, Roy, ed. Intelligent Information Systems: Progress and Prospects. Chichester, England: Ellis Horwood; 1986. 206-237. ISBN: 0-85312-896-0.

JACOB, R. 1987. Human-Computer Interaction. In: Shapiro, Stuart C., ed. Encyclopedia of Artificial Intelligence: Volume 1. New York, NY: John Wiley & Sons; 1987. 383-388. ISBN: 0-471-80748-6.

JACOBSON, CAROL E.; WITGES, SHIRLEY A., comps. 1986. Proceedings of the 2nd Conference on Computer Interfaces and Intermediaries for Information Retrieval; 1986 May 28-31; Boston, MA. Alexandria, VA: Office of Information Systems and Technology, Defense Technical Information Center; 1986. 371p. NTIS: AD A174-000-0. Available from: Office of Information Systems and Technology, Defense Technical Information Center, Cameron Station, Alexandria, VA 22304-6145.

JAKOBSON, GABRIEL. 1986. Toward Expert Database Front-Ends. See reference: JACOBSON, CAROL E.; WITGES, SHIRLEY A., comps. 317-318.

JÄRVELIN, KALERVO; REPO, AATTO J. 1984. A Taxonomy of Knowledge Work Support Tools. In: Flood, Barbara; Witiak, Joanne; Hogan, Thomas H., comps. 1984: Challenges to an Information Society: Proceedings of the American Society for Information Science (ASIS) 47th Annual Meeting: Volume 21; 1984 October 21-25; Philadelphia, PA. White Plains, NY: Knowledge Industry Publications, Inc.; 1984. 59-62. ISSN: 0044-7870; ISBN: 0-86729-115-X.

JOHNSON, GEORGE. 1986. Machinery of the Mind: Inside the New Science of Artificial Intelligence. New York, NY: Times Books; 1986. 336p. ISBN: 0-8129-1229-2.

JONES, KEVIN P., ed. 1983. Informatics 7: Intelligent Information Retrieval: Proceedings of a Conference Held by the Aslib Informatics Group and the Information Retrieval Group of the British Computer Society; 1983 March 22-23; Cambridge, England. London, England: Aslib; 1983. 149p. ISBN: 0-85142-187-5.

JONES, KEVIN P. 1984. The Effects of Expert and Allied Systems on Information Handling: Some Scenarios. Aslib Proceedings. 1984 May; 36(5): 213-217. ISSN: 0001-253X.

JONES, KEVIN P. 1985. Ten Years of Informatics. See reference: ASLIB. 263-307.

KANTOR, PAUL B. 1986. Information Retrieval Issues in the Design of Expert Systems. In: Hurd, Julie M., ed. ASIS '86: Proceedings of the American Society for Information Science (ASIS) 49th Annual Meeting: Volume 23; 1986 September 28-October 2; Chicago, IL. Medford, NJ: Learned Information, Inc.; 1986. 113-117. ISSN: 0044-7870; ISBN: 0-938734-14-8.

KAPLAN, S. JERROLD. 1984. The Industrialization of Artificial Intelligence: From By-Line to Bottom Line. AI Magazine. 1984 Summer; 5(2): 51–57. ISSN: 0738-4602.

KEHOE, CYNTHIA A. 1985. Interfaces and Expert Systems for Online Retrieval. Online Review. 1985 December; 9(6): 489–505. ISSN: 0309-314X.

KERSCHBERG, LARRY, ed. 1986. Expert Database Systems: Proceedings from the 1st International Workshop; 1984 October 24–27; Kiawah Island, SC. Menlo Park, CA: Benjamin/Cummings Publishing Company; 1986. 701p. ISBN: 0-8053-3270-7.

KINNELL, SUSAN K.; CHIGNELL, MARK H. 1987. Who's the Expert?: Conceptual Representation of Knowledge for End User Searching. In: Williams, Martha E.; Hogan, Thomas H., comps. Proceedings of the 8th National Online Meeting; 1987 May 5–7; New York, NY. Medford, NJ: Learned Information, Inc.; 1987. 237–243. ISBN: 0-938734-17-2.

KLEINBART, PAUL. 1985. Prolegomenon to "Intelligent" Thesaurus Software. Journal of Information Science. 1985; 11(2): 45–53. ISSN: 0165-5515.

KOLODNER, JANET L. 1983. Indexing and Retrieval Strategies for Natural Language Fact Retrieval. ACM Transactions on Database Systems. 1983 September; 8(3): 434–464. ISSN: 0362-5915.

KORFHAGE, ROBERT R. 1985. Intelligent Information Retrieval: Issues in User Modelling. In: Karna, Kamal N., ed. Expert Systems in Government Symposium. Washington, DC: IEEE Computer Society Press; 1985. 474–482. ISBN: 0-8186-8686-3.

KRAWCZAK, DEB; SMITH, PHILIP J.; SHUTE, STEVEN J. 1987. EP-X: A Demonstration of Semantically-Based Search of Bibliographic Databases. In: Yu, C. T.; Van Rijsbergen, C. J., eds. Proceedings of the Association for Computing Machinery, Special Interest Group on Information Retrieval (ACM SIGIR) 10th Annual International Conference on Research & Development in Information Retrieval; 1987 June 3–5; New Orleans, LA. New York, NY: ACM, Inc.; 1987. 263–271. ISBN: 0-89791-232-2.

KUHLEN, RAINER. 1984. Some Similarities and Differences Between Intellectual and Machine Text Understanding for the Purpose of Abstracting. In: Dietschmann, Hans J., ed. Representation and Exchange of Knowledge as a Basis of Information Processes: Proceedings of the 5th International Research Forum in Information Science (IRFIS 5); 1983 September 5–7; Heidelberg, Germany. New York, NY: North-Holland; 1984. 87–109. ISBN: 0-444-87563-8.

KURZWEIL, RAYMOND. 1985. What Is Artificial Intelligence Anyway? American Scientist. 1985 May–June; 73(3): 258–264. ISSN: 0003-0996.

LEBOWITZ, MICHAEL. 1983. Intelligent Information Systems. In: Kuehn, Jennifer J., ed. Research and Development in Information Retrieval: [Proceedings of the] Association for Computing Machinery, Special Interest Group on Information Retrieval (ACM SIGIR) 6th Annual International Conference; 1983 June 6–8; Bethesda, MD. New York, NY: ACM, Inc.; 1983. 25–30. ISBN: 0-89791-107-5.

LEBOWITZ, MICHAEL. 1986. An Experiment in Intelligent Information Systems: RESEARCHER. In: Davies, Roy, ed. Intelligent Information

Systems: Progress and Prospects. Chichester, England: Ellis Horwood; 1986. 127–149. ISBN: 0-85312-896-0.

LESK, MICHAEL. 1984. Programming Languages for Text and Knowledge Processing. In: Williams, Martha E., ed. Annual Review of Information Science and Technology: Volume 19. White Plains, NY: Knowledge Industry Publications, Inc. for the American Society for Information Science; 1984. 97–128. ISSN: 0066-4200; ISBN: 0-86729-093-5.

MACHLUP, FRITZ; MANSFIELD, UNA, eds. 1983. The Study of Information: Interdisciplinary Messages. New York, NY: John Wiley & Sons; 1983. 743p. ISBN: 0-471-88717-X.

MARCUS, RICHARD S. 1983. An Experimental Comparison of the Effectiveness of Computers and Humans as Search Intermediaries. Journal of the American Society for Information Science. 1983 November; 34(6): 381–404. ISSN: 0002-8231.

MARCUS, RICHARD S. 1986. Design Questions in the Development of Expert Systems for Retrieval Assistance. In: Hurd, Julie M., ed. ASIS '86: Proceedings of the American Society for Information Science (ASIS) 49th Annual Meeting: Volume 23; 1986 September 28–October 2; Chicago, IL. Medford, NJ: Learned Information, Inc.; 1986. 185–189. ISSN: 0044-7870; ISBN: 0-938734-14-8.

MAVOR, A. S.; KIDD, J. S.; VAUGHAN, W. S., JR. 1981. Cognitive Models of Scientific Work and Their Implications for the Design of Knowledge Delivery Systems. Annapolis, MD: W/V Associates; 1981 October. 46p. ERIC: ED 213 409.

MEADOW, CHARLES T.; HEWETT, THOMAS T.; AVERSA, ELIZABETH S. 1982a. A Computer Intermediary for Interactive Database Searching. I. Design. Journal of the American Society for Information Science. 1982 September; 33(5): 325–332. ISSN: 0002-8231.

MEADOW, CHARLES T.; HEWETT, THOMAS T.; AVERSA, ELIZABETH S. 1982b. A Computer Intermediary for Interactive Database Searching. II. Evaluation. Journal of the American Society for Information Science. 1982 November; 33(6): 357–364. ISSN: 0002-8231.

MICCO, H. MARY; SMITH, IRMA. 1986. Designing an Expert System for the Reference Function Subject Access to Information. In: Hurd, Julie M., ed. ASIS '86: Proceedings of the American Society for Information Science (ASIS) 49th Annual Meeting: Volume 23; 1986 September 28–October 2; Chicago, IL. Medford, NJ: Learned Information, Inc.; 1986. 204–210. ISSN: 0044-7870; ISBN: 0-938734-14-8.

MICHALSKI, RYSZARD S.; CARBONELL, JAIME G.; MITCHELL, TOM M., eds. 1983. Machine Learning: An Artificial Intelligence Approach. Palo Alto, CA: Tioga Publishing Company; 1983. ISBN: 0-935382-05-4.

MICHIE, DONALD. 1982. The Social Aspects of Artificial Intelligence. In: Michie, Donald. Machine Intelligence and Related Topics: An Information Scientist's Weekend Book. New York, NY: Gordon and Breach Science Publishers; 1982. 278–305. ISBN: 0-677-05560-9.

MICHIE, DONALD; JOHNSTON, RORY. 1984. The Creative Computer: Machine Intelligence and Human Knowledge. Harmondsworth, England: Penguin Books; 1984. 263p. ISBN: 0-14-022465-3.

MIKSA, FRANCIS L. 1985. Machlup's Categories of Knowledge as a Framework for Viewing Library and Information Science History. Journal of Library History. 1985 Spring; 20(2): 157-172. ISSN: 0275-3650.

MILLS, WILLIAM J. 1987. AI: Just How Scattered Is the Literature? An Online Investigation. SIGART Newsletter. 1987 January; 99: 24-26. Available from: ACM Order Dept., P.O. Box 64145, Baltimore, MD 21264.

MITEV, NATHALIE NADIA; WALKER, STEPHEN. 1985. Information Retrieval Aids in an Online Public Access Catalogue: Automatic Intelligent Search Sequencing. See reference: ASLIB. 215-226.

MOLHOLT, PAT. 1986. The Information Machine: A New Challenge for Librarians. Library Journal. 1986 October 1; 111(16): 47-52. ISSN: 0363-0277.

MONTGOMERY, CHRISTINE A. 1981. Where Do We Go from Here? In: Oddy, R. N.; Robertson, S. E.; Van Rijsbergen, C. J.; Williams, P. W., eds. Information Retrieval Research. London, England: Butterworths; 1981. 370-385. ISBN: 0-408-10775-8.

NEDOBITY, WOLFGANG. 1985. Terminology and Artificial Intelligence. International Classification. 1985; 12(1): 17-19. ISSN: 0340-0050.

NEWELL, ALLEN. 1983. Intellectual Issues in the History of Artificial Intelligence. In: Machlup, Fritz; Mansfield, Una, eds. The Study of Information: Interdisciplinary Messages. New York, NY: John Wiley & Sons; 1983. 187-227. ISBN: 0-471-88717-X.

NOWAK, ELZBIETA J.; SZABLOWSKI, BOGUMIŁ F. 1984. Expert Systems in Scientific Information Exchange. Journal of Information Science. 1984; 8(3): 103-111. ISSN: 0165-5515.

OBERMEIER, KLAUS K.; COOPER, LINDA E. 1984. Information Network Facility Organizing System (INFOS)—An Expert System for Information Retrieval. In: Flood, Barbara; Witiak, Joanne; Hogan, Thomas H., comps. 1984: Challenges to an Information Society: Proceedings of the American Society for Information Science (ASIS) 47th Annual Meeting: Volume 21; 1984 October 21-25; Philadelphia, PA. White Plains, NY: Knowledge Industry Publications, Inc.; 1984. 95-98. ISSN: 0044-7870; ISBN: 0-86729-115-X.

OPPENHEIM, CHARLES. 1986. Impact of New Technology on the Information Professional. International Journal of Micrographics & Video Technology. 1986; 5(1): 19-24. ISSN: 0743-9636.

PAICE, CHRIS. 1986. Expert Systems for Information Retrieval? Aslib Proceedings. 1986 October; 38(10): 343-353. ISSN: 0001-253X.

PARROTT, JAMES R. 1986. Expert Systems for Reference Work. Microcomputers for Information Management. 1986 September; 3(3): 155-171. ISSN: 0742-2342.

PENZIAS, ARNO. 1985. Technology in the Coming Century. In: Traub, Joseph F., ed. Cohabiting with Computers. Los Altos, CA: William Kaufmann, Inc.; 1985. 127-143. ISBN: 0-86576-079-9.

PIETILÄINEN, PIRKKO. 1983. Local Feedback and Intelligent Automatic Query Expansion. Information Processing & Management. 1983; 19(1): 51-58. ISSN: 0306-4573.

POLLITT, A. S. 1986. A Rule-Based System as an Intermediary for Searching Cancer Therapy Literature on MEDLINE. In: Davies, Roy, ed. Intelligent Information Systems: Progress and Prospects. Chichester, England: Ellis Horwood; 1986. 82-126. ISBN: 0-85312-896-0.

POLLITT, A. S. 1987. CANSEARCH: An Expert Systems Approach to Document Retrieval. Information Processing & Management. 1987; 23(2): 119–138. ISSN: 0306-4573.

RADA, ROY. 1987. Knowledge-Sparse and Knowledge-Rich Learning in Information Retrieval. Information Processing & Management. 1987; 23(3): 195–210. ISSN: 0306-4573.

RICH, ELAINE. 1984. The Gradual Expansion of Artificial Intelligence. Computer. 1984 May; 17(5): 4–12. ISSN: 0018-9162.

RICH, ELAINE. 1986. Users are Individuals: Individualizing User Models. In: Davies, Roy, ed. Intelligent Information Systems: Progress and Prospects. Chichester, England: Ellis Horwood; 1986. 184–201. ISBN: 0-85312-896-0.

RODGERS, KAY, comp. 1986. LC Science Tracer Bullet: Artificial Intelligence. Washington, DC: Science Reference Section, Science and Technology Division, Library of Congress; 1986 January. 17p. ERIC: ED 271 107. Also available from: Science Reference Section, Science and Technology Division, Library of Congress, 10 First Street SE, Washington DC 20540.

SAGER, NAOMI; KOSAKA, MICHIKO. 1983. A Database of Literature Organized by Relations. In: Dayhoff, Ruth E., ed. Proceedings of the 7th Annual Symposium on Computer Applications in Medical Care; 1983 October 23-26; Washington, DC. Silver Spring, MD: IEEE Computer Society Press; 1983. 692–695. ISSN: 0195-4210; ISBN: 0-8186-8503-4.

SALTON, GERARD. 1985. Some Characteristics of Future Information Systems. SIGIR (Association for Computing Machinery) Forum. 1985 Fall; 18(2–4): 28–39.

SALTON, GERARD. 1986a. Another Look at Automatic Text-Retrieval Systems. Communications of the ACM. 1986 July; 29(7): 648–656. ISSN: 0001-0782.

SALTON, GERARD. 1986b. On the Use of Knowledge-Based Processing in Automatic Text Retrieval. In: Hurd, Julie M., ed. ASIS '86: Proceedings of the American Society for Information Science (ASIS) 49th Annual Meeting: Volume 23; 1986 September 28–October 2; Chicago, IL. Medford, NJ: Learned Information, Inc.; 1986. 277–287. ISSN: 0044-7870; ISBN: 0-938734-14-8.

SCHANK, ROGER C.; CHILDERS, PETER. 1984. The Cognitive Computer: On Language, Learning, and Artificial Intelligence. Reading, MA: Addison-Wesley Publishing Company, Inc.; 1984. 268p. ISBN: 0-201-06443-X.

SCHANK, ROGER C.; KOLODNER, JANET L.; DEJONG, GERALD. 1981. Conceptual Information Retrieval. In: Oddy, R. N.; Robertson, S. E.; Van Rijsbergen, C. J.; Williams, P. W., eds. Information Retrieval Research. London, England: Butterworths; 1981. 94–116. ISBN: 0-408-10775-8.

SCHWARTZ, J. 1987. Limits of Artificial Intelligence. In: Shapiro, Stuart C., ed. Encyclopedia of Artificial Intelligence: Volume 1. New York, NY: John Wiley & Sons; 1987. 488–503. ISBN: 0-471-80748-6.

SHAPIRO, STUART C., ed. 1987. Encyclopedia of Artificial Intelligence. New York, NY: John Wiley & Sons; 1987. 2 volumes. ISBN: 0-471-80748-6.

SHORE, JOHN. 1985. The Sachertorte Algorithm and Other Antidotes to Computer Anxiety. New York, NY: Viking; 1985. 270p. ISBN: 0-670-80541-6.

SHOVAL, PERETZ. 1983. Knowledge Representation in Consultation Systems for Users of Retrieval Systems. In: Keren, C.; Perlmutter, L., eds. The Application of Mini- and Micro-Computers in Information, Documentation and Libraries: Proceedings of the International Conference on the Application of Mini- and Micro-Computers in Information, Documentation and Libraries; 1983 March 13–18; Tel Aviv, Israel. Amsterdam, The Netherlands: North-Holland; 1983. 631–643. ISBN: 0-444-86767-8.

SHOVAL, PERETZ. 1985. Principles, Procedures and Rules in an Expert System for Information Retrieval. Information Processing & Management. 1985; 21(6): 475–487. ISSN: 0306-4573.

SHOVAL, PERETZ. 1986. Comparison of Decision Support Strategies in Expert Consultation Systems. International Journal of Man–Machine Studies. 1986 February; 24(2): 125–139. ISSN: 0020-7373.

SIEGEL, ELLIOT R. 1982. Transfer of Information to Health Practitioners. In: Dervin, Brenda; Voigt, Melvin J. Progress in Communication Sciences: Volume III. Norwood, NJ: Ablex Publishing Corporation; 1982. 311–334. ISSN: 0163-5689; ISBN: 0-89391-081-3.

SIMON, HERBERT A. 1981. The Sciences of the Artificial. 2nd edition. Cambridge, MA: MIT Press; 1981. 247p. ISBN: 0-262-19193-8.

SLOCUM, JONATHAN; BICHTELER, JULIE; AMSLER, ROBERT. 1980. Inference Nets for Modeling Geoscience Reference Knowledge. In: Benenfeld, Alan R.; Kazlauskas, Edward John, eds. Communicating Information: Proceedings of the American Society for Information Science (ASIS) 43rd Annual Meeting: Volume 17; 1980 October 5–10; Anaheim, CA. White Plains, NY: Knowledge Industry Publications, Inc.; 1980. 183–185. ISSN: 0044-7870; ISBN: 0-914236-73-3.

SLOMAN, AARON. 1983. An Overview of Some Unsolved Problems in Artificial Intelligence. In: Jones, Kevin P., ed. Informatics 7: Intelligent Information Retrieval: Proceedings of a Conference Held by the Aslib Informatics Group and the Information Retrieval Group of the British Computer Society; 1983 March 22–23; Cambridge, England. London, England: Aslib; 1983. 3–14. ISBN: 0-85142-187-5.

SMITH, KAREN F. 1986. Robot at the Reference Desk? College & Research Libraries. 1986 September; 47(5): 486–490. ISSN: 0010-0870.

SMITH, LINDA C. 1980. Artificial Intelligence Applications in Information Systems. In: Williams, Martha E., ed. Annual Review of Information Science and Technology: Volume 15. White Plains, NY: Knowledge Industry Publications, Inc. for the American Society for Information Science; 1980. 67–105. ISSN: 0066-4200; ISBN: 0-914236-65-2.

SMITH, LINDA C. 1985. A Guide to Information Sources in Artificial Intelligence. Science & Technology Libraries. 1985 Spring; 5(3): 79–100. ISSN: 0194-262X.

SMITH, LINDA C. 1986a. Knowledge-Based Systems, Artificial Intelligence and Human Factors. In: Ingwersen, Peter; Kajberg, Leif; Pejtersen, Annelise Mark, eds. Information Technology and Information Use: Towards a Unified View of Information and Information Technology. London, England: Taylor Graham; 1986. 98–110. ISBN: 0-947568-06-9.

SMITH, LINDA C. 1986b. Machine Intelligence vs. Machine-Aided Intelligence As a Basis for Interface Design. See reference: JACOBSON, CAROL E.; WITGES, SHIRLEY A., comps. 107–112.

SMITH, LINDA C.; WARNER, AMY J. 1984. A Taxonomy of Representations in Information Retrieval System Design. Journal of Information Science. 1984; 8(3): 113–121. ISSN: 0165-5515.

SMITH, PHILIP J.; CHIGNELL, MARK. 1984. Development of an Expert System to Aid in Searches of the Chemical Abstracts. In: Flood, Barbara; Witiak, Joanne; Hogan, Thomas H., comps. 1984: Challenges to an Information Society: Proceedings of the American Society for Information Science (ASIS) 47th Annual Meeting: Volume 21; 1984 October 21–25; Philadelphia, PA. White Plains, NY: Knowledge Industry Publications, Inc.; 1984. 99–102. ISSN: 0044-7870; ISBN: 0-86729-115-X.

SOWIZRAL, HENRY A. 1985. Expert Systems. In: Williams, Martha E., ed. Annual Review of Information Science and Technology: Volume 20. White Plains, NY: Knowledge Industry Publications, Inc. for the American Society for Information Science; 1985. 179–199. ISSN: 0066-4200; ISBN: 0-86729-175-3.

SPARCK JONES, KAREN. 1983a. Intelligent Retrieval. In: Jones, Kevin P., ed. Informatics 7: Intelligent Information Retrieval: Proceedings of a Conference Held by the Aslib Informatics Group and the Information Retrieval Group of the British Computer Society; 1983 March 22–23; Cambridge, England. London, England: Aslib; 1983. 136–142. ISBN: 0-85142-187-5.

SPARCK JONES, KAREN. 1983b. Programmes in Advanced Information Technology. Journal of Documentation. 1983 June; 39(2): 85–87. ISSN: 0022-0418.

SPARCK JONES, KAREN. 1985. Issues in User Modelling for Expert Systems. In: Cohn, A. G.; Thomas, J. R., eds. Artificial Intelligence and Its Applications. Chichester, England: John Wiley & Sons; 1985. 183–195. ISBN: 0-471-91175-5.

SPARCK JONES, KAREN. 1987. Information Retrieval. In: Shapiro, Stuart C., ed. Encyclopedia of Artificial Intelligence: Volume 1. New York, NY: John Wiley & Sons; 1987. 419–421. ISBN: 0-471-80748-6.

STAUGAARD, ANDREW C., JR. 1987. Robotics and AI: An Introduction to Applied Machine Intelligence. Englewood Cliffs, NJ: Prentice Hall, Inc.; 1987. 373p. ISBN: 0-13-782269-3.

STEFIK, MARK. 1986. The Next Knowledge Medium. AI Magazine. 1986 Spring; 7(1): 34–46. ISSN: 0738-4602.

THIEL, ULRICH; HAMMWÖHNER, RAINER. 1987. Information Zooming: An Interaction Model for the Graphical Access to Text Knowledge Bases. In: Yu, C. T.; Van Rijsbergen, C. J., eds. Proceedings of the Association for Computing Machinery, Special Interest Group on Information Retrieval (ACM SIGIR) 10th Annual International Conference on Research & Development in Information Retrieval; 1987 June 3–5; New Orleans, LA. New York, NY: ACM, Inc.; 1987. 45–56. ISBN: 0-89791-232-2.

THOMPSON, ROGER H.; CROFT, W. BRUCE. 1985. An Expert System for Document Retrieval. In: Karna, Kamal N., ed. Expert Systems in Government Symposium. Washington, DC: IEEE Computer Society Press; 1985. 448–456. ISBN: 0-8186-8686-3.

THORNBURG, GAIL. 1986. Where Expert Systems Intersect with Information Retrieval: Issues and Applications. Urbana, IL: Intelligent

Systems Group, Department of Computer Science, University of Illinois at Urbana-Champaign; 1986 May. 26p. (UIUCDCS-F-86-956). Available from: Department of Computer Science, University of Illinois at Urbana-Champaign, Urbana, IL 61801.

TOLIVER, DAVID E. 1986. Whether and Whither Micro-Based Front Ends? See reference: JACOBSON, CAROL E.; WITGES, SHIRLEY A., comps. 225-234.

TONG, RICHARD M., APPELBAUM, LEE A.; ASKMAN, VICTOR N.; CUNNINGHAM, JAMES F. 1987. Conceptual Information Retrieval Using RUBRIC. In: Yu, C. T.; Van Rijsbergen, C. J., eds. Proceedings of the Association for Computing Machinery, Special Interest Group on Information Retrieval (ACM SIGIR) 10th Annual International Conference on Research & Development in Information Retrieval; 1987 June 3-5; New Orleans, LA. New York, NY: ACM, Inc.; 1987. 247-253. ISBN: 0-89791-232-2.

VAN RIJSBERGEN, C. J., ed. 1984. Research and Development in Information Retrieval: Proceedings of the British Computer Society (BCS) and Association for Computing Machinery (ACM) 3rd Joint Symposium; 1984 July 2-6; Cambridge, England. Cambridge, England: Cambridge University Press; 1984. 433p. ISBN: 0-521-26865-6.

VAUGHAN, W. S., JR.; MAVOR, ANNE S. 1981. Simulation of a Schema Theory-Based Knowledge Delivery System for Scientists. Annapolis, MD: W/V Associates; 1981 May. 133p. ERIC: ED 210 011.

VEANER, ALLEN B. 1983. Technical Services Research Needs for the 1990s. Library Resources & Technical Services. 1983 April/June; 27(2): 199-210. ISSN: 0024-2527.

VICKERY, ALINA. 1984. An Intelligent Interface for Online Interaction. Journal of Information Science. 1984 August; 9(1): 7-18. ISSN: 0165-5515.

VICKERY, ALINA, BROOKS, HELEN M. 1987. PLEXUS—The Expert System for Referral. Information Processing & Management. 1987; 23(2): 99-117. ISSN: 0306-4573.

VICKERY, ALINA, BROOKS, HELEN M.; VICKERY, B. C. 1986. An Expert System for Referral: The PLEXUS Project. In: Davies, Roy, ed. Intelligent Information Systems: Progress and Prospects. Chichester, England: Ellis Horwood; 1986. 154-183. ISBN: 0-85312-896-0.

VICKERY, B. C. 1986. Knowledge Representation: A Brief Review. Journal of Documentation. 1986 September; 42(3): 145-159. ISSN: 0022-0418.

VIGIL, PETER J. 1986. The Software Interface. In: Williams, Martha E., ed. Annual Review of Information Science and Technology: Volume 21. White Plains, NY: Knowledge Industry Publications, Inc. for the American Society for Information Science; 1986. 63-86. ISSN: 0066-4200; ISBN: 0-86729-209-1.

WALKER, DONALD E. 1981. The Organization and Use of Information: Contributions of Information Science, Computational Linguistics and Artificial Intelligence. Journal of the American Society for Information Science. 1981 September; 32(5): 347-363. ISSN: 0002-8231.

WALKER, DONALD E. 1986. Knowledge Resource Tools for Information Access. Future Generations Computer Systems. 1986; 2: 161-171. ISSN: 0167-739X.

WALKER, GERALDENE; JANES, JOSEPH W. 1984. Expert Systems as Search Intermediaries. In: Flood, Barbara; Witiak, Joanne; Hogan,

Thomas H., comps. 1984: Challenges to an Information Society; Proceedings of the American Society for Information Science (ASIS) 47th Annual Meeting: Volume 21; 1984 October 21–25; Philadelphia, PA. White Plains, NY: Knowledge Industry Publications, Inc.; 1984. 103–105. ISSN: 0044-7870; ISBN: 0-86729-115-X.

WARNER, AMY J. 1987. Natural Language Processing. In: Williams, Martha E., ed. Annual Review of Information Science and Technology: Volume 22. Amsterdam, The Netherlands: Elsevier Science Publishers B.V. for the American Society for Information Science; 1987. ISBN: 0-444-70302-0.

WATERMAN, DONALD A. 1986. A Guide to Expert Systems. Reading, MA: Addison-Wesley Publishing Company; 1986. 419p. ISBN: 0-201-08313-2.

WATERS, SAMUEL T. 1986. Answerman, the Expert Information Specialist: An Expert System for Retrieval of Information from Library Reference Books. Information Technology & Libraries. 1986 September; 5(3): 204–212. ISSN: 0730-9295.

WATTERS, C. R.; SHEPHERD, M. A.; ROBERTSON, W. 1987. Towards an Expert System for Bibliographic Retrieval: A Prolog Prototype. In: Yu, C. T.; Van Rijsbergen, C. J., eds. Proceedings of the Association for Computing Machinery, Special Interest Group on Information Retrieval (ACM SIGIR) 10th Annual International Conference on Research & Development in Information Retrieval; 1987 June 3–5; New Orleans, LA. New York, NY: ACM, Inc.; 1987. 272–281. ISBN: 0-89791-232-2.

WILKINSON, JULIA. 1986. Database in Artificial Intelligence. Online Review. 1986 October; 10(5): 307–315. ISSN: 0309-314X.

WILLIAMS, MARTHA E. 1986. Transparent Information Systems through Gateways, Front Ends, Intermediaries, and Interfaces. Journal of the American Society for Information Science. 1986 July; 37(4): 204–214. ISSN: 0002-8231.

WILLIAMS, P. W. 1985. The Design of an Expert System for Access to Information. In: Proceedings of the 9th Internationl Online Information Meeting; 1983 December 3–5; London, England. Oxford, England: Learned Information; 1985. 23–29. ISBN: 0-904933-50-4.

WINETT, SHEILA G.; FOX, EDWARD A. 1985. Information Retrieval Techniques in an Expert System for Foster Care. In: Karna, Kamal N., ed. Expert Systems in Government Symposium. Washington, DC: IEEE Computer Society Press; 1985. 391–396. ISBN: 0-8186-8686-3.

WINOGRAD, TERRY; FLORES, FERNANDO. 1986. Understanding Computers and Cognition: A New Foundation for Design. Norwood, NJ: Ablex Publishing Corp.; 1986. 207p. ISBN: 0-89391-050-3.

WINSTON, PATRICK H.; PRENDERGAST, KAREN A., eds. 1984. The AI Business: The Commercial Uses of Artificial Intelligence. Cambridge, MA: MIT Press; 1984. 324p. ISBN: 0-262-23117-4.

YAGHMAI, N. SHAHLA; MAXIN, JACQUELINE A. 1984. Expert Systems: A Tutorial. Journal of the American Society for Information Science. 1984 September; 35(5): 297–305. ISSN: 0002-8231.

YANNAKOUDAKIS, E. J.; FAWTHROP, D. 1983. An Intelligent Spelling Error Corrector. Information Processing & Management. 1983; 19(2): 101–108. ISSN: 0306-4573.

ZARRI, GIAN PIERO. 1985. Interactive Information Retrieval: An Artificial Intelligence Approach to Deal with Biographical Data. See reference: ASLIB. 101–119.

3 Natural Language Processing

AMY J. WARNER
University of Wisconsin, Madison

INTRODUCTION

This review discusses recent developments in computerized natural language processing. Previous *ARIST* chapters on this topic have appeared since the first *ARIST* volume in 1966, the most recent being that by BECKER, in 1981. *ARIST* chapters on related topics within the past five years include those by AMSLER (machine-readable dictionaries) and TUCKER & NIRENBURG (machine translation).

The title of this chapter was deliberately changed from "Automated Language Processing," the title used in previous *ARIST* volumes, to "Natural Language Processing" to reflect a narrower scope, which is necessary to cover the subject thoroughly. Natural language processing is that area of research and application that explores how natural language that is entered into a computer system can be manipulated and stored in a form that preserves certain aspects of the original (M. HARRIS). The underlying organization of the data should incorporate information at various levels—morphological, grammatical, contextual, and so forth. Input and output may be in the form of single sentences or sentence fragments, or in connected text. Further, language can be entered and retrieved in spoken or written (keyed) form, but this chapter discusses written (keyed) form only.

Natural language processing may be put to various uses, among them natural language interfaces to databases and expert systems, text understanding, text generation, and machine translation. Within information retrieval, natural language processing is particularly relevant in the design of natural language interfaces and in the semantic analysis of documents. This chapter

Annual Review of Information Science and Technology (ARIST), Volume 22, 1987
Martha E. Williams, Editor
Published for the American Society for Information Science (ASIS)
by Elsevier Science Publishers B.V.

covers literature on all of these except machine translation, which was the subject of a recent *ARIST* chapter (TUCKER & NIRENBURG), and information retrieval, where some excellent reviews of the major trends and issues already exist (DOSZKOCS; SALTON & MCGILL).

This chapter could have been organized around the technical aspects of natural language processing applications. However, in gathering and assimilating the recent literature on natural language processing, it became clear that the literature reflects various issues and problems being faced by research and development in the area, and this review is organized around those points. These issues provide an overall organization for discussion of specific systems, how they relate to one another, and how they form part of the overall trends in natural language processing.

This review covers English-language literature published between 1982 and 1986. There is an admitted bias toward U.S. publications, but close attention was paid to foreign contributions, in particular the many conferences and journals that have a strong international component. Finally, the documents in the bibliography were selected because they represent the important trends and issues discussed.

MAIN THEMES IN NATURAL LANGUAGE PROCESSING

Two interrelated issues that underlie natural language processing (NLP) and that run through the literature are: 1) to what extent does NLP require understanding of human cognition (as manifested in linguistic ability) for the creation of useful products; and 2) to what extent is knowledge processing required by natural language processing systems.

Simulation vs. Engineering

Natural language processing reflects an interest in basic research and in applied systems. If a natural language processing system attempts to simulate human cognition, it is called a cognitive approach; otherwise, the working program or system is an end in itself and is characterized by a strong engineering approach (HAYES).

The cognitive approach assumes that the information processing theory of cognition is valid (reviewed by SIMON). Proponents of this theory claim that human mental processes behave like a computer. Within natural language processing, the computer thus becomes a tool for building models of human language production and comprehension. Literature reporting on research that is steeped in the cognitive approach emphasizes the psychological validity of the proposed system. One of the strongest articulations of the cognitive approach is the knowledge representation formalism known as conceptual dependency theory and the systems based on that theory (SCHANK & CHILDERS). More detailed information on some of the specific systems can be found in LEBOWITZ, DEJONG, and LEHNERT ET AL. Appeals to psychological reality have also been made by MARCUS (1980) for his deterministic parser.

In contrast, GARVIN explains that the cognitive approach is useful only when simulation of the human activity is seen as an end in itself. In applied endeavors, he favors a strong engineering approach that takes advantage of the unique capabilities of computers. Garvin explains what he sees as a relative lack of progress in the field by noting the negative attitude currently plaguing applied research. This attitude seems to favor using theories that seem to psychologically "fit" without a true consideration of their conceptual appropriateness to the application. The engineering approach is nonetheless evident in the restricted subject domains of natural language processing systems and in the system designs that are intended to foster transportability to other subject domains and applications.

There is a relationship between the cognitive and engineering approaches in natural language processing. The use of theoretical models in system design has already been noted. There are also well-documented engineering solutions, such as augmented transition networks, which were originally proposed as a solution to the intractability of transformational grammar in parsing human language and were only later shown to have psychological validity (WILKS & SPARCK JONES). However, there remains the unresolved issue of how far applied natural language processing can go without using advances in the study of theoretical models of language comprehension and production. In contrast to Garvin, the book of readings edited by BARA & GUIDA posits that progress in the design of high-performance natural language systems depends on using concomitant strides in the theoretical domain.

Knowledge Representation

WALTZ notes that a major realization of the 1970s was that knowledge representation formalisms are of central importance in natural language processing. Two of the most crucial issues in knowledge representation are what should be represented and how it should be represented.

WINOGRAD (1981) and WINOGRAD & FLORES point out the shift in what natural language systems represent. Winograd notes the need for natural language systems to represent *facts* about linguistic structures, about the correspondence between linguistic structures and the world, and about the cognitive structures of people, as well as the need to treat utterances as *acts*, stressing the notion that speakers are entering into an interaction pattern, committing both participants to future actions, some linguistic and others not.

M. HARRIS claims that purely syntactic structures do not represent knowledge since knowledge representation requires some internal meaning structure to be created. However, WEISCHEDEL takes a broader view, classifying the knowledge required by a system into the traditional components: morphological and phonetic, syntactic, semantic, and pragmatic. He also notes various important issues, such as how much and what kinds of knowledge need to be exploited in understanding certain constructions and at what point in the process that knowledge should be exploited.

The formal mechanisms and structures for representing knowledge are covered in a number of texts. M. HARRIS gives a general overview, while

CULLINGFORD focuses on a knowledge structure that is an amalgam of conceptual dependency, common-sense algorithms, and preference semantics. BARR & FEIGENBAUM sketch the broad division of knowledge representation systems into procedural and declarative representations and give brief accounts of the major schemes, including logic, semantic networks, case frames, semantic primitives, and scripts. GRISHMAN presents an overview of the major structures and formalisms for representing syntactic and semantic knowledge.

Sophisticated natural language systems need large knowledge bases. Work is under way to decide the principles on which the knowledge base could be constructed (HOBBS) and also to automate the construction process (FREY ET AL.).

LEXICAL ANALYSIS

The lexical phase of a natural language processing system involves the mapping between the input or output stream and the lexical items to be manipulated by the system (M. HARRIS). Various issues have surfaced in the literature on this aspect of natural language processing, including word- vs. morpheme-based lexicons, the role of machine-readable dictionaries in natural language processing, and the importance of the lexicon in the natural language processing system.

Word-based approaches involve retaining all word variants in the lexicon. The Linguistic String Project (LSP) (SAGER) uses this approach since it involves preclassification of text words to avoid incorrect analyses. LEHRBERGER describes TAUM-METEO, a system for automated translation of weather reports from English to French, which uses a word-based approach since the reports are highly telegraphic with less morphological variation than in other systems. WEHRLI discusses the tradeoffs of word-based and morpheme-based approaches and advocates a relational word-based system in which morphology is dealt with in the relations among lexical items.

On the other hand, considerable attention has been given to the principles of affixation in various languages and how these may be exploited in natural language processing systems. Morpheme-based approaches are particularly preferable over word-based approaches in cases where the system designer wishes to account for linguistic productivity and/or idiosyncrasy. BYRD ET AL. describe one portion of a project that is studying derivational affixation in English, seeking to discover linguistically significant generalizations in English affixation that can then be used for automated word recognition. KOKTOVA describes a morphemic analysis of Czech technical texts wherein the burden is entirely on the algorithm so that there is no need for a dictionary of etymological irregularities. In contrast to the previous two studies, which describe language-dependent analyses, KOSKENNIEMI describes a morphological analysis that is purported to be language independent. BYRD describes a morpheme-based system that can deal with coinages by placing all idiosyncratic information into the dictionary and deriving systematic information from word structure.

In purely theoretical natural language systems, the lexicon has typically been small and has been generated by hand. As systems designers have given more attention to practical applications, they have explored ways to automate the creation and maintenance of the lexicon. Specifically, there have been investigations of how large machine-readable dictionaries can be used to automate lexicon creation.

MARKOWITZ ET AL. describe an automated lexicon builder that uses a machine-readable version of *Webster's Seventh Collegiate Dictionary.* Their system focuses on certain key terms and phrases that signal semantic relationships such as set membership, taxonomy, and selectional (contextual) restrictions. ALSHAWI ET AL. describe research on the development of an automated link between the *Longman Dictionary of Contemporary English* and a natural language processing system, outlining the process of restructuring the information in the dictionary and the use of grammar codes and word sense definitions in the PATR-II parsing system.

In some natural language processing systems, the lexicon has taken on a central importance because much of the information necessary for processing has been placed there, resulting in fewer grammar rules. In these systems, much of the processing is driven by information contained in the lexical component. FLICKINGER ET AL. describe the central role played by the lexicon in the Head-driven Phrase Structure Grammar (HPSG), where lexical entries are represented as frames in a hierarchical structure. SMALL & RIEGER advance the hypothesis that language processing should not start with rules that describe general syntactic and semantic regularities but with a system that can handle irregularities; this is embodied in the Word Expert Parser (WEP), in which processing of linguistic input is controlled by passing information among internally complex "experts," one per linguistic unit (word or morpheme).

PARSING

A central component of natural language processing systems is the parser, a computational process that takes individual sentences or connected texts and converts them to some representational structure useful for further processing (WILKS & SPARCK JONES). The importance of parsing is illustrated by the many books and texts devoted exclusively to it.

WINOGRAD (1983) describes syntactic (grammatical) algorithms, including context-free grammars, transformational grammars, augmented transition network grammars, and feature and function grammars, and emphasizes their formal properties; applied issues are also noted, including points related to: 1) completeness (full vs. partial) of analysis; 2) the basic framework (e.g., augmented transition networks); 3) form of assigned structures (deep vs. surface structures); 4) handling of ambiguous input (parallel parsing vs. backtracking, determinism); 5) syntactic coverage (including more difficult phenomena such as complex noun phrases and conjunctions); 6) domain specificity; and 7) system engineering (including efficiency).

The book edited by KING (1983a) emphasizes linguistic theory, wherein the development of parsing procedures is seen as a reaction to many of the problems associated with trying to apply various linguistic theories directly to the computational analysis of sentences. The book edited by SPARCK JONES & WILKS stresses trends in natural language processing over the past decade, including the trend toward phrase structure grammars and deterministic parsing and toward the closer integration of syntax and semantics.

The contributors to DOWTY ET AL. react to two theoretical themes in the parsing problem: 1) the psychological reality of various parsing mechanisms (verified through carefully controlled experiments), and 2) the formal properties that parsers must possess, including how much power is required to describe human languages adequately and the ability of parsers to make significant linguistic generalizations.

The Power Issue

An early attempt at computer parsing of human language drew from linguistic theory, particularly from the syntactic theory known as transformational grammar. The experiences and problems associated with this attempt are reviewed by KING (1983b). Transformational theory within linguistics serves only to define the acceptable sentences of a language, and the use of its rules to analyze (and also generate) sentences is problematic. A later syntactic formalism called the Augmented Transition Network (ATN) (explained and reviewed by R. JOHNSON) was seen as a reaction to the computational intractability of transformational grammars (WILKS & SPARCK JONES) but is equally powerful (i.e., can analyze the same linguistic structures).

ATNs have proved very successful in natural language processing and are used extensively in many systems. Many versions of implemented ATNs are discussed in the literature that extend, refine, and improve the structure, particularly in order to incorporate semantics into the originally syntactic model. SHAPIRO describes a generalization of the ATN framework that allows the same ATN grammar both to analyze and to generate sentences from semantic networks; this is an improvement on the previous situation in which inconsistencies in the semantics of the two components meant that the same ATN interpreter could not be used for both. CHRISTALLER & METZING describe a system based on cascaded ATNs, originally introduced to allow feedback with the semantic interpretation process. They describe its use in analyzing discourse properties, where it should be possible to use the most appropriate knowledge representations and processes for specific tasks in the dialog grammar.

One reason for using transformational grammar and ATNs was the belief that other grammatical formalisms such as finite state grammars and phrase structure grammars are inadequate for treating natural languages. The literature reflects a reevaluation of that assumption. GAZDAR argues that context-free phrase structure grammars: 1) can deal with long-distance dependencies (such as subject–predicate agreement); 2) can capture significant generalizations; and 3) can support a semantics (i.e., there exists a semantic theory that can interpret the syntactic structures). He embodies these claims in Generalized Phrase Structure Grammar (GPSG), the basics of which are spelled out in GAZDAR ET AL. Other formalisms investigated include Tree Adjoining Grammar (JOSHI), Lexical Functional Grammar (KAPLAN & BRESNAN), and Functional Unification Grammar (KAY, 1985).

These formalisms are important in natural language processing for two reasons other than purely mathematical interest. First is the issue of whether there can be a clear relationship between human linguistic processing mechanisms and computational parsing; this issue is addressed by CRAIN & FODOR, who provide experimental evidence that GPSG may be psychologically accurate. Second is the issue of computational complexity—i.e., if a formalism can be shown to be computationally intractable, then it becomes less attractive in any applied domain; RISTAD addresses this issue for GPSG, BERWICK (1982) does the same for Lexical Functional Grammar, and RITCHIE (1986) does so for Functional Unification Grammar. The application of many of these grammatical formalisms has also been investigated, including GPSG (RAMSAY), Lexical Functional Grammar (REYLE & FREY), Tree Adjoining Grammar (MCDONALD & PUSTEJOVSKY, 1985b), and Functional Unification Grammar (KAY, 1984).

Syntax vs. Semantics

A much-debated and still-unresolved issue within natural language processing involves the roles to be played by syntax and semantics in the parsing process. Some researchers have begun to explore the involvement of other information (e.g., pragmatic) as well.

The exploration of the roles of syntax and semantics is a common theme in the literature. SCHANK & BIRNBAUM discuss the relationships among memory, meaning, and syntax and make the point that meaning and world knowledge are often crucial even at the earliest points in the process of understanding language. They refer to this idea as the integrated processing hypothesis, wherein syntax plays little or no role in the parsing process. In contrast, MARCUS (1984) shows how "semantics mainly" and "semantics only" natural language analyzers are not adequate to cover the full range of language people use. He further claims that the process of understanding human language is to determine the meanings of utterances but that syntactic structures appear to be a necessary stop along the way.

There has been a trend in syntax-based parsing toward making algorithms that incorporate more semantics, but syntax nevertheless retains its primacy. CHARNIAK (1983b) argues for syntax and semantics that operate in parallel as well as another semantic component called "compositional semantics," which creates the logical form of the sentence. MELLISH (1983) discusses in general terms a framework he calls incremental semantic processing. He claims that it can include strong syntactic and semantic modules but the modules are given much more flexibility to make progress on their own even when input is ambiguous.

Arguments favoring a weaker role for syntax are also common. Conceptual dependency theory is one of the more well known articulations of this position. Here, syntax is used only in parsing when absolutely necessary in what is primarily a process involving meaning and memory structures (SCHANK & BIRNBAUM). CATER describes a system that delivers a conceptual dependency representation by using an expectation-based mechanism that uses various syntactic and semantic cues to aid the parser in choosing between parallel competing analyses. He uses explicit syntactic information only to group input words into smaller constituents and then uses the semantic and syntactic preferences of nouns and verbs to guide the parsing. GERSHMAN describes another general framework based on conceptual dependency for building dictionary-based expectation-driven conceptual analyzers. It uses both bottom-up and top-down analyses, going through the sentence from left to right to find a conceptualization that forms the backbone of the meaning representation of the sentence (bottom-up process); then it uses the predictions that come with this general framework to analyze the rest of the sentence in a top-down manner.

Other parsing mechanisms with strong semantic components are discussed in the literature. Case frames are used for parsing by HAYES ET AL., who discuss the strengths and weaknesses of this approach for their parser called Plume. Case frames are also used by SHIMAZU ET AL., who describe their implementation in a semantic analyzer of the Japanese language.

Coping with Ambiguity

Another common theme in the literature centers around how to handle ambiguity in ways that are both computationally efficient and psychologically plausible. Earlier parsing mechanisms had used parallelism or backtracking. WINOGRAD (1983) notes the characteristics, advantages, and disadvantages of both approaches. More recently, determinism has become a major issue in parsing theory and practice.

The issue of determinism was proposed by MARCUS (1980) who claims that his PARSIFAL parser makes linguistically significant generalizations and is psychologically accurate. PARSIFAL's key feature is the claim that no backtracking or parallel analyses need to be carried out on sentences and that it fails only in cases of obvious psychological complexity. BRISCOE supports the claim that determinism is an essential goal for a psychologically interesting parser but provides evidence that Marcus's use of delayed processing instead of backtracking or parallelism is psychologically and linguistically implausible. More importantly, Briscoe demonstrates that it is not possible to provide a structural definition of garden-path sentences,[1] a contention supported by experimental evidence supplied by CRAIN & STEEDMAN.

Although the psychological reality of Marcus's deterministic parser has been questioned, determinism is still an important issue within natural language processing. RITCHIE (1983) describes the PIDGIN notation (the high-level language Marcus designed for writing his grammar) in enough detail

---

[1] These are sentences in which human subjects in psycholinguistic experiments appear to analyze a sentence incorrectly and then go back and correct an original parsing—e.g., "The granite rocks during an earthquake."

for it to be implemented. Others have explored the utility of determinism in dealing with a variety of constructions. MILNE shows that a deterministic parser can easily resolve certain classes of lexical ambiguity, particularly that concerning part of speech; KOSY describes how a deterministic parser can process conjunctions efficiently; and BERWICK (1983) describes an extension that broadens the syntactic coverage of Marcus's formalism. CHARNIAK (1983a) introduces another parser based on the Marcus formalism, which is "semigrammatical" in the sense that it can accept sentences that do not fit the grammar while noting how the sentences are deviant. CARTER & FREILING describe a small implementation of a parser called PARSER (Deterministic PARSER), which is intended to reduce the complexity of deterministic grammars.

Parsing Difficult Constructions

Specific types of constructions are known to be highly ambiguous, including complex noun phrases, temporal constructions, and constructions involving conjunctions, quantification, and anaphora. Approaches to these phenomena address several issues: 1) the means for incorporating analyses of these constructions as part of the formalism that describes the rest of the language; 2) the means for constraining competing choices between analyses; and 3) the types of linguistic and world knowledge required to handle these constructions.

The problems of interpreting noun phrases are described by SPARCK JONES (1985), who argues that handling complex noun phrases involves extensive inference from other types of information, including semantic information and world knowledge. MELLISH (1985) focuses on noun phrases in the context of problems associated with early semantic analysis. In the face of local uncertainty in semantic analysis of certain types of noun phrases, he develops a system of incremental reference evaluation wherein representation of partial information can be exploited so that the system can continue reliably with the analysis through the uncertainties. FININ describes a continuation of work carried out on the UNCLE system, in which the various interpretations of nominal compounds in a technical domain are filtered through the use of discourse context.

Parsing conjunctions involves problems in building systems with enough power to handle all of the phenomena involved in these coordinate expressions but with enough constraints to reject ungrammatical constructions. It is also desirable to achieve this in linguistically and computationally uncomplicated ways. BOGURAEV deals with an extension of the ATN interpreter that can deal with conjunctions. FONG & BERWICK describe a PROLOG grammar for conjunctions and show that a different syntactic structure—one not based on tree structures—can improve the parsing problem for conjunctions; its advantages are that it can analyze a number of related constructions using the same approach, it is completely reversible (i.e., can be used to generate as well as to interpret sentences), and it is comparable with other systems in efficiency. They compare their system with the logic grammar formalism called Modifier Structure Grammar (MSG) introduced by DAHL & MCCORD, claiming that theirs is more modular and just as efficient. LESMO & TORASSO describe how conjunctions are parsed in the FIDO system (A Flexible Interface for Database Operations), stressing that the overall system is designed so that conjunctions can be accommodated easily. KOSY describes the parser in NEXUS (Non-Expert Understanding System), which uses a deterministic device to analyze conjunctions and appears to be faster than other parsers described in the literature.

Quantification is another complex problem for parsers. Quantified expressions are highly ambiguous since there are complex relations between nominal (noun) and verbal (verb) constituents. BUNT describes a method of dealing with these in the TENDUM dialog system. Quantification is accounted for efficiently since the analysis generates the logical representations of only those interpretations that are relevant in a given domain of discourse. SAINT-DIZIER describes a system that parses and generates French sentences; a set of rewrite rules builds all the different relevant quantifier configurations, and the correct configuration is then selected according to contextual information.

Anaphora poses problems in natural language processing since it is necessary to identify the entity referred to by some potentially ambiguous expression, such as a pronoun. HIRST surveys the literature on anaphora in natural language understanding, creating a useful catalog of anaphoric expressions, summarizing the ways various systems deal with anaphora, and advocating the use of discourse theme (the subject central to the ideas expressed in a text) in resolving anaphora. This view has been taken up more recently by JOHNSON & KLEIN, who present a reformulation of an earlier discourse-representation model that can handle both inter- and intrasentential anaphora.

Another difficult problem in natural language processing is how to account for temporal information. HAFNER and DE ET AL. deal with how to incorporate such a facility in handling database queries, stressing that it must be accounted for in the underlying database as well as in the interface design. HARPER & CHARNIAK examine five criteria for judging the semantic representations that handle temporal constructions in English.

NATURAL LANGUAGE PROCESSING SYSTEM ISSUES

Robustness

Research and development in natural language processing has been oriented toward producing systems with greater depth of analysis and flexibility. A completely robust system would successfully process ungrammatical or partial input, novel language including metaphor, the context of sentences or texts, and the goals and plans of the participants in a cooperative dialog. It would do all these things in a completely unrestricted semantic domain. Because natural language processing systems have not come anywhere near this point, various compromises are made, including less ambitious attempts at robustness.

The definition of robustness adopted by SELFRIDGE focuses on the ability to cope with partial or ungrammatical input, to correct misunderstanding, and to incorporate changes entered by the user. He describes the MURPHY system, which is used to question and direct a robot assembly system and achieves some robustness by using conceptual dependency representations and the principle of integrated processing.

RIESBECK deals with the general requirements for Realistic Language Comprehension (RLC) systems. He focuses on the need to create systems that can deal with real texts rather than designer-created text. He presents a system architecture that he claims can begin to deal both with malformed input and with the internal inadequacies of the system itself. Any new structure built is tested for reasonability in the domain and modified if something is wrong and there is a known way to reinterpret it.

The ability of a natural language interface to continue processing and yield a correct output when given partial or ungrammatical input is a desirable feature. WEISCHEDEL & SONDHEIMER propose a system of meta-rules as a standard framework for handling ill-formed input. These meta-rules, which correspond to patterns of ill-formedness, are used to diagnose a problem as a violated rule of the normal processing and then relax the violated rule to allow processing to continue. They propose that this general framework will work for cases of lexical, syntactic, semantic, and pragmatic ill-formedness. CARBONELL & HAYES classify different types of grammatical deviations at both the lexical and sentential levels and discuss recovery strategies for transition-network, pattern-matching, and case-frame approaches to natural language analysis. They only touch on the notion that ill-formedness may be due to ungrammatical dialog rather than ungrammatical linguistic structures. In contrast, CARBERRY notes that a sentence may be syntactically and semantically well formed yet violate the pragmatic rules of the world model. She presents a context-based strategy for dealing with these cases, in which "suggestion heuristics" propose revisions to the ill-formed query based on a model of the dialog and "selection heuristics" are then used to evaluate these suggestions based on semantic and relevance criteria.

Ill-formedness is also accounted for in text processing systems. JENSEN ET AL. describe the processing of syntactically ill-formed input in the EPISTLE system, which addresses the problems of grammar and style checking of texts. Central to its purely syntactic strategy is the technique known as "fitted" parsing, where the fitting procedure begins after the syntactic parse has been applied but has failed to produce an acceptable interpretation. GRANGER describes a somewhat different application in the handling of "scruffy text" in the NOMAD system, which provides a well-formed English translation of unedited Navy ship-to-shore messages. Processing of ill-formed input is seen as a necessary part of the overall problems of processing text, and general error-correction processes operate on syntax, semantics, and pragmatics.

Sublanguages

Generally speaking, every document pertains to some subject, and each subject has a somewhat specialized language of its own. Since each of these "languages" is just a part of the overall language, each is called a sublanguage. There is a sublanguage of molecular biology, a sublanguage of astronomy, and so forth. Sublanguages are simpler than the whole language because each is characterized by a small set of linguistic and pragmatic patterns—i.e., characteristic constructions. At present, some natural language processing systems are being built to process text using particular small sublanguages to reduce the size of the operations that must be coded and carried out. At the same time, more theoretical work is being done to pinpoint the characteristics and patterns of various sublanguages in areas as diverse as aviation hydraulics, cooking recipes, weather synopses and stock market reports (KITTREDGE).

FITZPATRICK ET AL. assess the relationship of a sublanguage to the language as a whole, pondering the basic question of whether sublanguages can be considered as relatively independent syntactic systems with internal consistency or whether they must be described in terms of the whole language. They conclude that there are sublanguage-specific constraints, which, if isolated and characterized correctly, can result in an accurate and simple grammar. In contrast, Kittredge discusses the characteristics of sublanguages themselves. He tries to assess the unique properties of sublanguages, such as their status as closed linguistic systems and specialized subject domains as well as the variations among them.

A number of researchers have also addressed sublanguage issues in the practical domain. MONTGOMERY & GLOVER describe a sublanguage used by space analysts for reporting and evaluating past, present, and future space events. The existence of two discourse levels—an event-reporting level and a meta-level of commentary on the events—prompts them to ask whether there are two sublanguages involved.

One long-standing project on the characterization of sublanguages for both practical and theoretical purposes is New York University's Linguistic String Project, described by SAGER, HIRSCHMAN & SAGER, and FRIEDMAN. It uses a precise sublanguage description to convert hospital records into a structured format, which can then be used in various information retrieval applications—e.g., the production of summary reports and question answering. Also in the medical domain, DUNHAM describes the syntactic features of medical diagnostic statements. He makes the case that since the sublanguage is telegraphic, rich in complex noun phrases, and often lacking verbs, syntactic information plays a lesser role than semantics and pragmatics in guiding the processing.

WALKER & AMSLER investigate the application of knowledge resources to the processing of natural language texts. Specifically, they use the semantic codes found in the *Longman Dictionary of Contemporary English* to determine the subject domains for texts from the New York Times News Service wire. By determining the semantic codes associated with each word in the text, accumulating the frequencies for these senses, and then ordering the list of categories in terms of frequency, the overall subject matter of the text can be identified.

Transportability

Since NLP systems can now operate only in restricted domains, one of the greatest problems is how to best use restricted-domain techniques in new domains. Work is under way to separate the domain-independent from the domain-dependent aspects of systems and to find the most effective ways to add the domain-specific information of a new application to the system. A number of authors have discussed their experiences in transporting a particular system to a new domain. MARSH & FRIEDMAN discuss their experiences in transporting the Linguistic String Project (LSP) from a medical to a Navy domain, stressing the features of LSP that make it transportable as well as the syntactic and semantic similarities between the Navy and medical domains. They also offer a number of requirements for portable natural language processing: a grammar that is broad in coverage, a modular system design, and a consistent linguistic theory. THOMPSON & THOMPSON discuss the technical issues and possible solutions in making the ASK (A Simple Knowledgeable) system for structuring and manipulating database information transportable, including both new subjects and new capabilities (e.g., transportability to new natural languages, new programming languages, or new computer families). They stress the need to address the transportability issue at all stages of design and implementation.

In discussing particular systems, many researchers have stressed the need for modular design in achieving transportability. To achieve modular design, domain dependencies (i.e., information specific to that subject area) must be isolated.

GRISHMAN ET AL. (1983) explore the problem of isolating domain dependencies within two systems, describing the domain information schema they use to capture the domain-dependent information. They then describe how these are used in the two systems, claiming that information about the structure of information in a domain rather than specific facts about a domain can be adequate. In a related work, GRISHMAN ET AL. (1986) address the problems of identifying selectional constraints, which are domain specific, each time a new subject domain needs to be described; they then outline a semiautomated procedure for achieving this.

EPSTEIN describes PRE (Purposefully Restricted English), a restricted English database query language, which is based on the philosophy that an explicitly minimalist approach to natural language processing can facilitate transportability. He contrasts his system with those that seek full coverage of the English language within a given domain, claiming that transportability cannot be achieved in such systems. The PRE system achieves transportability through addition of domain-specific information contained in a few tables.

Other systems attempt to achieve transportability while accommodating a much less restrictive query capability. HAFNER & GODDEN describe Datalog, an experimental natural language query system that uses a three-level architecture to facilitate transportability, showing how this feature can be achieved by changing only one "layer" of the system's knowledge in each case. SOPENA describes the USL (User Specialty Languages) system, an applications-independent natural language interface to a relational database system than can be used in various natural languages. It has been designed so that only the grammar and the structural vocabulary need to be changed to shift from one language to another. BATES & BOBROW (1983) describe the development of a facility for natural language access to various computer databases and database systems; it achieves transportability largely through the RUS parsing system, a general English parser based on the ATN framework.

To allow for commercial application of transportable systems, it must be feasible to customize them to new domains. Thus, whereas the preceding group of studies emphasized the design characteristics that made their systems transportable, the following studies emphasize the means by which various systems acquire new domain-specific knowledge in a customization process. GROSZ describes TEAM (Transportable English Access Data Manager), in which domain- and database-dependent information are obtained from an interactive dialog with database experts. DAMERAU describes the TQA (Transformational Question Answering) system, which can be customized by a database administrator at various points in its multilayered structure, including the lexicon, the grammar rules, the semantic interpreter, and the output formatter. BALLARD & TINKHAM describe the Layered Domain Class (LDC) System. It consists of a learning component through which the user provides information about the structure of the domain and the language to be used in describing it; the user can later modify this information. This project has since evolved into the TELI (Transportable English-Language Interface) project, described by BALLARD & STUMBERGER. IZUMIDA ET AL. provide details of the Japanese language interface called KID (Knowledge-based Interface to Databases). It is based on a modular design, with all domain-dependent knowledge contained in a world model, and an associated world model editor that provides various levels of user interfaces to aid in the customization process for various levels of user sophistication.

Dealing with Context

Another topic in NLP is how to process natural language within a context— i.e., a sentence embedded in a dialog. Context makes a crucial contribution to the understanding of most natural language expressions. Context is taken into account in undertaking dialogs with users who are interacting with databases and expert systems as well as in the understanding, generating, and summarizing of texts.

Cooperative dialog with a database or expert system. Within interactive natural language processing systems, the emphasis has been on developing flexible, cooperative systems that not only provide answers but can also provide advice, anticipate needs, and prevent or clear up misconceptions. Some database fact-retrieval systems resolve ambiguity and ill-formed input. However, more can be done by using context to define parameters in user models (WAHLSTER & KOBSA) and to deduce user plans and goals (ALLEN & LITMAN). An open issue here is whether capabilities for this added sophistication can come solely from the interface. FININ ET AL. discuss how the underlying reasoning system and knowledge base will have to be structured to support context processing.

COHEN ET AL. sketch the general outline of a new system architecture that would support a cooperative dialog between person and machine. REICHMAN describes techniques for recognizing change of focus in a dialog and for understanding the flow of discourse; these techniques are applied in a computer program aimed at tracking the conversational flow of dialog participants.

The trend to accommodate cooperative dialog focuses on many of the capabilities noted by Finin et al. One of these is clearly the need for the system to be able to do more than retrieve facts in response to a query. There have been various attempts to enable systems to explain their actions. Aspects of explanation that have received attention include accuracy (i.e., they correctly specify their reasoning) (NECHES ET AL.; SWARTOUT), clarity (HASLING ET AL; KUKICH), and appropriateness to the subject and level of difficulty of the domain for the user (MCKEOWN ET AL.). A related capability that has been explored is that of paraphrase, wherein new versions of users' questions are generated and displayed to give the user a chance to correct any possible misunderstandings (MCKEOWN, 1983; SPARCK JONES & BOGURAEV).

Finin et al. also make the point that a system and a user may hold different world views and do not realize it or fail to do anything about it. GAL & MINKER use a system of integrity constraints in their natural language interface; these are used to generate cooperative responses when the user is operating under the misconception that a relationship exists between objects or classes in the database when they cannot according to the system. The ROMPER (Responding to Object-Related Misconceptions) system (MCCOY) can respond to misconceptions involving an object's classification or attributes. SPIRIT (System for Plan Inference that Reasons about Invalidities Too) (POLLACK) can resolve misconceptions involving the inappropriateness of a user's plans in relation to the system's plans.

In addition to the ability to resolve misconceptions, it should also be possible for a system to avoid misconstrual by anticipating that the user may misinterpret an utterance and thus clarify it in advance. JOSHI ET AL. state that if a system has reason to believe that its planned response might mislead the user, then it must modify its response. Their model identifies and avoids potentially misleading responses by acknowledging types of "informing behavior" usually expected of an expert.

Much of the flexibility in generating cooperative responses comes from the ability of the system to model the user and his role in the conversation with the system (WAHLSTER & KOBSA). Since users rarely state their beliefs and goals directly, systems that are to be cooperative and flexible must be able to generate assumptions about them from their inputs. SIDNER presents a model of the interpretation of what a speaker means that takes into account the speaker's goals, the discourse context, the speaker's knowledge of the hearer's capacities, and the hearer's assumptions. The ARGOT system (ALLEN) incorporates an important component of natural language dialog, which is the recognition that there are multiple goals underlying utterances. This phenomenon is also explored in the KAMP system (APPELT). A related topic is the need to deal with subdialogs and topic change, phenomena addressed by LITMAN & ALLEN.

User modeling is especially important if the dialog system is to be used by a heterogeneous group (WAHLSTER & KOBSA). In its hotel reservation application, HAM-ANS (HAMburg Application-oriented Natural language

System) (MORIK, 1985) can derive assumptions about the user along several dimensions using collections of frequently occurring characteristics of users (stereotypes).

Text generation. Natural language generation is another area of research in which attention has been given to connected linguistic units rather than to sentences as discrete elements. MCKEOWN (1986) reviews the areas in which language generation is relevant, including question answering systems, communicating with expert systems, and text summarization. She also notes the important distinctions between language interpretation and generation that make language generation a research area in its own right. In automating the language interpretation process, one must determine how a speaker's options are limited at various points whereas in the generation process, the task is to formulate the means for selecting among valid options.

MCKEOWN (1985) also considers a model for text generation, which has been used successfully in many such systems. It involves the separation of the generation process into strategic and tactical components, corresponding to the distinction between deciding what to say and how to say it. This division of the problem into conceptual and linguistic domains makes it possible to study one or the other in isolation or to study the interaction between them.

MCKEOWN (1986) states that much earlier work in text generation concentrated on the tactical component. More recently, study of the strategic component has been undertaken by APPELT, whose KAMP (Knowledge and Modalities Planner) system takes high-level descriptions of speakers' goals and then plans both linguistic and nonlinguistic actions. MANN (1984) has studied a large corpus of texts and then embodied his findings in Rhetorical Structure Theory. In this theory of text structure each region of text has a central nucleus and a number of related satellites. This study can be contrasted with Mann's earlier work on the Penman system, in which he addressed primarily tactical problems (MANN, 1983).

In her TEXT system, MCKEOWN (1986) formalizes strategies as schema with a graph representation and then uses a grammar and dictionary to transform this message into actual text. MCDONALD & PUSTEJOVSKY (1985a) have extended the MUMBLE system through the addition of a process called attachment that mediates between content planning and linguistic realization; it can accommodate variations in prose style by allowing the attachment process to choose among various alternatives. VAUGHAN & MCDONALD propose a model of text generation with revision, noting that there are stylistic decisions to make as well as conceptual and linguistic ones; the revision model they propose is implemented at the stylistic level since alterations to the conceptual component would alter the meaning of the text.

Text summarization. MCKEOWN (1986) notes text summarization as one of the noninteractive applications of text generation. The system uses an internal formal representation of the text and selects information to include in the summary. LEHNERT presents an approach to summarization based on plot units; these can be used to locate the central concepts of a narrative by examining structural features of graphs and then generating a good summary. TAIT describes a summarizing system, called Scrabble, which uses conceptual dependency structures to incorporate unexpected information in the input text into its summaries. These two approaches operate from a high level of conceptual representation to summarize stories. In contrast, MARSH ET AL. describe a multilevel approach to incorporating several sources of knowledge into a linguistic analysis and production rule system that generates one-line summaries of Navy messages.

Text understanding. Work continues in the area of text understanding, where the structures of texts are analyzed and where the context provided by structures facilitates sentence comprehension. Much of this work has focused on conceptual analysis, which skims the text and extracts its gist. This is characteristic of the FRUMP (Fast Reading Understanding and Memory Program) program (DEJONG), which skims and summarizes news wire stories; its main principle of design is that text analysis can benefit from pragmatic knowledge at all levels of analysis. The Integrated Partial Parser (IPP) (LEBOWITZ) also works in this generic way, using high-level memory structures. In contrast, the BORIS system (LEHNERT ET AL.) attempts to understand everything it reads and to as great a depth as possible; it does this through an integrated parsing mechanism that incorporates diverse sources of knowledge in a single concept-driven process.

Nonliteral Uses of Language

Natural language processing systems have typically stressed the manipulation and understanding of literal language. This is because figurative uses of language are highly idiosyncratic since words can be combined to form concepts that are not easily understood as a simple combination of the component parts. Work is under way to understand the use of metaphor and to accommodate it in natural language processing systems.

An understanding of the basic workings of metaphors would allow natural language processing systems to accommodate them in a systematic manner. CARBONELL proposes that there are only a few general metaphors in English, making the problem of understanding many metaphors a task of recognition rather than reconstruction. In contrast, WEINER focuses on the comprehension of metaphors, attempting to formulate a computational model whose central component is a knowledge representation system incorporating elements that appear to be a part of the comprehension process.

Other researchers have addressed the more immediate goals of accommodating metaphor within specific systems or theories. DYER & ZERNIK discuss the RINA system, which can handle figurative phrases, incorporating these new terms into a phrasal lexicon by acquiring their meanings through context. FASS & WILKS discuss how metaphor might be accommodated within their semantics-based processing framework, called preference semantics.

INTO THE MARKETPLACE

Recently natural language processing has been moving from the research domain into the commercial sector. HARRIS & DAVIS outline a number of reasons for this phenomenon, including the use of computers by many people who are not skilled in their operations, the overall increase in installed computers within the commercial sector, the entry of computers into new environments (e.g., factories), and the growth in complexity of noncomputerized fields.

DAVIS describes the introduction of a number of different kinds of commercial products: those that translate a user's query into a database query language, those that target the microcomputer market, and other systems that model the domain. Names and pertinent data for recent products are given by GEVARTER. T. JOHNSON gives recent facts about existing companies and products as well as who is using them and how they are being used.

Database front ends are early products that have been marketed successfully; the most well-known is probably INTELLECT, produced by Artificial Intelligence Corp. (AIC) (L. HARRIS). Other well-known systems include Symantec's Q & A (HENDRIX) and Microrim's Clout 2 (WILLIAMSON). There is also an interest in producing systems that can help the user formulate a problem (WEBBER & FININ). SHWARTZ and SCHANK & SLADE investigate the needs for producing advisory systems, claiming that these are potentially the most useful user services. The activities within the commercialization effort thus mirror the research sector, where there is a distinct division between those who claim that natural language access to databases has outlived its usefulness (SPARCK JONES, 1984) and those who say it has not (BATES).

Since natural language systems are entering the marketplace, it is necessary to evaluate them for their potential usefulness. BATES & BOBROW (1984) address the suitability of natural language systems to various applications and discuss general ways to judge them, including their linguistic coverage, their level of inferencing, their ability to handle difficult constructions, and the degree of control given to the user. Guida and Mauri also address performance evaluation, first in a formal and quantitative way (GUIDA & MAURI, 1984) and then under the general criteria applied to software—efficiency, correctness, reliability, and adequacy (GUIDA & MAURI, 1986). Evaluation needs to take into account the requirements of customers, a consideration addressed by MORIK (1985) in her market inquiry on German natural language systems.

CONCLUSIONS

This chapter has surveyed several key topics in the current literature on natural language processing, by reviewing examples that reflect important issues within the field. The overall trend has been for researchers to develop systems that are more robust by accounting for more types of knowledge that interact in increasingly complex ways; this includes traditional linguistic knowledge and pragmatic knowledge as well. Systems are beginning to use contextual knowledge and eventually may demonstrate conversational capabilities that rival those of a human.

Although the goal is to build systems that can simulate the performance of a human expert (even in a limited domain), it is unclear whether or not the mechanisms required to do so must emulate human cognition.

In the end, the production of useful systems may not require the actual simulation of human competence or performance. On the other hand, it may be that humans possess information processing representations and mechanisms that can be directly emulated by machine; it is also possible that only a portion of human capabilities can be emulated. Two general questions thus arise: 1) can fully robust natural language systems be designed; 2) if they can, how much of their eventual success depends on good engineering and how much on the knowledge of human cognitive processes, and how can these be rendered computationally tractable?

BIBLIOGRAPHY

ALLEN, JAMES F. 1983. ARGOT: A System Overview. In: Cercone, Nick, ed. Computational Linguistics. Oxford, England: Pergamon Press; 1983. 97–109. (Rodin, E. Y., ed. International Series in Modern Applied Mathematics and Computer Science). ISBN: 0-08-030253-X.

ALLEN, JAMES F.; LITMAN, DIANE. 1986. Plans, Goals, and Language. Proceedings of the IEEE. 1986 July; 74(7): 939–947. ISSN: 0018-9219.

ALSHAWI, HIYAN; BOGURAEV, BRAN; BRISCOE, TED. 1985. Towards a Dictionary Support Environment for Real Time Parsing. See reference: ASSOCIATION FOR COMPUTATIONAL LINGUISTICS. 1985a. 171–178.

AMSLER, ROBERT A. 1984. Machine-Readable Dictionaries. In: Williams, Martha E., ed. Annual Review of Information Science and Technology: Volume 19. White Plains, NY: Knowledge Industry Publications; 1984. 161–209. ISSN: 0066-4200; ISBN: 0-86729-093-5.

APPELT, DOUGLAS E. 1985. Planning English Sentences. London, England: Cambridge University Press; 1985. 171p. (Joshi, Aravind K., ed. Studies in Natural Language Processing). ISBN: 0-521-30115-7.

ASSOCIATION FOR COMPUTATIONAL LINGUISTICS. 1984. Proceedings of the 10th International Conference on Computational Linguistics and the Association for Computational Linguistics 22nd Annual Meeting; 1984 July 2-6; Stanford, CA. Morristown, NJ: Association for Computational Linguistics; 1984. 561p. Available from: Donald E. Walker (ACL), Bell Communications Research, 445 South Street, MRE 2A379, Morristown, NJ 07960 USA.

ASSOCIATION FOR COMPUTATIONAL LINGUISTICS. 1985a. Proceedings of the European Chapter of the Association for Computational Linguistics 2nd Conference; 1985 March 27–29; Geneva, Switzerland. Morristown, NJ: Association for Computational Linguistics; 1985. 276p. Available from: Donald E. Walker (ACL), Bell Communications Research, 445 South Street, MRE 2A379, Morristown, NJ 07960 USA.

ASSOCIATION FOR COMPUTATIONAL LINGUISTICS. 1985b. Proceedings of the Association for Computational Linguistics 23rd Annual Meeting; 1985 July 8-12; Chicago, IL. Morristown, NJ: Association for Computational Linguistics; 1985. 332p. Available from: Donald E. Walker (ACL), Bell Communications Research, 445 South Street, MRE 2A379, Morristown, NJ 07960 USA.

ASSOCIATION FOR COMPUTATIONAL LINGUISTICS. 1986. Proceedings of the Association for Computational Linguistics 24th Annual Meeting; 1986 June 10-13; New York, NY. Morristown, NJ: Association for

Computational Linguistics; 1986. 270p. Available from: Donald E. Walker (ACL), Bell Communications Research, 445 South Street, MRE 2A379, Morristown, NJ 07960 USA.

BALLARD, BRUCE; STUMBERGER, DOUGLAS. 1986. Semantic Acquisition in TELI: A Transportable, User-Customized Natural Language Processor. See reference: ASSOCIATION FOR COMPUTATIONAL LINGUISTICS. 1986. 20–29.

BALLARD, BRUCE W.; TINKHAM, NANCY L. 1984. A Phrase-Structured Grammatical Framework for Transportable Natural Language Processing. Computational Linguistics. 1984 April–June; 10(2): 81–96. ISSN: 0362-613X.

BARA, BRUNO G.; GUIDA, GIOVANNI, eds. 1984. Computational Models of Natural Language Processing. Amsterdam, The Netherlands: North-Holland; 1984. 327p. (Fundamental Studies in Computer Science; Volume 9). ISBN: 0-444-87598-0.

BARR, AVRON; FEIGENBAUM, EDWARD A., eds. 1981. The Handbook of Artificial Intelligence: Volume 1. Stanford, CA: HeurisTech Press; 1981. 409p. (pp. 230–231). ISBN: 0-86576-005-5.

BATES, MADELEINE. 1984. There Still is Gold in the Database Mine. See reference: ASSOCIATION FOR COMPUTATIONAL LINGUISTICS. 1984. 184–185.

BATES, MADELEINE; BOBROW, ROBERT J. 1983. Information Retrieval Using a Transportable Natural Language Interface. In: Kuehn, Jennifer J., ed. Research and Development in Information Retrieval: Proceedings of the ACM/SIGIR 6th Annual International Conference; 1983 June 6–8; Bethesda, MD. New York, NY: Association for Computing Machinery (ACM); 1983. 81–86. ISBN: 0-89791-107-5.

BATES, MADELEINE; BOBROW, ROBERT J. 1984. Natural Language Interfaces: What's Here, What's Coming, and Who Needs It. See reference: REITMAN, WALTER, ed. 179–194.

BECKER, DAVID. 1981. Automated Language Processing. In: Williams, Martha E., ed. Annual Review of Information Science and Technology: Volume 16. White Plains, NY: Knowledge Industry Publications for the American Society for Information Science; 1981. 113–138. ISSN: 0066-4200; ISBN: 0-914236-90-3.

BERWICK, ROBERT C. 1982. Computational Complexity of Lexical-Functional Grammar. American Journal of Computational Linguistics. 1982 July-December; 8(3–4): 97–109. ISSN: 0362-613X.

BERWICK, ROBERT C. 1983. A Deterministic Parser With Broad Coverage. See reference: BUNDY, ALAN, ed. 710–712.

BOGURAEV, B. K. 1983. Recognising Conjunctions within the ATN Framework. See reference: SPARCK JONES, KAREN; WILKS, YORICK, eds. 39–45.

BRISCOE, E. J. 1983. Determinism and Its Implementation in PARSIFAL. See reference: SPARCK JONES, KAREN; WILKS, YORICK, eds. 61–68.

BUNDY, ALAN, ed. 1983. IJCAI-83: Proceedings of the 8th International Joint Conference on Artificial Intelligence; 1983 August 8–12; Karlsruhe, West Germany. Los Altos, CA: William Kaufmann; 1983. 1206p. ISBN: 0-86576-064-0.

BUNT, HARRY. 1984. The Resolution of Quantificational Ambiguity in the TENDUM System. See reference: ASSOCIATION FOR COMPUTATIONAL LINGUISTICS. 1984. 130–133.

BYRD, ROY J. 1983. Word Formation in Natural Language Processing Systems. See reference: BUNDY, ALAN, ed. 704-706.

BYRD, ROY J.; KLAVANS, JUDITH L.; ARONOFF, MARK; ANSHEN, FRANK. 1986. Computer Methods for Morphological Analysis. See reference: ASSOCIATION FOR COMPUTATIONAL LINGUISTICS. 1986. 120-127.

CARBERRY, SANDRA M. 1984. Understanding Pragmatically Ill-Formed Input. See reference: ASSOCIATION FOR COMPUTATIONAL LINGUISTICS. 1984. 200-206.

CARBONELL, JAIME G. 1982. Metaphor: An Inescapable Phenomenon in Natural Language Comprehension. See reference: LEHNERT, WENDY G.; RINGLE, MARTIN H., eds. 415-434.

CARBONELL, JAIME G.; HAYES, PHILIP J. 1984. Coping With Extragrammaticality. See reference: ASSOCIATION FOR COMPUTATIONAL LINGUISTICS. 1984. 437-443.

CARTER, ALAN W.; FREILING, MICHAEL J. 1984. Simplifying Deterministic Parsing. See reference: ASSOCIATION FOR COMPUTATIONAL LINGUISTICS. 1984. 239-242.

CATER, A. 1983. Request-Based Parsing with Low-Level Syntactic Recognition. See reference: SPARCK JONES, KAREN; WILKS, YORICK, eds. 141-147.

CHARNIAK, E. 1983a. A Parser with Something for Everyone. See reference: KING, MARGARET, ed. 1983a. 117-149.

CHARNIAK, E. 1983b. Parsing, How to. See reference: SPARCK JONES, KAREN; WILKS, YORICK, eds. 156-163.

CHRISTALLER, T.; METZING, D. 1983. Parsing Interactions and a Multi-Level Parser Formalism Based on Cascaded ATNs. See reference: SPARCK JONES, KAREN; WILKS, YORICK, eds. 46-60.

COHEN, PHILIP R.; PERRAULT, C. RAYMOND; ALLEN, JAMES F. 1982. Beyond Question Answering. See reference: LEHNERT, WENDY G.; RINGLE, MARTIN H., eds. 245-274. ISBN: 0-89859-191-0.

CRAIN, STEPHEN; FODOR, JANET DEAN. 1985. How Can Grammars Help Parsers? See reference: DOWTY, DAVID R.; KARTTUNEN, LAURI; ZWICKY, ARNOLD, eds. 94-128.

CRAIN, STEPHEN; STEEDMAN, MARK. 1985. On Not Being Led Up the Garden Path: The Use of Context by the Psychological Syntax Processor. See reference: DOWTY, DAVID R.; KARTTUNEN, LAURI; ZWICKY, ARNOLD M., eds. 320-358.

CULLINGFORD, RICHARD E. 1986. Natural Language Processing: A Knowledge-Engineering Approach. Totowa, NJ: Rowman & Littlefield; 1986. 406p. ISBN: 0-8476-7358-8.

DAHL, VERONICA; MCCORD, MICHAEL C. 1983. Treating Coordination in Logic Grammars. American Journal of Computational Linguistics. 1983 April-June; 9(2): 69-91. ISSN: 0362-613X.

DAMERAU, FRED J. 1985. Problems and Some Solutions in Customization of Natural Language Database Front Ends. ACM Transactions on Office Information Systems. 1985 April; 3(2): 165-184. ISSN: 0734-2047.

DAVIS, DWIGHT B. 1983. Communicating with Computers in English: The Emergence of Natural-Language Processing. Mini-Micro Systems. 1983 October; 16(11): 153-154, 156, 158, 163. ISSN: 0364-9342.

DE, SURANJAN; PAN, SHUH-SHEN; WHINSTON, ANDREW B. 1985. Natural Language Query Processing in a Temporal Database. Data and Knowledge Engineering. 1985; 1: 3-15. ISSN: 0169-023X.

DEJONG, GERALD. 1982. An Overview of the FRUMP System. See reference: LEHNERT, WENDY G.; RINGLE, MARTIN H., eds. 149-176.

DOSZKOCS, TAMAS E. 1986. Natural Language Processing in Information Retrieval. Journal of the American Society for Information Science. 1986 July; 37(4): 191-196. ISSN: 0002-8231.

DOWTY, DAVID R.; KARTTUNEN, LAURI; ZWICKY, ARNOLD M., eds. 1985. Natural Language Parsing: Psychological, Computational, and Theoretical Perspectives. London, England: Cambridge University Press; 1985. 413p. (Studies in Natural Language Processing). ISBN: 0-521-26203-8.

DUNHAM, GEORGE. 1986. The Role of Syntax in the Sublanguage of Medical Diagnostic Statements. See reference: GRISHMAN, RALPH, KITTREDGE, RICHARD, eds. 175-194.

DYER, MICHAEL G.; ZERNIK, URI. 1986. Encoding and Acquiring Meanings for Figurative Phrases. See reference: ASSOCIATION FOR COMPUTATIONAL LINGUISTICS. 1986. 106-111.

EPSTEIN, SAMUEL S. 1985. Transportable Natural Language Processing through Simplicity—The PRE System. ACM Transactions on Office Information Systems. 1985 April; 3(2): 107-120. ISSN: 0734-2047.

FASS, DAN; WILKS, YORICK. 1983. Preference Semantics, Ill-Formedness, and Metaphor. American Journal of Computational Linguistics. 1983 July-December; 9(3-4): 178-187. ISSN: 0362-613X.

FININ, TIMOTHY W. 1986. Constraining the Interpretation of Nominal Compounds in a Limited Context. See reference: GRISHMAN, RALPH; KITTREDGE, RICHARD, eds. 163-173.

FININ, TIMOTHY W., JOSHI, ARAVIND K.; WEBBER, BONNIE LYNN. 1986. Natural Language Interactions with Artificial Experts. Proceedings of the IEEE. 1986 July; 74(7): 921-938. ISSN: 0018-9219.

FITZPATRICK, EILEEN; BACHENKO, JOAN; HINDLE, DON. 1986. The Status of Telegraphic Sublanguages. See reference: GRISHMAN, RALPH; KITTREDGE, RICHARD, eds. 39-51.

FLICKINGER, DANIEL; POLLARD, CARL; WASOW, THOMAS. 1985. Structure Sharing in Lexical Representation. See reference: ASSOCIATION FOR COMPUTATIONAL LINGUISTICS. 1985b. 262-267.

FONG, SANDIWAY; BERWICK, ROBERT. 1985. New Approaches to Parsing Conjunctions Using Prolog. See reference: ASSOCIATION FOR COMPUTATIONAL LINGUISTICS. 1985b. 118-126.

FREY, WERNER; REYLE, UWE; ROHRER, CHRISTIAN. 1983. Automatic Construction of a Knowledge Base by Analysing Texts in Natural Language. See reference: BUNDY, ALAN, ed. 727-729.

FRIEDMAN, CAROL. 1986. Automatic Structuring of Sublanguage Information: Application to Medical Narrative. See reference: GRISHMAN, RALPH; KITTREDGE, RICHARD, eds. 85-102.

GAL, ANNIE; MINKER, JACK. 1985. A Natural Language Database Interface that Provides Cooperative Answers. In: The Engineering of Artificial Intelligence Applications: 2nd Conference on Artificial Intelligence Applications; 1985 December 11-13; Miami, FL. Washington, DC; IEEE Computer Society Press; 1985. 352-357. ISBN: 0-8186-0688-6.

GARVIN, PAUL L.   1985.   The Current State of Language Data Processing.
In: Yovits, Marshall C., ed.   Advances in Computers: Volume 24.   Orlando,
FL: Academic Press; 1985.   217-275.   ISBN: 0-12-012124-7.

GAZDAR, G.   1983.   NLs, CFLs and CF-PSGs.   See reference: SPARCK
JONES, KAREN; WILKS, YORICK, eds.   81-93.

GAZDAR, GERALD; KLEIN, EWAN; PULLUM, GEOFFREY; SAG, IVAN.
1985.   Generalized Phrase Structure Grammar.   Cambridge, MA:
Harvard University Press; 1985.   276p.   ISBN: 0-674-34456-1.

GERSHMAN, ANATOLE V.   1982.   A Framework for Conceptual Analyzers.
See reference: LEHNERT, WENDY G.; RINGLE, MARTIN H., eds.
177-197.

GEVARTER, WILLIAM B.   1984.   Artificial Intelligence, Expert Systems,
Computer Vision, and Natural Language Processing.   Park Ridge, NJ:
Noyes Publications; 1984.   226p.   ISBN: 0-8155-0994-4.

GRANGER, RICHARD H.   1983.   The NOMAD System: Expectation-Based
Detection and Correction of Errors During Understanding of Syntactically
and Semantically Ill-Formed Text.   American Journal of Computational
Linguistics.   1983 July-December; 9(3/4): 188-196.   ISSN: 0362-613X.

GRISHMAN, RALPH.   1986.   Computational Linguistics: An Introduction.
Cambridge, England: Cambridge University Press; 1986.   193p.   (Joshi,
Aravind K., ed.   Studies in Natural Language Processing).   ISBN: 0-521-
32502-1.

GRISHMAN, RALPH; HIRSCHMAN, L.; FRIEDMAN, C.   1983.   Isolating
Domain Dependencies in Natural Language Interfaces.   In: Proceedings
of the Conference on Applied Natural Language Processing; 1983
February 1-3; Santa Monica, CA.   Menlo Park, CA: Association for
Computational Linguistics; 1983.   46-53.

GRISHMAN, RALPH; HIRSCHMAN, LYNETTE; NHAN, NGO THANH.
1986.   Discovery Procedures for Sublanguage Selectional Patterns: Initial
Experiments.   Computational Linguistics.   1986 July-September; 12(3):
205-215.   ISSN: 0362-613X.

GRISHMAN, RALPH; KITTREDGE, RICHARD, eds.   1986.   Analyzing
Language in Restricted Domains: Sublanguage Description and Processing.   Hillsdale, NJ: Lawrence Erlbaum Associates; 1986.   246p.   ISBN:
0-89859-620-3.

GROSZ, BARBARA J.   1983.   TEAM. A Transportable Natural Language
Interface System.   In: Proceedings of the Conference on Applied Natural
Language Processing; 1983 February 1-3; Santa Monica, CA.   Menlo
Park, CA: Association for Computational Linguistics; 1983.   39-45.

GUIDA, GIOVANNI; MAURI, GIANCARLO.   1984.   A Formal Basis for
Performance Evaluation of Natural Language Understanding Systems.
Computational Linguistics.   1984 January-March; 10(1): 15-30.   ISSN:
0362-613X.

GUIDA, GIOVANNI; MAURI, GIANCARLO.   1986.   Evaluation of Natural
Language Processing Systems: Issues and Approaches.   Proceedings of
the IEEE.   1986 July; 74(7): 1026-1035.   ISSN: 0018-9219.

HAFNER, CAROLE D.   1985.   Semantics of Temporal Queries and
Temporal Data.   See reference: ASSOCIATION FOR COMPUTATIONAL
LINGUISTICS.   1985b.   1-8.

HAFNER, CAROLE D.; GODDEN, KURT.   1985.   Portability of Syntax
and Semantics in Datalog.   ACM Transactions on Office Information
Systems.   1985 April; 3(2): 141-164.   ISSN: 0734-2047.

HARPER, MARY P.; CHARNIAK, EUGENE. 1986. Time and Tense in English. See reference: ASSOCIATION FOR COMPUTATIONAL LINGUISTICS. 1986. 3–9.

HARRIS, LARRY R. 1984. Natural Language Front Ends. In: Winston, Patrick H.; Prendergast, Karen A., eds. The AI Business: The Commercial Uses of Artificial Intelligence. Cambridge, MA: The M.I.T. Press; 1984. 149–161. ISBN: 0-262-23117-4.

HARRIS, LARRY R.; DAVIS, DWIGHT B. 1986. Artificial Intelligence Enters the Marketplace. Toronto, Canada: Bantam Books; 1986. 193p. ISBN: 0-553-34293-2.

HARRIS, MARY DEE. 1985. Introduction to Natural Language Processing. Reston, VA: Reston Publishing Company; 1985. 368p. (p. 95, p. 283). ISBN: 0-8359-3253-2.

HASLING, D.; CLANCEY, W.; RENNELS, G. 1984. Strategic Explanations for a Diagnostic Consultation System. International Journal of Man-Machine Studies. 1984 January; 20(1): 3–20. ISSN: 0020-7373.

HAYES, PATRICK. 1978. Trends in Artificial Intelligence. International Journal of Man-Machine Studies. 1978 May; 10(3): 295–299. ISSN: 0020-7373.

HAYES, PHILIP J.; ANDERSEN, PEGGY M.; SAFIER, SCOTT. 1985. Semantic Caseframe Parsing and Syntactic Generality. See reference: ASSOCIATION FOR COMPUTATIONAL LINGUISTICS. 1985b. 153–160.

HENDRIX, GARY G. 1986. The Story of Q & A. See reference: ASSOCIATION FOR COMPUTATIONAL LINGUISTICS. 1986. p. 2.

HIRSCHMAN, LYNETTE; SAGER, NAOMI. 1982. Automatic Information Formatting of a Medical Sublanguage. In: Kittredge, Richard; Lehrberger, John, eds. Sublanguage: Studies of Language in Restricted Semantic Domains. Berlin, W. Germany: Walter de Gruyter; 1982. 27–80. (Posner, Roland, ed. Foundations of Communication: Library Edition). ISBN: 3-11-008244-6.

HIRST, GRAEME. 1981. Anaphora in Natural Language Understanding: A Survey. Berlin, W. Germany: Springer-Verlag; 1981. 128p. (Goos, G.; Hartmanis, J., eds. Lecture Notes in Computer Science; 119). ISBN: 3-540-10858-0.

HOBBS, JERRY R. 1984. Building a Large Knowledge Base for a Natural Language System. See reference: ASSOCIATION FOR COMPUTATIONAL LINGUISTICS. 1984. 283–286.

IZUMIDA, YOSHIO; ISHIKAWA, HIROSHI; YOSHINO, TOSHIAKI; HOSHIAI, TADASHI; MAKINOUCHI, AKIFUMI. 1985. A Natural Language Interface Using a World Model. See reference: ASSOCIATION FOR COMPUTATIONAL LINGUISTICS. 1985a. 205–212.

JENSEN, K.; HEIDORN, G. E.; MILLER, L. A.; RAVIN, Y. 1983. Parse Fitting and Prose Fixing: Getting a Hold on Ill-Formedness. American Journal of Computational Linguistics. 1983 July–December; 9(3-4): 147–160. ISSN: 0362-613X.

JOHNSON, MARK; KLEIN, EWAN. 1986. Discourse, Anaphora and Parsing. In: COLING-86: Proceedings of the 11th International Conference on Computational Linguistics; 1986 August 25–29; Bonn, West Germany. Morristown, NJ: Association for Computational Linguistics; 1986. 669–675.

JOHNSON, R. 1983. Parsing with Transition Networks. See reference: KING, MARGARET, ed. 1983a. 59–72.

JOHNSON, TIM. 1986. NLP Takes Off. Datamation. 1986 January 15; 32(2): 91-93. ISSN: 0011-6963.

JOSHI, ARAVIND K. 1985. Tree Adjoining Grammars: How Much Context-Sensitivity Is Required to Provide Reasonable Structural Descriptions? See reference: DOWTY, DAVID R.; KARTTUNEN, LAURI; ZWICKY, ARNOLD M., eds. 206-250.

JOSHI, ARAVIND, WEBBER, BONNIE; WEISCHEDEL, RALPH. 1984. Living Up to Expectations: Computing Expert Responses. In: AAAI-84: Proceedings of the American Association for Artificial Intelligence (AAAI) National Conference on Artificial Intelligence; 1984 August 6-10; Austin, TX. Los Altos, CA: William Kaufmann; 1984. 169-175. ISBN: 0-86576-080-2.

KAPLAN, RONALD; BRESNAN, JOAN. 1982. Lexical-Functional Grammar: A Formal System for Grammatical Representation. In: Bresnan, Joan, ed. The Mental Representation of Grammatical Relations. Cambridge, MA: M.I.T. Press; 1982. 173-281. ISBN: 0-262-02518-7.

KAY, MARTIN. 1984. Functional Unification Grammar: A Formalism for Machine Translation. See reference: ASSOCIATION FOR COMPUTATIONAL LINGUISTICS. 1984. 75-78.

KAY, MARTIN. 1985. Parsing in Functional Unification Grammar. See reference: DOWTY, DAVID R.; KARTTUNEN, LAURI, ZWICKY, ARNOLD M., eds. 251-278.

KING, MARGARET, ed. 1983a. Parsing Natural Language. London, England: Academic Press; 1983. 308p. ISBN: 0-12-408280-7.

KING, MARGARET. 1983b. Transformational Parsing. See reference: KING, MARGARET, ed. 1983a. 19-34.

KITTREDGE, RICHARD. 1982. Variation and Homogeneity of Sublanguages. In: Kittredge, Richard; Lehrberger, John, eds. Sublanguage: Studies of Language in Restricted Semantic Domains. Berlin, W. Germany: Walter de Gruyter; 1982. 107-137. (Posner, Roland, ed. Foundations of Communication: Library Edition). ISBN: 3-11-008244-6.

KOKTOVA, EVA. 1985. Towards a New Type of Morphemic Analysis. See reference: ASSOCIATION FOR COMPUTATIONAL LINGUISTICS. 1985a. 179-186.

KOSKENNIEMI, KIMMO. 1983. Two-Level Morphological Analysis. See reference: BUNDY, ALAN, ed. 683-685.

KOSY, DONALD W. 1986. Parsing Conjunctions Deterministically. See reference: ASSOCIATION FOR COMPUTATIONAL LINGUISTICS. 1986. 78-82.

KUKICH, KAREN. 1985. Explanation Structures in XSEL. See reference: ASSOCIATION FOR COMPUTATIONAL LINGUISTICS. 1985b. 228-237.

LEBOWITZ, MICHAEL. 1983. Memory-Based Parsing. Artificial Intelligence. 1983 November; 21(4): 363-404. ISSN: 0004-3702.

LEHNERT, WENDY G. 1983. Narrative Complexity Based on Summarization Algorithms. See reference: BUNDY, ALAN, ed. 713-716.

LEHNERT, WENDY G.; DYER, MICHAEL G.; JOHNSON, PETER N.; YANG, C. J.; HARLEY, STEVE. 1983. BORIS—An Experiment in In-Depth Understanding of Narratives. Artificial Intelligence. 1983 January; 20(1): 15-62. ISSN: 0004-3702.

LEHNERT, WENDY G.; RINGLE, MARTIN H., eds. 1982. Strategies for Natural Language Processing. Hillsdale, NJ: Lawrence Erlbaum Associates, 1982. 533p. ISBN: 0-89859-191-0.

LEHRBERGER, JOHN. 1982. Automatic Translation and the Concept of Sublanguage. In: Kittredge, Richard; Lehrberger, John, eds. Sublanguage: Studies of Language in Restricted Semantic Domains. Berlin, W. Germany: Walter de Gruyter; 1982. 81–106. (Posner, Roland, ed. Foundations of Communication: Library Edition). ISBN: 3-11-008244-6.

LESMO, LEONARDO; TORASSO, PIETRO. 1985. Analysis of Conjunctions in a Rule-Based Parser. See reference: ASSOCIATION FOR COMPUTATIONAL LINGUISTICS. 1985b. 180–187.

LITMAN, DIANE J.; ALLEN, JAMES F. 1984. A Plan Recognition for Clarification Subdialogues. See reference: ASSOCIATION FOR COMPUTATIONAL LINGUISTICS. 1984. 302–311.

MANN, WILLIAM C. 1983. An Overview of the Penman Text Generation System. In: AAAI-83: Proceedings of the American Association for Artificial Intelligence (AAAI) National Conference on Artificial Intelligence; 1983 August 22–26; Washington, DC. Los Altos, CA: William Kaufmann; 1983. 261–265. ISBN: 0-86576-080-2.

MANN, WILLIAM C. 1984. Discourse Structures for Text Generation. See reference: ASSOCIATION FOR COMPUTATIONAL LINGUISTICS. 1984. 367–375.

MARCUS, MITCHELL P. 1980. A Theory of Syntactic Recognition for Natural Language. Cambridge, MA: MIT Press; 1980. 335p. ISBN: 0-262-13149-8.

MARCUS, MITCHELL P. 1984. Some Inadequate Theories of Human Language Processing. In: Bever, Thomas G.; Carroll, John M.; Miller, Lance A., eds. Talking Minds: The Study of Language in Cognitive Science. Cambridge, MA: MIT Press; 1984. 253–278. ISBN: 0-262-02181-1.

MARKOWITZ, JUDITH; AHLSWEDE, THOMAS; EVENS, MARTHA. 1986. Semantically Significant Patterns in Dictionary Definitions. See reference: ASSOCIATION FOR COMPUTATIONAL LINGUISTICS. 1986. 112–119.

MARSH, ELAINE; FRIEDMAN, CAROL. 1985. Transporting the Linguistic String Project System from a Medical to a Navy Domain. ACM Transactions on Office Information Systems. 1985 April; 3(2): 121–140. ISSN: 0734-2047.

MARSH, ELAINE; HAMBURGER, HENRY; GRISHMAN, RALPH. 1984. A Production Rule System for Message Summarization. In: AAAI-84: Proceedings of the American Association for Artificial Intelligence (AAAI) National Conference on Artificial Intelligence; 1984 August 6–10; Austin, Texas. Los Altos, CA: William Kaufmann; 1984. 243–246.

MCCOY, KATHLEEN F. 1986. The ROMPER System: Responding to Object-Related Misconceptions Using Perspective. See reference: ASSOCIATION FOR COMPUTATIONAL LINGUISTICS. 1986. 97–105.

MCDONALD, DAVID; PUSTEJOVSKY, JAMES D. 1985a. A Computational Theory of Prose Style for Natural Language Generation. See reference: ASSOCIATION FOR COMPUTATIONAL LINGUISTICS. 1985a. 187–193.

MCDONALD, DAVID D.; PUSTEJOVSKY, JAMES D. 1985b. TAG's as a Grammatical Formalism for Generation. See reference: ASSOCIATION FOR COMPUTATIONAL LINGUISTICS. 1985b. 94–103.

MCKEOWN, KATHLEEN R. 1983. Paraphrasing Questions Using Given and New Information. American Journal of Computational Linguistics. 1983 January–March; 9(1): 1–10. ISSN: 0362-613X.

MCKEOWN, KATHLEEN R. 1985. Text Generation: Using Discourse Strategies and Focus Constraints to Generate Natural Language Text. London, England: Cambridge University Press; 1985. 246p. (Joshi, Aravind K., ed. Studies in Natural Language Processing). ISBN: 0-521-30116-5.

MCKEOWN, KATHLEEN R. 1986. Language Generation: Applications, Issues, and Approaches. Proceedings of the IEEE. 1986 July; 74(7): 961–968. ISSN: 0018-9219.

MCKEOWN, KATHLEEN R.; WISH, M.; MATTHEWS, K. 1985. Tailoring Explanations for the User. In: Joshi, Aravind, ed. IJCAI-85: Proceedings of the 9th International Joint Conference on Artificial Intelligence; 1985 August 18-24; Los Angeles, CA. Los Altos, CA: William Kaufmann; 1985. 794–798. ISBN: 0-86576-064-0.

MELLISH, C. S. 1983. Incremental Semantic Interpretation in a Modular Parsing System. See reference: SPARCK JONES, KAREN; WILKS, YORICK, eds. 148–155.

MELLISH, C. S. 1985. Computer Interpretation of Natural Language Descriptions. Chichester, England: Ellis Horwood Ltd.; 1985. 182p. (Campbell, John, ed. Ellis Horwood Series in Artificial Intelligence). ISBN: 0-85312-828-6.

MILNE, ROBERT. 1986. Resolving Lexical Ambiguity in a Deterministic Parser. Computational Linguistics. 1986 January–March; 12(1): 1–12. ISSN: 0362-613X.

MONTGOMERY, CHRISTINE A.; GLOVER, BONNIE. 1986. A Sublanguage for Reporting and Analysis of Space Events. See reference: GRISHMAN, RALPH; KITTREDGE, RICHARD, eds. 129–161.

MORIK, KATHARINA. 1984. Customers' Requirements for Natural Language Systems: Results of an Inquiry. International Journal of Man-Machine Studies. 1984 October; 21(4): 401–414. ISSN: 0020-7373.

MORIK, KATHARINA. 1985. User Modelling, Dialog Structure, and Dialog Strategy in HAM-ANS. See reference: ASSOCIATION FOR COMPUTATIONAL LINGUISTICS. 1985a. 268–273.

NECHES, R.; SWARTOUT, R.; MOORE, J. 1985. Explainable (and Maintainable) Expert Systems. In: Joshi, Aravind, ed. IJCAI-85: Proceedings of the 9th International Joint Conference on Artificial Intelligence; 1985 August 18-24; Los Angeles, CA. Los Altos, CA: William Kaufmann; 1985. 382–389. ISBN: 0-86576-064-0.

POLLACK, MARTHA E. 1986. A Model of Plan Inference that Distinguishes Between the Beliefs of Actors and Observers. See reference: ASSOCIATION FOR COMPUTATIONAL LINGUISTICS. 1986. 207–214.

RAMSAY, ALLAN. 1985. Effective Parsing with Generalized Phrase Structure Grammar. See reference: ASSOCIATION FOR COMPUTATIONAL LINGUISTICS. 1985a. 57–61.

REICHMAN, RACHEL. 1985. Getting Computers to Talk Like You and Me: Discourse, Context, Focus and Semantics (An ATN Model). Cambridge, MA: MIT Press; 1985. 221p. ISBN: 0-262-18118-5.

REITMAN, WALTER, ed. 1984. Artificial Intelligence Applications for Business. Norwood, NJ: Ablex; 1984. 343p. ISBN: 0-89391-220-4.

REYLE, UWE; FREY, WERNER. 1983. A PROLOG Implementation of Lexical Functional Grammar. See reference: BUNDY, ALAN, ed. 693–695.

RIESBECK, CHRISTOPHER K. 1982. Realistic Language Comprehension. See reference: LEHNERT, WENDY G.; RINGLE, MARTIN H., eds. 37–54.

RISTAD, ERIC SVEN. 1986. Computational Complexity of Current GPSG Theory. See reference: ASSOCIATION FOR COMPUTATIONAL LINGUISTICS. 1986. 30–39.

RITCHIE, GRAEME D. 1983. The Implementation of a PIDGIN Interpreter. See reference: SPARCK JONES, KAREN; WILKS, YORICK, eds. 69–80.

RITCHIE, GRAEME D. 1986. The Computational Complexity of Sentence Derivation in Functional Unification Grammar. In: COLING-86: Proceedings of the 11th International Conference on Computational Linguistics; 1986 August 25–29; Bonn, West Germany. Morristown, NJ: Association for Computational Linguistics; 1986. 584–586.

SAGER, NAOMI. 1981. Natural Language Information Processing: A Computer Grammar of English and Its Applications. Reading, MA: Addison-Wesley; 1981. 399p. ISBN: 0-201-06769-2.

SAINT-DIZIER, PATRICK. 1985. Handling Quantifier Scoping Ambiguities in a Semantic Representation of Natural Language Sentences. In: Dahl, Veronica; Saint-Dizier, Patrick, eds. Natural Language Understanding and Logic Programming: Proceedings of the 1st International Workshop on Natural Language Understanding and Logic Programming; 1984 September 18–20; Rennes, France. Amsterdam, The Netherlands: North-Holland; 1985. 49–63. ISBN: 0-444-87714-2.

SALTON, GERARD; MCGILL, MICHAEL J. 1983. Introduction to Modern Information Retrieval. New York, NY: McGraw-Hill; 1983. 448p. (pp. 257–302). ISBN: 0-07-054484-0.

SCHANK, ROGER C.; BIRNBAUM, LAWRENCE. 1984. Memory, Meaning and Syntax. In: Bever, Thomas G.; Carroll, John M.; Miller, Lance A., eds. Talking Minds: The Study of Language in Cognitive Science. Cambridge, MA: MIT Press; 1984. 209–251. ISBN: 0-262-02181-1.

SCHANK, ROGER C.; CHILDERS, PETER. 1984. The Cognitive Computer: On Language, Learning, and Artificial Intelligence. Reading, MA: Addison-Wesley; 1984. 268p. ISBN: 0-201-06446-4.

SCHANK, ROGER C.; SLADE, STEPHEN. 1984. Advising Systems. See reference: REITMAN, WALTER, ed. 249–265.

SELFRIDGE, MALLORY. 1986. Integrated Processing Produces Robust Understanding. Computational Linguistics. 1986 April–June; 12(2): 89–106. ISSN: 0362-613X.

SHAPIRO, STUART C. 1982. Generalized Augmented Transition Network Grammars for Generation from Semantic Networks. American Journal of Computational Linguistics. 1982 January–March; 8(1): 12–25. ISSN: 0362-613X.

SHIMAZU, AKIRA; NAITO, SYOZO; NOMURA, HIROSATO. 1983. Japanese Language Semantic Analyzer Based on an Extended Case Frame Model. See reference: BUNDY, ALAN, ed. 717–719.

SHWARTZ, STEVEN P. 1984. Natural Language Processing in the Commercial World. See reference: REITMAN, WALTER, ed. 235–248.

SIDNER, CANDACE L. 1983. What the Speaker Means: The Recognition of Speakers' Plans in Discourse. In: Cercone, Nick, ed. Computational Linguistics. Oxford, England: Pergamon Press; 1983. 71–82. (Rodin, E. Y., ed. International Series in Modern Applied Mathematics and Computer Science). ISBN: 0-08-030253-X.

SIMON, HERBERT A. 1981. Information Processing Models of Cognition. Journal of the American Society for Information Science. 1981 September; 32(5): 364-377. ISSN: 0002-8231.

SMALL, STEVEN; RIEGER, CHUCK. 1982. Parsing and Comprehending with Word Experts (A Theory and its Realization). See reference: LEHNERT, WENDY G.; RINGLE, MARTIN H., eds. 89-147.

SOPENA, LUIS DE. 1983. Natural Language Grammars for an Information System. In: Kuehn, Jennifer J., ed. Research and Development in Information Retrieval: ACM/SIGIR 6th Annual International Conference; 1983 June 6-8; Bethesda, MD. New York, NY: Association for Computing Machinery; 1983. 75-80. ISBN: 0-89791-107-5.

SPARCK JONES, KAREN. 1984. Natural Language and Databases, Again. See reference: ASSOCIATION FOR COMPUTATIONAL LINGUISTICS. 1984. 182-183.

SPARCK JONES, KAREN. 1985. Compound Noun Interpretation Problems. In: Fallside, Frank; Woods, William A., eds. Computer Speech Processing. Englewood Cliffs, NJ: Prentice Hall; 1985. 363-381. ISBN: 0-13-163841-6.

SPARCK JONES, KAREN; BOGURAEV, BRAN K. 1984. A Natural Language Front End to Data Bases with Evaluative Feedback. In: Gardarin, George; Gelenbe, Erol, eds. New Applications of Data Bases. London, England: Academic Press; 1984. 159-182. ISBN: 0-12-275550-2.

SPARCK JONES, KAREN; WILKS, YORICK, eds. 1983. Automatic Natural Language Parsing. Chichester, England: Ellis Horwood Ltd.; 1983. 208p. (Campbell, John, ed. Ellis Horwood Series in Artificial Intelligence). ISBN: 0-85312-621-6.

SWARTOUT, WILLIAM R. 1985. Knowledge Needed for Expert System Explanation. In: Wojcik, Anthony S., ed. Proceedings of the American Federation of Information Processing Societies (AFIPS) Conference; 1985 July 15-18; Chicago, IL. Reston, VA: AFIPS Press; 1985. 93-98. ISBN: 0-88283-046-5.

TAIT, J. I. 1985. Generating Summaries Using a Script-Based Language Analyzer. In: Steels, Luc; Campbell, J. A., eds. Progress in Artificial Intelligence. Chichester, England: Ellis Horwood; 1985. 312-318. (Campbell, John, ed. Ellis Horwood Series in Artificial Intelligence). ISBN: 0-470-20171-1.

THOMPSON, BOZENA HENISZ; THOMPSON, FREDERICK B. 1985. ASK Is Transportable in Half a Dozen Ways. ACM Transactions on Office Information Systems. 1985 April; 3(2): 185-203. ISSN: 0734-2047.

TUCKER, ALLEN B., JR.; NIRENBURG, SERGEI. 1984. Machine Translation: A Contemporary View. In: Williams, Martha E., ed. Annual Review of Information Science and Technology: Volume 19. White Plains, NY: Knowledge Industry Publications; 1984. 129-160. ISSN: 0066-4200; ISBN: 0-86729-093-5.

VAUGHAN, MARIE M.; MCDONALD, DAVID D. 1986. A Model of Revision in Natural Language Generation. See reference: ASSOCIATION FOR COMPUTATIONAL LINGUISTICS. 1986. 90-96.

WAHLSTER, WOLFGANG; KOBSA, ALFRED. 1986. Dialogue-Based User Models. Proceedings of the IEEE. 1986 July; 74(7): 948-960. ISSN: 0018-9219.

WALKER, DONALD E.; AMSLER, ROBERT A. 1986. The Use of Machine-Readable Dictionaries in Sublanguage Analysis. See reference: GRISHMAN, RALPH; KITTREDGE, RICHARD, eds. 69-83.

WALTZ, DAVID L. 1982. The State of the Art in Natural Language Understanding. See reference: LEHNERT, WENDY G.; RINGLE, MARTIN H., eds. 3-32.

WEBBER, BONNIE LYNN; FININ, TIM. 1984. In Response: Next Step in Natural Language Interaction. See reference: REITMAN, WALTER, ed. 211-234.

WEHRLI, ERIC. 1985. Design and Implementation of a Lexical Data Base. See reference: ASSOCIATION FOR COMPUTATIONAL LINGUISTICS. 1985a. 146-153.

WEINER, E. JUDITH. 1984. A Knowledge Representation Approach to Understanding Metaphors. Computational Linguistics. 1984 January-March; 10(1): 1-14. ISSN: 0362-613X.

WEISCHEDEL, RALPH M. 1986. Knowledge Representation and Natural Language Processing. Proceedings of the IEEE. 1986 July; 74(7): 905-920. ISSN: 0018-9219.

WEISCHEDEL, RALPH M.; SONDHEIMER, NORMAN K. 1983. Meta-Rules as a Basis for Processing Ill-Formed Input. American Journal of Computational Linguistics. 1983 July-December; 9(3-4): 161-177. ISSN: 0362-613X.

WILKS, Y.; SPARCK JONES, K. 1983. Introduction: A Little Light History. See reference: SPARCK JONES, KAREN; WILKS, YORICK, eds. 11-21.

WILLIAMSON, MICKEY. 1985. Artificial Intelligence for Microcomputers: The Guide for Business Decision Makers. New York, NY: Brady Communications; 1985. 184p. ISBN: 0-89303-483-5.

WINOGRAD, TERRY. 1981. What Does It Mean to Understand Language? In: Norman, Donald A., ed. Perspectives on Cognitive Science. Norwood, NJ: Ablex; 1981. 231-263. ISBN: 0-89391-071-6.

WINOGRAD, TERRY. 1983. Language as a Cognitive Process. Volume I: Syntax. Reading, MA: Addison-Wesley; 1983. 640p. ISBN: 0-201-08571-2.

WINOGRAD, TERRY; FLORES, FERNANDO. 1986. Understanding Computers and Cognition: A New Foundation for Design. Norwood, NJ: Ablex; 1986. 207p. ISBN: 0-89391-050-3.

4   Retrieval Techniques

NICHOLAS J. BELKIN
Rutgers University

W. BRUCE CROFT
University of Massachusetts

INTRODUCTION

Retrieval Techniques

This review is concerned with a specific aspect of research and development (R&D) in information retrieval (IR) systems—that is, the means for identifying, retrieving, and/or ranking texts (or text surrogates or portions of texts), in a collection of texts, that might be relevant to a given query (or useful for resolving a particular problem). In particular, retrieval techniques address the issue of comparing a representation of a query with representations of texts for the above purpose. Although we necessarily discuss different means of representation, our focus is on different techniques for comparison and not on the generation of the representations. Figure 1 indicates the situation with which we are concerned and shows that different representations allow different retrieval techniques without necessarily specifying them.

Our limitation of the topic of this review also means that a number of techniques used for retrieval, or discussed in the IR literature as having to do with retrieval, will be discussed only as aspects of simpler, or basic matching techniques. For instance, in the context of Figure 1, it is clear that "feedback" techniques are not different retrieval techniques but rather methods

The authors would like to thank Barbara Kwaśnik for her invaluable efforts, bibliographic, technical and otherwise supportive, without which this document could not have been produced.

Annual Review of Information Science and Technology (ARIST), Volume 22, 1987
Martha E. Williams, Editor
Published for the American Society for Information Science (ASIS)
by Elsevier Science Publishers B.V.

Figure 1   Retrieval technique situation

for enhancing the query or request model, which can then be used with various techniques for eventual or subsequent comparison. Thus, rather than discussing feedback on its own as a retrieval technique, we mention it as a representation mechanism for the various specific retrieval techniques.

Limits on This Review

Although there has been no review in *ARIST* specifically and solely on retrieval techniques as we have defined them, there are two that we consider precursors to this one--namely the chapters by MCGILL & HUITFELDT and by BOOKSTEIN (1985). The former deals with research in information retrieval in general and covers retrieval techniques quite well. It thus provides a starting point for our review. The latter deals with the use of probability and fuzzy set theories in IR and thus also with their related retrieval techniques. Because Bookstein's review is so recent, we do not review in depth all of the post-1980 material he covers. Instead we comment in general on probability and fuzzy set dependent techniques, and provide detailed discussion only of selected work covered by Bookstein, and of material published since his review. Thus, our review covers the R&D literature on retrieval techniques since 1980 except for that dealing with probability theory and fuzzy set theory, which is primarily post-1985.

Our review also omits research concerned primarily with representational issues (whether of query or of text); rather it concentrates on work concerned explicitly with different models for actually comparing query representation and text representation. We also do not treat research on file organization or efficiency of storage and retrieval, concentrating rather on the logic of retrieval. This review covers text retrieval only; it excludes pattern recognition, image matching, numerical representation, etc. as well as chemical structure searching.

In keeping with previous *ARIST* practice, we try to be reasonably comprehensive within these limits in our bibliography, but in the text we discuss at some length selected items as examples of particular approaches. Thus, some items in the bibliography, although read by us, are not cited in the text.

Aims of This Review

This seems to be an especially appropriate time to review R&D in retrieval techniques. The general situation in this field can be characterized as follows. There are a few techniques actually used in large-scale operational IR systems (namely, Boolean or string searching), and there appears to be some general dissatisfaction with them. These techniques have become established more through practice than theory. At the same time, there are a number of quite different techniques, usually with a strong theoretical basis, that have been used almost exclusively in experimental settings (e.g., probabilistic retrieval), but these are now well developed in many of their aspects. These "experimental" techniques, when compared with the "operational" techniques in a controlled setting, have almost always performed better on standard measures and often *very* much better. Since these two types of techniques are now well established in their respective settings, why has the experimental experience had so little effect on the operational environment? We attempt to deal with this question on a technical rather than social level.

A related issue is that although there appears to be a general feeling that retrieval techniques can be classed broadly as experimental or operational, the basis for this classification is nothing more than an historical accident of custom or use. To deal with the issues involved in understanding retrieval techniques and the associated problems in IR systems, a more principled and detailed classification seems necessary. One of its aims would be to establish meaningful relationships among the classes of objects involved. We propose such a classification and base the structure of our review on it.

For some time a number of researchers in the field have noted that all retrieval techniques perform better for some queries than for others. In the cases where techniques have been compared with one another on a micro level, it appears that differences in performance on specific queries are masked by evaluation on cumulated results, and that especially, even when specific techniques have been shown to perform more poorly than another technique overall, they may have done much better on at least some individual queries. Thus, are some techniques better for some kinds of queries than others? Although there has not been much research on this issue, there is sufficient experience now to think about an approach to this issue.

Finally, integrated information systems with vastly different problem contexts, databases, and so on within them are now being either constructed or contemplated (e.g., office automation systems, integrated IR and database management systems (DBMS)). Given the experience discussed above, will multiple retrieval techniques be necessary within such systems, and if so, on what bases should they be chosen and used and how can they be integrated within a single system design? With these questions in mind, we hope to provide a framework for work already done and perhaps to provide some guidelines to research on these issues.

Organization of the Review

On the basis of the objectives, the rest of this chapter is structured as follows. First, we present a classification of retrieval techniques and use this classification as the basis for discussing specific R&D in retrieval techniques since about 1980. We then discuss comparative performance studies and the relationship of request representation to retrieval techniques; this leads to the issues of choosing appropriate retrieval techniques and using multiple techniques. The next section deals with architectures for integrated information systems, including so-called "expert" intermediaries, especially from the point of view of problems of retrieval techniques. We conclude with a general discussion of the current status of R&D in retrieval techniques, identification of the issues and problems that appear to be most crucial, and identification of research directions that seem to be the most pressing as well as those that seem most likely to be fruitful in developing retrieval techniques for truly effective information systems.

A CLASSIFICATION OF RETRIEVAL TECHNIQUES

We have defined a retrieval technique as a technique for comparing the query with the document representations. We can further classify retrieval techniques in terms of the characteristics of the retrieved set of documents and the representations that are used. Some techniques do not fall naturally into only a single category in this classification, and others are hybrids of techniques from different categories, but the scheme is useful for discussing the broad distinctions among retrieval techniques. Figure 2 gives a diagrammatic view of the classification. The first distinction that we make among retrieval techniques is whether the set of retrieved documents contains only documents whose representations are an exact match with the query or a partial match with the query. For a partial match, the set of retrieved documents will include also those that are an exact match with the query.

The next level of the classification distinguishes between retrieval techniques that compare the query with individual document representatives and techniques that use a representation of documents that emphasizes connections to other documents in a network. In this category, individual documents are retrieved, but the retrieval is based on connections to other documents and

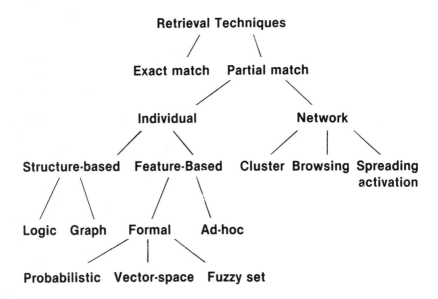

Figure 2 A classification of retrieval techniques

not solely on the contents of an individual document. In the network category, we identify the subcategories of cluster-based searches, searches based on browsing a network of documents, and spreading activation searches.

The individual category breaks down into retrieval techniques that use a feature-based representation of queries and documents and techniques that use a structure-based representation. In a feature-based representation, queries and documents are represented as sets of features, such as index terms. Features can be weighted and can represent more complex entities in the text than single words. The structure-based category is divided into representations based on logic, that is, those in which the meaning of queries and documents are represented using some formal logic, and on representations that are similar to graphs, in which documents and queries are represented by graph-like structures composed of nodes and edges connecting these nodes. Such graphs can be produced by natural language processing (e.g., semantic nets and frames) or statistical techniques.

The feature-based category includes techniques based on formal models (including the vector space model, probabilistic model, fuzzy set model and others) and techniques based on ad-hoc similarity measures. In the following sections, we discuss the techniques that make up these categories in more detail.

EXACT MATCH TECHNIQUES

Exact match retrieval techniques are those that require that the request model be contained, precisely as represented in the query formulation, within the text representation. Implemented as Boolean, full-text, or string searching, this is the retrieval technique in current use in most of the large operational IR systems. The disadvantages of this type of technique are well known and well documented and a variety of aids such as thesauri are required to achieve reasonable performance. In the simple case exact match searching: 1) misses many relevant texts whose representations match the query only partially; 2) does not rank retrieved texts; 3) cannot take into account the relative importance of concepts either within the query or within the text; 4) requires complicated query logic formulation; and 5) depends on the two representations being compared having been drawn from the same vocabulary. BOOK-STEIN (1985) mentions several other undesirable characteristics of Boolean retrieval. Note that we do not consider the provision of "wild card" strings to be true partial matching.

Given its many and obvious objections, why does exact match searching remain the paradigm for operational systems? The traditional answers are that the investment in current systems is so great that changing them is economically unfeasible, that alternative techniques are untested in large-scale environments, and that the results of alternative techniques are not sufficiently better even in experimental environments to justify any changes. A more significant argument, accepted it seems by all parties, is that the structures of Boolean statements represent important aspects of user's queries or problems.

Research in exact match retrieval techniques in the period covered has dealt with all of the problems mentioned above to some extent and with one additional problem: that of efficiency of searching for strings. The major efforts in the logic of exact match retrieval have been in making it less exact, in taking into account relative importance, and in achieving sensible ranking rules. The problem of difficulty of use has been approached mainly via interface design, without reference to the underlying technique, while the vocabulary problem has been considered primarily, although not exclusively, one of request-model elaboration. The efficiency problem has been dealt with by techniques such as file organization (e.g., ARNOW ET AL.), specialized hardware (e.g., CARROLL ET AL.; HOLLAAR, 1984), and text compression (e.g., MOHAN & WILLETT). As far as exact match retrieval techniques themselves are concerned, there has really been no research during the period covered by this review. That is, the only exact match logics that are available are Boolean or simple string matching, and no one has suggested any new exact matching logic. Thus our discussion devolves to brief mention of some attempts to modify exact match searching.

It is possible to relax some constraints on exact match searching by the expedient of specifying parts of the string to be matched which can be ignored. Truncation (that is, ignoring endings of words after some point) and so-called "wild-card" searching are examples of this approach. Since there is really no way to make exact retrieval techniques less exact within their own logic, all attempts to do so have necessarily resulted in hybrids of exact techniques and partial match techniques. Since partial match techniques automatically produce retrieved text rankings, so do these hybrids. Similarly, all attempts to take into account the relative importance of aspects of query and text have relied on partial match techniques of one sort or another. We therefore do not attempt to separate these from one another.

Perhaps the most interesting approach to the question of extending the logic of exact matching has been the recent effort to deal with Boolean techniques as a special case of either vector or probabilistic models. Salton and his colleagues have developed an extended vector model, and CROFT (1986) has proposed a method for making use of the relations established in a Boolean query within a probabilistic search model. For more detail on these approaches, see BOOKSTEIN (1985) and the sections on the "Vector Space Model" and "Probabalistic Model" below.

The vocabulary and efficiency approaches to modifying exact match techniques are more properly seen as attempts to deal with problems common to various retrieval techniques. Thus, for instance, Fox's relational thesaurus for Boolean retrieval (FOX, 1987b) is equally relevant for probabilistic or vector retrieval as are suggestions for a "user thesaurus" (e.g., BATES). Some methods have been developed within operational IR systems to support request-model elaboration. The ZOOM facility on ESA-IRS, for instance, allows frequency-ranked display of terms associated with a retrieved document set (INGWERSEN). All such work deals with the problem of reconciling the query vocabulary with the document or with the index vocabulary in matching and, as such, is relevant to any retrieval technique that faces this issue.

PARTIAL MATCH TECHNIQUES

Individual, Feature-Based Techniques

Techniques in this category are used to compare queries with documents represented as sets of features or index terms. The document representatives are derived from the text of the document either by manual or automated indexing. Similarly, the query terms can either be derived from a query expressed in natural language, or an indexing vocabulary can be used directly for specifying queries. The retrieval techniques used do not depend on the indexing method, and the relative merits of automated and manual indexing are not discussed here. For details on automated indexing techniques and comparisons with manual indexing, see SALTON (1986a), SALTON & MCGILL, and SPARCK JONES.

Features can represent single words, stems, phrases, or concepts and can have weights associated with them. The weights are typically derived from the way the feature is used in an individual document, such as a within document frequency weight, or the way it is used in the document collection, such as the inverse document frequency weight (SALTON & MCGILL, SPARCK JONES).[1] The interpretation of the weights, the way they are used, and the way they are calculated depend on the particular retrieval technique and the underlying retrieval model selected.

In the following discussion of feature-based retrieval techniques, we assume that a document has a representation consisting of a vector of terms $(d_1, d_2, \ldots d_m)$, where d_i indicates the presence or absence of term i and has the value 1 or 0. Weights associated with these terms are introduced as needed. A query has a similar representation, with q_i referring to the ith query term. Techniques that deviate from the purely feature-based approach but are strongly related to it, such as the partial match Boolean query techniques, are also discussed in this section.

Formal. These retrieval techniques are based on formal models of document retrieval and indexing. In this review, we concentrate on the major modeling approaches that have been used for information retrieval: vector space, probabilistic, and fuzzy set. The reader is referred to BOOKSTEIN (1985), ROBERTSON (1977b), SALTON (1979), and VAN RIJSBERGEN (1979) for further discussions of models of information retrieval. Bookstein's 1985 review, in particular, has described the probabilistic and fuzzy set models in detail, and the presentation in this paper emphasizes the retrieval techniques based on those models, rather than theoretical aspects.

Vector space model. In the vector space model, documents and queries are vectors in an n-dimensional space, where each dimension corresponds to an index term. The model has intuitive appeal and has formed the basis of a

---

[1] The within-document frequency is the number of times an index term (usually a word stem) occurs in the document text (usually the abstract). The inverse document frequency is the inverse of the relative frequency of a term in the collection. The logarithm of this ratio is often used, giving a weight of log N/n, where N is the number of documents in the collection and n is the number of documents that contain the term.

large part of IR research, including the SMART system (SALTON, 1968; SALTON & MCGILL). Although this was one of the first models proposed, modifications to it are still appearing (WONG & RAGHAVAN; WONG ET AL.). From the point of view of retrieval techniques, there have been very specific recommendations about how the model should be applied in operational systems (SALTON, 1981; 1986a). The recommended retrieval process is as follows:

1. Term weights are calculated using a combination of the normalized within-document frequency (*tf*) and the inverse document frequency (*idf*). This *tf.idf* weight can be calculated for document terms either as part of the retrieval process or, less accurately, when the document is indexed.
2. Terms that have poor "discrimination value" (terms that are not useful for distinguishing among documents) are replaced by terms representing thesaurus classes for low-frequency terms and phrases for high-frequency terms. Discrimination values are calculated by observing the document space before and after assignment of a term. If the documents tend to move together after assignment (as measured by the average pairwise similarity), the term is a poor discriminator.
3. Documents are ranked in decreasing order of similarity to the query as measured by the cosine correlation (intuitively retrieving those documents closest to the query in vector space). This is calculated by formula 1:

$$\sum d_i q_i \left/ \sqrt{\sum d_i^2 q_i^2} \right. \qquad (1)$$

where d_i is the *tf.idf* weight.

There are some important points to make about this retrieval technique that also apply to other techniques. First, the exact formula for calculating term weights, such as *tf.idf*, may vary from one system or set of experiments to another. For example, it is common in calculating the *idf* weight to normalize the document collection frequency with the maximum collection frequency rather than simply the number of documents in the collection; this is done to expand the range of *idf* weights that result. Similarly, in some experiments the *tf* weight is normalized with the maximum within-document frequency, and in others it is not normalized. These details can have a significant impact on the effectiveness of a system, and it is essential to check their validity.

Second, when similarity measures are used, the query is not directly compared with every document in the collection to produce a ranking. Typically, an inverted file is used to exclude documents that have no terms in common with the query. Refinements of this technique have been devised to reduce search time further (BUCKLEY & LEWIT; CROFT & PARENTY; SMEATON & VAN RIJSBERGEN, 1981).

Finally, although the term weighting done in steps 1 and 2 is often regarded as part of the indexing process, it can also be done during retrieval. The weights used follow directly from the retrieval models described in the next section, and because the collection is dynamic, they can be more accurately calculated during retrieval. The identification of important relationships among words and the expansion of terms to include thesaurus classes can also be part of retrieval since it is those relationships and terms that are relevant to a particular query that should be identified (CROFT & THOMPSON, 1987).

An important extension of the retrieval techniques based on the vector space model is extended Boolean retrieval (FOX, 1983; SALTON, 1983, 1985; SALTON & VOORHEES; SALTON ET AL., 1983b; SALTON ET AL., 1985). This technique overlaps the structure-based category because it uses structured (Boolean) queries. That is, the query is formulated with index terms and the Boolean operators AND, OR, and NOT. The prevalence of systems that use Boolean queries has led researchers to consider the problem of producing ranked output from Boolean specifications. In this approach to the problem, a similarity measure is defined that ranks documents, giving precedence to those that match all or part of the Boolean specification. For example, consider a situation in which document 1 contains terms A and B, document 2 contains terms B and C, and the query is A AND (B OR C). Ignoring the effect of term weights, the standard cosine correlation would rank documents 1 and 2 equally because both have two query terms. The extended Boolean similarity measure would rank document 1 higher because it matches the Boolean query specification. The addition of term weights makes the calculation more complex, but the general effect is the same.

Research based on the vector space model has led to other techniques, such as relevance feedback, clustering, and document space modification (SALTON, 1968). These techniques are discussed in other sections.

Probabilistic model. The version of the probabilistic model that is discussed most often in research papers was introduced by ROBERTSON & SPARCK JONES and VAN RIJSBERGEN (1979). Other forms of the model are discussed by BOOKSTEIN (1985), but these have not contributed significantly to the development of new retrieval techniques. The advantages of the probabilistic model are the insights it gives into techniques that have been used in previous research and that it is a powerful framework for developing new techniques.

The retrieval techniques based on the probabilistic model are very similar to those developed from the vector space model. The basic aim is to retrieve documents in order of their probability of relevance to the query (ROBERT-SON, 1977a). If we assume that document term weights are either 1 or 0 and that terms are independent of each other, this can be shown to be achieved by ranking documents according to Formula 2:

$$\sum d_i \, q_i \qquad\qquad (2)$$

where q_i is a weight equal to $\log pr_i$ $(1\text{-}pnr_i)/pnr_i$ $(1\text{-}pr_i)$ where pr_i is the probability that term i occurs in the relevant set of documents, and pnr_i is the probability that term i occurs in the nonrelevant set of documents.

The problem in applying this ranking function is to estimate the probabilities in the query term weights. Experimental results have shown that pnr is best estimated from the entire collection (HARPER & VAN RIJSBERGEN). This means that the formula, $pnr_i = n_i/N$ (where n_i is the number of documents that contain term i and N is the number of documents), is a reasonable estimate. If a set of relevant documents is available (for example, after feedback), the best estimate for pr_i is r_i/R, where r_i is the number of relevant documents that contain term i and R is the total number of relevant documents. This is another case in which details are important. The actual estimation formula that provides the best results is $(r_i + 0.5)/(R + 1)$ (SPARCK JONES & WEBSTER). When r_i is 0, this estimate is too high and a value of 0.05 for pr_i is used. If a set of relevant documents is not available, as in an initial search, the retrieval technique suggested by the simple probabilistic model is approximately equivalent to using the idf weight in Formula 2 (CROFT & HARPER).

The probabilistic model can be extended to use within-document frequency information (CROFT, 1983b). In this case, the term weight used in Formula 2 is $ts.idf$, where ts is a term significance weight that measures the importance of a term to a particular document. The ts weight is best estimated with the normalized within-document frequency. This form of the ranking function is virtually identical to that developed from the vector space model (CROFT, 1984). Another approach to incorporating term weights is described by FUHR.

A number of proposals have been made to remove the term independence assumption of the probabilistic model. HARPER & VAN RIJSBERGEN, VAN RIJSBERGEN (1977), and YU ET AL. (1979; 1983) describe retrieval techniques that involve calculating correlations between terms (or term dependencies) in the document collection. These dependencies are then used to expand queries and change estimates of the relevance of documents. It has been shown that if there is sufficient information about the occurrence of terms in relevant documents, these retrieval techniques could significantly improve effectiveness. In practice, however, estimation problems make it difficult to realize any of the potential benefits (VAN RIJSBERGEN ET AL.; YU ET AL., 1983).

Another approach is to identify important dependencies in the query and to use the presence of those dependencies to modify the document scores according to a probabilistic model that assumes term dependence (CROFT, 1986). This avoids the calculation of dependencies that are not used in queries and identifies those dependencies most likely to affect retrieval effectiveness. The retrieval technique based on this model modifies a ranking produced with Formula 2 and $ts.idf$ weights by adding a correction factor to a document score for each set of dependent terms that the document contains. The dependencies used by this technique can be identified using groups of query terms joined with the Boolean AND, thereby allowing structured queries to be used with a probabilistic retrieval model.

The maximum entropy approach suggested by COOPER and COOPER & HUIZINGA can be interpreted (as it has been by them) as effectively simulating Boolean relations within a statistical retrieval environment. For details on this work, see BOOKSTEIN (1985).

Fuzzy set. A fuzzy set approach to information retrieval has been discussed in many papers (BOOKSTEIN, 1985). The main contribution of this work in terms of retrieval techniques has been the integration of Boolean queries with ranking techniques. This integration is limited, however, when compared with extended Boolean retrieval based on the vector space model or the use of term dependencies in probabilistic models.

Ad hoc. A number of similarity measures for comparing queries and documents have been proposed in the literature (MCGILL ET AL.; SALTON, 1968). Many of them were developed in the context of numerical taxonomy (SNEATH & SOKAL). Similarity measures typically consist of a measure of the overlap of the query and document sets of terms normalized by the size of the sets involved. For example, Dice's coefficient is $2(Q \cap D)/(|Q| + |D|)$ for queries and documents represented as sets of unweighted index terms. Although these measures are similar to those described in the last section (we do not consider the cosine correlation to be *ad hoc*), they are not based on a particular model of document retrieval. Thus, there is no means for comparing measures apart from exhaustive evaluations such as that done by McGill et al. These types of evaluations are never conclusive, and a more appropriate motivation for using a particular technique for ranking documents is to base the choice on a well-founded retrieval model. Small differences in weights can lead to significant differences in results, and approaching the design of similarity measures in an ad-hoc manner can lead to a confusing collection of results.

Individual, Structure-Based Techniques

In this category of retrieval techniques, either the query or the documents or both are represented by more complicated structures than the sets of terms used in feature-based techniques. We have already encountered retrieval techniques designed to deal with Boolean queries, although these were still primarily feature-based. The types of techniques described here typically rely on a much richer representation of the knowledge in the subject domain covered by the documents and queries. This domain knowledge can be regarded as a more complex form of the thesaurus information found in many systems.

Logic. It is theoretically possible to represent the information conveyed by the text in documents as sentences in a formal logic. For example, the statement, "DEC sells computers," could be represented in first-order predicate calculus as (sells dec computer). Similarly, the statement, "If a company sells computers, it is financially viable," could be represented as (forall (x) (if (sells x computer) (viable x))). More complex sentences require more complex logic representations (CHARNIAK & MCDERMOTT). Given a logic representation of document content, a query in the same logic could be answered by inference using the rules associated with that logic. For example, the query (viable ?) can be answered by forward chaining (a form of *modus ponens*) from the sentences given above. This approach to information retrieval has been studied by WALKER & HOBBS and SIMMONS and is related to the natural language research of people like Schank (SCHANK; SCHANK & ABELSON). Simmons has represented a portion of the *Handbook of Artificial Intelligence* (COHEN & FEIGENBAUM) in logic and retrieves answers to queries in this domain (SIMMONS). The critical problem with this approach is the translation of the text into logic. In current experimental systems, this is done manually.

VAN RIJSBERGEN (1986a) has proposed a framework for information retrieval based on logic. He describes retrieval as a process of determining if a query (expressed in logic) can be inferred from a document's content (expressed in logic). In many cases, this inference cannot be made directly because information is missing in the document; in these cases the inference is uncertain. This framework can be used to describe other models of retrieval and may lead to further insights but currently has not produced new retrieval techniques.

The notion of uncertain inference is also the basis of the RUBRIC system (TONG ET AL.). A part of this system provides standard full-text document retrieval. Queries, however, are represented as rules that describe how pieces of evidence in the document text can be used to infer the relevance of the document. Numbers are attached to the rules to represent the certainty of the statement. For example, a query may be represented by the rules

> "information" AND "retrieval" → information-retrieval (0.6)
> 'information" ADJACENT "retrieval" →
>   information-retrieval (0.9)
> "probabilit" → probabilistic-model (0.5)
> "information-retrieval" AND "probabilistic-model" →
>   probabilistic-information-retrieval (0.9)

If a particular document contains the sentence, "Retrieval of information with high probability of relevance is desirable," the rules above will be used to infer that the document is about information retrieval with a certainty of .6, probabilistic models with a certainty of .5, and probabilistic information retrieval with a certainty of .45 (.5 x .9). This type of rule-based representation has also been used for thesaurus information to infer concepts that are related to terms in a query (CROFT & THOMPSON, 1987; SHOVAL).

Graph. A number of structures fall into this general category. The general characteristic of a graph-like representation is a set of nodes and edges (or links) connecting these nodes. Specific examples include semantic nets and frames (CHARNIAK & MCDERMOTT), which are typically produced by natural language processing. Simpler network structures can also be produced by statistical techniques, such as those used in the ASK (anomalous states of knowledge) project (BELKIN ET AL., 1982; BELKIN & KWASNIK). Retrieval techniques in this category must look for similarities in the structures of query and document graphs. This similarity can be used directly to determine if a document should be retrieved or to modify a document ranking.

BELKIN & KWASNIK, for instance, describe the identification of regions of interest in graphs generated by a co-occurrence analysis of narrative "problem statements" (ASK representations). These areas are identified as specific kinds of structures within the graph, such as groups of highly interconnected nodes at high association strengths or two such groups weakly connected with one another. The structural nature of the ASK graph is then used to determine first the terms that will be matched against the database of documents, then the ranking of the retrieved documents. The first retrieval stage takes little account of the relationships of the terms within the document structures (computed in the same way as the ASK structures), using instead the ASK structure to identify terms that either must or may appear in the text structures. In the second stage, the candidate set of retrieved documents is then ranked according to how well each satisfies desirable criteria of term position, importance, and relationship to other terms as established by rules associated with the structures identified in the ASK representation. This retrieval method has not been tested in a formal experiment, but it appears to have some possibilities of providing a way to use graphical representations to choose different retrieval strategies. It may also be of use in specifying term dependencies that can then be used by other retrieval techniques.

The general method of first making a rough pass at the text collection, omitting graph matching of any sort, and then using the graphs to order the retrieved set is typical of graph- and structure-based techniques. The reason for this is that graph matching of any sort is computationally difficult, and searching the entire database is almost always an intractable problem. Another approach is to use very general characteristics of the graphs to define the search space and then progressively to refine the search. This has been used successfully in similarity searching for chemical structures (e.g., WILLETT ET AL.).

Network

Cluster. A cluster is a group of documents whose contents are similar. A particular clustering method gives a more detailed definition of a cluster and provides a technique for generating them. The use of clustering for information retrieval was a major topic in the SMART project (SALTON, 1968). The approach used was to form a cluster hierarchy using an ad-hoc clustering technique. The cluster hierarchy was formed by dividing documents into a few large clusters, dividing these clusters into smaller clusters, and so on. A top-down search of the cluster hierarchy is performed by comparing (using a similarity measure) the query to cluster representatives of the top-level (largest) clusters, choosing the best clusters, comparing the query with representatives of lower-level clusters within these clusters, and so on until a ranked list of lowest-level clusters is produced. The documents in the top-ranked clusters are then ranked individually for presentation to the user. A cluster representative can be generated in various ways, but in general it represents the average properties of documents in the cluster.

Jardine and Van Rijsbergen also used a top-down search of a cluster hierarchy, with the difference that the hierarchy was produced using a formal clustering method (single-link) and clusters were retrieved in their entirety without individual document ranking (JARDINE & VAN RIJSBERGEN). The cluster hypothesis was introduced as a basis for using cluster searches to improve retrieval effectiveness relative to ranking individual documents.

CROFT (1980) described a probabilistic model of cluster searching and introduced the bottom-up retrieval technique. Here the query is compared with representatives of the lowest-level clusters directly, and documents in the top-ranked clusters are retrieved. The emphasis on small, well-defined clusters has led to the development of retrieval techniques based on the generation of the document's nearest neighbors (CROFT & PARENTY; GRIFFITHS ET AL., 1986). A document's nearest neighbors are those most similar to it, and a cluster of nearest neighbors is very similar to the lowest-level single-link clusters (WILLETT, 1984b). Given a network of documents connected to their nearest neighbors, it is possible to generate clusters and their representatives at search time with considerable storage savings (CROFT & PARENTY).

A retrieval technique that has strong similarities to those based on nearest neighbors is Goffman's indirect retrieval method (GOFFMAN). This technique has been used in some recent research (BADRAN). Other recent research on clustering techniques has compared the relative effectiveness and efficiency of different types of clusters (VOORHEES).

Browsing. If the documents, terms, and other bibliographic information are represented in the system as a network of nodes and connections, the user can browse through this network with system assistance. Browsing is an interesting retrieval technique in that it places less emphasis on query formulation than do other techniques and relies heavily on the immediate feedback provided by user browsing decisions. The THOMAS system (ODDY) uses index terms as starting points in a simple network of documents and terms. Through dialog with the user, the system uses the network to build a model of the user's information need that includes relevant documents found during the process. The browsing component of the I^3R system (CROFT & THOMPSON, 1987) contains nodes that represent documents, index terms, domain knowledge, authors, and journals. The links represented include indexing information, thesaurus information, nearest neighbors, citations, and authorship. The system makes browsing recommendations based on the number and types of connections between and among documents but allows users to choose any path.

Other research in browsing concentrates on the use of visual representations of the document database to acquire information from the user interactively (FREI & JAUSLIN).

Spreading activation. Spreading activation is a retrieval technique that has some similarities to browsing. A query is used to "activate" parts of a network that describes the contents of documents and how they are related to each other. In the simplest case, the query would activate index term nodes that are connected to document nodes and other terms. In more knowledge-intensive networks, the links and nodes represent concepts from the subject domain and how they relate to each other as well as the documents that contain those concepts (COHEN & KJELDSEN; RAU). From the "start nodes" provided by the query, other nodes connected to those nodes are in turn "activated" (hence, the term "spreading activation"). Criteria, such as threshold values that decrease as the activation propagates through the network or rules about the reasonableness of the inference implied by using a particular link, are used to control the spread of activation. Activation can converge on particular document nodes from a number of links. These highly activated nodes are retrieved.

In a simple network of documents and terms, the documents that have a high level of activation after the first links from the query nodes are followed will be those documents that have a high number of terms in common with the query. If the activation spreads to other terms connected to those documents and then to other documents, the documents retrieved in this second phase will be similar to those found by a cluster search based on nearest neighbors. When the activation is refined using inference rules and more link types, it is more difficult to relate the retrieved documents to those found with conventional techniques. The retrieval technique in this case is more similar to structure-based techniques.

FEEDBACK METHODS

Relevance feedback techniques are not considered retrieval techniques by our criteria. Rather they are used to refine the request model, which is then used for another search. Feedback techniques are, however, an extremely important part of ensuring that a document retrieval system will be effective.

These techniques were primarily developed in the context of feature-based retrieval (SALTON, 1968) although the principles apply to any retrieval technique. The main part of relevance feedback is the adjustment of weights associated with query terms. This adjustment is done on the basis of term occurrences in the documents identified by the user as relevant. The probabilistic model has a particularly strong motivation behind this weight adjustment in that the identified relevant documents provide a sample to estimate the pr values. The query (or request model) can also be changed by the addition of new terms from relevant documents. Some control is needed over the number of new terms added, and it seems that the most reliable method is to have users identify interesting terms in relevant documents.

Other types of modifications based on feedback are possible, such as adding term dependencies identified in relevant documents or modifying the "document space" (document indexing) to make relevant documents more similar to the queries. Attempts have also been made to use adaptive mechanisms to select retrieval techniques appropriate for a particular query (CROFT & THOMPSON, 1984). There have also been some attempts to use this type of information for feedback in operational systems, presenting it to the user, who then adjusts the query manually (e.g., INGWERSEN).

RELATIVE PERFORMANCE OF RETRIEVAL TECHNIQUES

Comparative Performance Studies

The evaluation of IR systems has been a major topic of research for a number of years (VAN RIJSBERGEN, 1979). Although there have been and continue to be problems with the evaluations that have been done, these results, together with theoretical results derived from underlying models, provide valuable information about the relative performance of retrieval techniques. Again, we do not argue the relative merits of automated vs. manual indexing in this paper. In a recent article, SALTON (1986a) summarizes these arguments and other results. In making the comparisons in this section, we shall not discuss the particular effectiveness measures used. When a difference is described as significant, we are following the generally accepted "rule of thumb" guidelines of at least 10 percent increases in recall and precision.

The first important result is that all available evidence points to the superiority of partial match techniques over exact match techniques (in particular, see SALTON ET AL., 1983b). Although there are problems with making direct comparisons between the sets of retrieved documents, it appears that the difference in effectiveness is significant. For feature-based retrieval, the evidence indicates that the best performance is provided by the probabilistic retrieval strategy incorporating term significance weights or its equivalent, the $tf.idf$/cosine correlation combination. This retrieval technique uses a simple similarity measure (the inner product) and the effectiveness is due entirely to the index term weights used. Results that indicate superiority of one technique over another in this context (e.g., SALTON, 1986b) can be interpreted in terms of the estimates used for the weights. The use of good estimates is the major factor in obtaining the best performance from these techniques.

The use of term dependencies to modify document rankings can also improve performance but only if the dependencies are accurately identified by the user or natural language processing techniques (CROFT, 1986). The same problems occur with the extended Boolean retrieval technique (SALTON & VOORHEES). Techniques that rely on identifying dependencies in document collections independently of a particular query do not seem to give significant improvements (VAN RIJSBERGEN ET AL.; YU ET AL., 1983). The automated use of thesaurus information to expand queries appears to be effective but only if the terms expanded and the type of thesaurus information used are tightly controlled. Relevance feedback can give very good results even when few relevant documents are identified (SPARCK JONES & WEBSTER).

Cluster-based searches can achieve levels of performance that are similar to individual feature-based searches (CROFT, 1980; GRIFFITHS ET AL., 1986) but in general they tend to be better for high-precision results. The primary advantage of cluster searching, however, is that it retrieves different relevant documents than, say, a $tf.idf$ search, and for some queries it works much better (CROFT & HARPER; GRIFFITHS ET AL., 1986). Cluster-based retrieval, therefore, is a good alternative technique to the individual feature-based method. Systems have been designed to allow both retrieval techniques to be used simultaneously or for cluster searches to be used when other techniques fail (CROFT & THOMPSON, 1987).

Although the techniques described so far can achieve reasonable levels of performance and can be implemented efficiently in operational systems, there is still a lot of room for improvement in terms of absolute performance. To obtain much higher levels of performance, it is apparently necessary to consider knowledge-intensive techniques such as structure-based retrieval or spreading activation. The problem is that because these techniques are knowledge-intensive, they are difficult to implement and have been tested only with very small collections of documents in very specific domains. Techniques that use some form of natural language processing to construct representations of document and query content have been studied in IR for some time (e.g., SPARCK JONES, 1974). Experiments with these techniques have never achieved significant performance benefits, often because the information derived from natural language processing was used in inappropriate ways. Hybrid systems (CROFT & LEWIS; SPARCK JONES & TAIT; TONG ET AL.) that combine knowledge-intensive techniques with efficient full-text retrieval or ranking strategies appear to have significant promise. The RUBRIC approach seems particularly suited to users who are prepared to spend a lot of effort in constructing queries and would not be appropriate in a general environment. Cohen's recent paper (COHEN & KJELDSEN) provides a detailed evaluation of a spreading activation technique that achieved good results. The problem of translating text into the representations used, however, remains unsolved.

Relationship of the Request Model to Retrieval Techniques

It has probably become clear that it is difficult to separate the retrieval technique from the form of representation used in the request model. There is, however, a clear message from the evaluations performed in IR research. Whatever retrieval technique is used, the quality of the results depends almost entirely on the accuracy of the information in the request model. More sophisticated retrieval techniques can use more detailed request models but constructing these models requires more user effort. It is this process of an intermediary's interaction with the user to formulate the query that is the heart of current retrieval systems, and it is also crucial for more advanced retrieval systems. This obvious fact has led to a lot of research in expert intermediary systems (e.g., BELKIN ET AL., 1983; CROFT & THOMPSON, 1987; MARCUS). These systems engage the user in a dialog with a variety of facilities to acquire a detailed request model. The systems also assist the user in evaluating the retrieved documents and, in some cases, in selecting retrieval techniques.

Predicting Appropriate Retrieval Techniques

A number of experimental results have indicated that although different retrieval techniques appear to have similar results, they often retrieve different relevant documents for the same query. The use of alternative content representations, such as citation information, can also result in the retrieval of different documents. Different techniques also vary in their performance for different queries. If the best results from different techniques for individual queries could be selected, very high performance would be possible (CROFT & THOMPSON, 1984; GRIFFITHS ET AL., 1986). The problem is then to identify which technique (and representation) is appropriate for a particular query. Unfortunately, growing evidence shows that this is extremely difficult if not impossible (e.g., MCCALL & WILLETT). One solution is to design systems to use alternative strategies (such as probabilistic, cluster, and browsing) and alternative representations (such as index terms, citations, and semantic nets) (BELKIN ET AL., 1982; CROFT & THOMPSON, 1987). The selection of a particular strategy and representation can be guided by rules in consultation with the user. Given such a system, retrieval can be viewed as a form of plausible inference (VAN RIJSBERGEN, 1986a) or gathering of evidence about the relevance of documents from various sources.

ARCHITECTURES AND TECHNIQUES
FOR INTEGRATED SYSTEMS

It seems, therefore, that there are strong arguments either for using specific retrieval techniques for specific kinds of queries—on the assumption that some techniques are more appropriate for some queries than for others—or for using many retrieval techniques on a single query in the hope that the combination will result in a satisfactory response when no single technique is adequate. The latter position does not really require any specific kind of system architecture or design to be implemented other than having representations that allow the various techniques to be used. On the other hand, the task of choosing a specific technique for specific circumstances presents a more significant problem.

The SIRE system (KOLL ET AL.) is an example of a system that provides several retrieval techniques that can be used singly or in combination for any specific query. In this system the user chooses the technique. Systems that automatically choose the technique or that have special techniques or data structures associated with specific kinds of queries or problems present are more problematic. This approach has arisen in two basic contexts. One is the context associated with so-called "expert" information systems in which it is assumed that different techniques will be required for different kinds of queries; the other is in office automation and/or integration of DBM and IR systems. Here different data types or structures seem to imply the need for different techniques, and there are many different user populations with presumably different data needs. We discuss some approaches to these problems below.

"Expert" Information Intermediaries and Systems

Several "expert" or "intelligent" information systems have been proposed that have assumed that multiple retrieval strategies will be necessary to respond to different query types. We have already discussed one such proposal, the ASK approach (BELKIN ET AL., 1982). It is based on the hypothesis that there will be different categories of anomalous states of knowledge that will require different retrieval strategies. The architecture of this proposal was never fully specified, although it was clear that an elaborate request model would be necessary to do the correct request classification. BELKIN & KWASNIK suggest some strategies and note how they can be chosen according to structural representations of the user's ASK. For details of the kinds of techniques, and the types of characteristics of ASKs, see the section on *Graphs* above. Here, it is sufficient to say that the system design requires a representation of the texts in structural terms, as well as of the queries, and that the alternative retrieval techniques are chosen in a rule-based stepwise manner, operating directly on the text representations.

There have been several suggestions for intelligent information systems based on a "distributed expert" model, in which a number of functions (e.g., building up a request model, building a user model, and choosing a retrieval technique) are isolated as separate processes that communicate with one another. BELKIN ET AL. (1983) suggested the logical design for such a system, and CROFT & THOMPSON (1987) and FOX (1987a) have built prototypes of similar systems. Although the details of these systems differ, they share some significant characteristics in terms of what functions are specified and how they are related. From our point of view, they all assume that different retrieval strategies will be necessary for different user situations, and therefore a major function of their retrieval strategy or technique expert is to choose one of several techniques available, based on information provided by the other experts about the user. Croft and Thompson have available in their system both probabilistic and cluster searching as well as a separate browsing component that the user can choose to instigate, and Fox has implemented separate *p*-norm search (a form of extended Boolean retrieval—see FOX, 1983) and browsing experts. As yet there is no strong reasoning capability in any of these systems for choosing one or another retrieval technique, but the general architecture of separation of functions seems to be useful in this regard. This work in general has progressed on the assumption that the more elaborate the request model, the more likely that: 1) any retrieval technique will work well, and 2) that a most appropriate technique can be selected. These assumptions have yet to be rigorously tested.

Other intelligent interface or system designs have paid less attention to incorporating multiple retrieval techniques, relying instead on elaboration of the request model through interaction (e.g., MARCUS) or through natural language access to the system (e.g., GUIDA & TASSO) in order to use a single technique already in place. KRAWCZAK ET AL. use a hierarchical representation of the knowledge associated with a particular domain (in this case environmental pollution), in order to guide this interaction.

Multimedia and "Integrated" Information Systems

For some time there has been active research in attempting to integrate DBM and IR systems into a single system or model (e.g., SCHEK). The impetus for such work has been twofold. First, the two models offer different capabilities for, and have different problems in, data management and query formulation; these might complement one another in various ways if they were incorporated into one system. Second, it has been suggested that the two systems might provide different retrieval techniques and interfaces appropriate to different classes of users or to the same users with different problems. CROFT (1982) provides an overview of these positions and suggests that knowledge-based or "expert" systems be included in such an amalgamation.

Proposals for integration have generally taken the form of putting an IR interface on top of a DBM system (DBMS), thus effectively translating the IR-based query into a DBMS "exact match" retrieval technique. This technique may be useful for managing very large databases efficiently, but as a multiple architecture it leaves something to be desired. A different approach is exemplified by the HAM-ANS (HAMburg- Application-Oriented Natural language System) project (HOEPPNER ET AL.), in which a single interface is used to access three different kinds of databases: factual, document, and "knowledge." Although the project did not actually manage to integrate this access entirely, the system was meant to be able to choose an appropriate database (and therefore its appropriate retrieval technique) according to characteristics of the user's query.

Research in office automation has also led to proposals for new architectures for information systems (VAN RIJSBERGEN, 1986c). Here the problem is twofold. First, the documents in the office environment are of many types, and even within a single type, they are often multimedia (i.e., made up of mixtures of text, images, data, tables, voice, and so on). Standard DBMS retrieval techniques are inadequate in this environment. Second, the range of users and uses of such systems is very broad, meaning that it might be necessary to identify particular uses with specific retrieval techniques or at least types of documents. The typical response in the first case has been to define strictly the general office document or object and its parts. One can then presumably identify from the query what kind of document, or what part of the document, is desired and search accordingly (e.g., CHRISTODOULAKIS; HARPER ET AL.). This approach has led to proposals for architectures that include DBMS and conventional IR techniques and sometimes others, such as browsing (for a review of such systems, see SMEATON & VAN RIJSBERGEN, 1986). In the case of the second issue, very little has been proposed since up to now there has been little systematic investigation of the users on which to base an architecture.

Overall, little seems to have been done to attack the problem of how, in principle, to choose the techniques that should be incorporated in such systems, although the work on document types is promising. Even less seems to have been done on how to design a system that will choose an appropriate technique automatically- i.e., without direction from the user.

CURRENT STATUS AND FUTURE RESEARCH
IN IR TECHNIQUES

The Situation Now

One can conclude from this review that there is a disquieting disparity between the results of research on IR techniques, which demonstrate fairly conclusively, on both theoretical and empirical grounds, the inadequacy of the exact match paradigm for effective information retrieval and the status of operational IR systems, which use almost exclusively just one exact match technique—Boolean logic. This, of course, is not exactly news. What is perhaps new is the directions in which IR technique research is going. In the past, most research in retrieval techniques was rather far removed from operational environments and to some extent even from operational constraints. The past several years seem to show movement in this research in three directions; two respond to some extent to what may have been problems in acceptance of research results; the third seems to offer promise in greatly improving retrieval performance.

First there has been a good deal of work on relating partial match techniques to exact matching, as in extended Boolean searching and the use of Boolean-derived dependencies in probabilistic searching. This can be seen as a response to the demands of the operational environment. Similarly, there are at least a few studies being carried out of partial-match techniques in operational environments (e.g., ROBERTSON ET AL.), which are explicitly designed to attend to the criticism that such techniques have never been demonstrated to be worthwhile in large systems. There are also several microcomputer-based systems that use partial match techniques in operational, albeit smaller, environments—e.g., the SIRE system (KOLL ET AL.).

The second response seems to be the realization that no one technique will be adequate for all purposes and that either a mix of techniques or a principled choice of techniques is required to improve IR system performance. Thus, there is perhaps less in-fighting among the various camps and more willingness to accept the usefulness of particular techniques in specific circumstances. This tendency may have been encouraged by the special problems of the office automation environment and by the many results indicating that although different techniques perform similarly, they retrieve different relevant documents.

The third new direction we note is increasingly complex representations of the request or user's problem. As noted, although retrieval techniques are not the same as representations, the techniques one can use are determined by the representation. The more complex the representation, one might think, the more kinds of retrieval techniques are possible. Much, although not all, of the work of this type has had its roots in the knowledge representation schemes associated with artificial intelligence research. In general, this work seems to have arisen because of an increased understanding of the importance of the request model (or understanding of the user) to all of IR.

Open Problems

Some issues that have been raised by current research in IR techniques and some that have been around for a while have become particularly important. These open problems will need to be resolved before there will be great improvements in IR system performance, but at least some IR technique research is now attacking them.

First, and extremely important, is our understanding of the gaps between perfect, optimal, and current performance of IR systems. We know more or less what current performance is (although we may not be happy with the evaluation measures). We do not know, however, what perfect performance is now that it is generally accepted that the ideal of all and only the relevant documents is not the goal for all users at all times. Nor do we have any understanding of what optimal performance might be other than a widespread feeling that none of the current techniques seems able to achieve it alone and that current performance therefore is suboptimal. This, of course, is an assumption, which we also hope is valid. Thus, the issues that need to be seriously addressed are: 1) what is perfect performance and why is it perfect; 2) what is optimal performance and why is it optimal; and 3) how can optimal performance be achieved.

There is a growing feeling that it is inappropriate to treat all queries with just one technique and that multiple techniques could be used on single queries. This might be a way to approach the problem of optimal performance, but first some questions must be answered. For example, different retrieval techniques seem to operate differently on queries. We need to discover what aspects of the request model the various techniques capture or reflect. Then we can begin to learn how to choose techniques appropriate to different user situations and decide whether and how to apply several techniques to the same request.

Also we need some ideas about how to develop new retrieval techniques. If the previous questions have been answered, at least to some extent, it may be possible to suggest techniques that directly approach the issue of optimal performance or that respond to aspects of the user's problem that other techniques do not but that theory indicates are important. This, of course, will depend on the development of representations that will allow such techniques to be useful.

There is now some reason for optimism because, as this review has shown, not only is there dissatisfaction with current retrieval techniques, but significant work is at least beginning to attack all of the questions we have raised. Thus, we expect to see much more work in the near future on the relationship between technique and request model, on the testing of experimental techniques in operational environments, on the specification and implementation of new architectures for multiple retrieval techniques, and especially on methods and effects of request model elaboration and their use in the development of new retrieval techniques.

BIBLIOGRAPHY

APPELRATH, HANS-JURGEN. 1985. Die Erweiterung von DB- und IR-Systemen zu Wissensbasierten Systemen [The Extension of Data Base and Information Retrieval Systems to Knowledge Based Systems]. Nachrichten für Dokumentation (West Germany). 1985; 36(1): 13-21. ISSN: 0027-7436.

ARNOW, DAVID M.; TENENBAUM, AARON M.; WU, CONNIE. 1985. P-Trees: Storage Efficient Multiway Trees. See reference: ASSOCIATION FOR COMPUTING MACHINERY. 111-121.

ASSOCIATION FOR COMPUTING MACHINERY (ACM). SPECIAL INTEREST GROUP ON INFORMATION RETRIEVAL (SIGIR). 1985. Research and Development in Information Retrieval: [Proceedings of the] 8th Annual International ACM SIGIR Conference; 1985 June 5-7; Montreal, Canada. New York, NY: ACM, Inc.; 1985. 288p. Available from: ACM Order Department, P.O. Box 64145, Baltimore, MD 21264 (Order no. 606850). ISBN: 0-89791-159-8.

BADRAN, ODETTE MAROUN. 1985. An Alternative Search Strategy to Improve Information Retrieval. In: Parkhurst, Carol A., ed. ASIS '85: Proceedings of the American Society for Information Science (ASIS) 48th Annual Meeting; 1985 October 20-24; Las Vegas, NV. White Plains, NY: Knowledge Industry Publications, Inc. for ASIS; 1985. 137-140. ISSN: 0044-7870; ISBN: 0-86729-176-1; CODEN: PAISDQ.

BARBI, E.; CALVO, F.; PERALE, C.; SIROVICH, F.; TURINI, F. 1984. A Conceptual Approach to Document Retrieval. In: Ellis, Clarence, ed. [Proceedings of the] 2nd Association for Computing Machinery, Special Interest Group on Office Automation (ACM-SIGOA) Conference on Office Information Systems; 1984 June 25-27; Toronto, Canada. New York, NY: ACM, Inc.; 1984. 219-226. (Published as Vol. 5(1-2) of SIGOA Newsletter). Available from: ACM Order Department; P.O. Box 64145, Baltimore, MD 21264 (Order no. 611840). ISBN: 0-89791-140-7.

BARR, AVRON; FEIGENBAUM, EDWARD A., eds. 1981. The Handbook of Artificial Intelligence: Volume 1. Los Altos, CA: William Kaufmann, Inc.; 1981. 409p. ISBN: 0-86576-004-7.

BATES, MARCIA J. 1986. Subject Access in Online Catalogs: A Design Model. Journal of the American Society for Information Science. 1986 November; 37(6): 357-376. ISSN: 0002-8231; CODEN: AISJB6.

BELKIN, N. J.; KWASNIK, B. H. 1986. Using Structural Representations of Anomalous States of Knowledge for Choosing Document Retrieval Strategies. See reference: RABITTI, FAUSTO, ed. 11-22.

BELKIN, N. J.; ODDY, R. N.; BROOKS, H. M. 1982. ASK for Information Retrieval: Part I. Background and Theory; Part II: Results of a Design Study. Journal of Documentation. 1982 June; September; 38(2-3): 61-71; 145-164. ISSN: 0022-0418.

BELKIN, N. J.; SEEGER, T.; WERSIG, G. 1983. Distributed Expert Problem Treatment as a Model for Information System Analysis and Design. Journal of Information Science. 1983; 5: 153-167. ISSN: 0165-5515.

BELL, D. A. 1985. An Architecture for Integrating Data, Knowledge, and Information Bases. In: Informatics 8: Advances in Intelligent Retrieval: Proceedings of a Conference Jointly Sponsored by Aslib, the Aslib Informatics Group, and the Information Retrieval Specialist Group of the British Computer Society; 1985 April 16-17; Wadham College, Oxford,

England. London, England: Aslib, The Association for Information Management, Information House; 1985. 240–257. Available from: Aslib, 26–27 Boswell Street, London WC1N3JZ, England. ISBN: 0-85142-195-4.

BILLER, HORST. 1983. On the Architecture of a System Integrating Data Base Management and Information Retrieval. In: Salton, Gerard; Schneider, Hans-Jochen, eds. Research and Development in Information Retrieval: Proceedings [of the 5th International Conference of the Association for Computing Machinery, Special Interest Group on Information Retrieval (ACM-SIGIR)]; 1982 May 18–20; Berlin. Berlin, West Germany: Springer-Verlag; 1983. 80–97. (Lecture Notes in Computer Science: 146). ISBN: 3-540-11978-7 (Berlin); ISBN: 0-387-11978-7 (New York).

BISWAS, GAUTAM; SUBRAMANIAN, VISWANATH; BEZDEK, JAMES C. 1985. A Knowledge Based System Approach to Document Retrieval. In: The Engineering of Knowledge-Based Systems: [Proceedings of the] 2nd Conference on Artificial Intelligence Applications; 1985 December 11–13; Miami Beach, FL. Washington, DC: Institute of Electrical and Electronics Engineers (IEEE) Computer Society Press; 1985. 455–460. Available from: IEEE Service Center, 445 Hoes Lane, Piscataway, NJ 08854 (Catalog no. 85CH2215-2). ISBN: 0-8186-0688-6; LC: 85-62321.

BISWAS, GAUTAM; SUBRAMANIAN, VISWANATH; MARQUES, M. M.; BEZDEK, JAMES C. 1985. A Document Retrieval System Using a Fuzzy Expert Systems Approach. In: Proceedings of the International Conference on Cybernetics and Society; 1985 November 12–15; Tucson, AZ. New York, NY: Institute of Electrical and Electronics Engineers, Inc.; 1985. 126–130. (IEEE Catalog no. 85CH2253-3). ISSN: 0360-8913.

BIVINS, KATHLEEN T.; ERIKSSON, LENNART. 1982. Reflink: A Microcomputer Information Retrieval and Evaluation System. Information Processing & Management. 1982; 18(3): 111–116. ISSN: 0306-4573; CODEN: IPMADK.

BLAIR, DAVID C. 1984. The Data-Document Distinction in Information Retrieval. Communications of the ACM. 1984 April; 27(4): 369–374. ISSN: 0001-0782.

BLAIR, DAVID C. 1986a. Full Text Retrieval: Evaluation and Implications. International Classification (West Germany). 1986; 13(1): 18–23. ISSN: 0340-0050.

BLAIR, DAVID C. 1986b. Indeterminacy in the Subject Access to Documents. Information Processing & Management. 1986; 22(2): 229–241. ISSN: 0306-4573; CODEN: IPMADK.

BLAIR, DAVID C.; MARON M. E. 1985. An Evaluation of Retrieval Effectiveness for a Full-Text Document-Retrieval System. Communications of the ACM. 1985 March; 28(3): 289–299. ISSN: 0001-0782.

BOOKSTEIN, ABRAHAM. 1980. Fuzzy Requests: An Approach to Weighted Boolean Searches. Journal of the American Society for Information Science. 1980 July; 31(4): 240–247. ISSN: 0002-8231; CODEN: AISJB6.

BOOKSTEIN, ABRAHAM. 1982. Recent Developments in the Theory of Information Retrieval. 1982 December. 28p. (NTIS Report TRITA-LIB-6019; Based on a seminar presented at the Royal Institute of Technology Library [Sweden], 1982 November 10). Also published as ISSN: 0346-9042.

BOOKSTEIN, ABRAHAM. 1983. Explanation and Generalization of Vector Models in Information Retrieval. In: Salton, Gerard; Schneider, Hans-Jochen, eds. Research and Development in Information Retrieval: Proceedings [of the 5th International Conference of the Association for Computing Machinery, Special Interest Group on Information Retrieval (ACM-SIGIR)]; 1982 May 18-20; Berlin. Berlin, West Germany: Springer-Verlag; 1983. 118-132. (Lecture Notes in Computer Science: 146). ISBN: 3-540-11978-7 (Berlin); ISBN: 0-387-11978-7 (New York).

BOOKSTEIN, ABRAHAM. 1985. Probability and Fuzzy-Set Applications to Information Retrieval. In: Williams, Martha E., ed. Annual Review of Information Science and Technology: Volume 20. White Plains, NY: Knowledge Industry Publications, Inc. for the American Society for Information Science; 1985. 117-151. ISSN: 0066-4200; ISBN: 0-86729-175-3; LC: 66-25096; CODEN: ARISBC.

BOSE, PRASANTA K.; RAJINIKANTH, M. 1985. KARMA: Knowledge-Based Assistant to a Database System. In: The Engineering of Knowledge-Based Systems: [Proceedings of the] 2nd Conference on Artificial Intelligence Applications; 1985 December 11-13; Miami Beach, FL. Washington, DC: Institute of Electrical and Electronics Engineers (IEEE) Computer Society Press; 1985. 467-472. Available from: IEEE Service Center, 445 Hoes Lane, Piscataway, NJ 08854 (Catalog no. 85CH2215-2). ISBN: 0-8186-0688-6; LC: 85-62321.

BROOKS, H. M. 1983. Information Retrieval and Expert Systems-Approaches and Methods of Development. In: Informatics 7: Intelligent Information Retrieval. London, England: Aslib; 1983. 65-75. ISBN: 0-85142-187-5.

BUCKLEY, CHRIS. 1985. Implementation of the SMART Information Retrieval System. Ithaca, NY: Cornell University. Department of Computer Science; 1985 May. 37p. (Technical Report TR 85-686). Available from: Chris Buckley, Department of Computer Science, Cornell University, Ithaca, NY 14853.

BUCKLEY, CHRIS; LEWIT, ALAN F. 1985. Optimization of Inverted Vector Searches. See reference: ASSOCIATION FOR COMPUTING MACHINERY. 97-110.

BUELL, DUNCAN A. 1985. A Problem in Information Retrieval with Fuzzy Sets. Journal of the American Society for Information Science. 1985; 36(6): 398-401. ISSN: 0002-8231; CODEN: AISJB6.

CARROLL, DAVID M.; POGUE, CHRISTINE A.; WILLETT, PETER. 1987. Bibliographic Pattern Matching Using the ICL Distributed Array Processor. Journal of the American Society for Information Science. 1987. (In press). ISSN: 0002-8231; CODEN: AISJB6.

CHARNIAK, EUGENE; MCDERMOTT, DREW. 1985. Introduction to Artificial Intelligence. Reading, MA: Addison-Wesley; 1985. 701p. ISBN: 0-201-11945-5.

CHRISTODOULAKIS, S. 1984. Framework for the Development of an Experimental Mixed-Mode Message System. See reference: VAN RIJSBERGEN, C. J., ed. 1984. 1-20.

CLEVELAND, DONALD B.; CLEVELAND, ANA D.; WISE, OLGA B. 1984. Less than Full-Text Indexing Using a Non-Boolean Searching Model. Journal of the American Society for Information Science. 1984; 35(1): 19-28. ISSN: 0002-8231; CODEN: AISJB6.

COHEN, PAUL R.; FEIGENBAUM, EDWARD A. eds. 1982. The Handbook of Artificial Intelligence: Volume 3. Los Altos, CA: William Kaufmann, Inc.; 1982. 639p. ISBN: 0-86576-007-1.

COHEN, PAUL R.; KJELDSEN, RICK. 1987. Information Retrieval by Constrained Spreading Activation in Semantic Networks. Information Processing & Management. 1987. (In press). ISSN: 0306-4573; CODEN: IPMADK.

COOPER, WILLIAM S. 1983. Exploiting the Maximum Entropy Principle to Increase Retrieval Effectiveness. Journal of the American Society for Information Science. 1983; 34(1): 31-39. ISSN: 0002-8231; CODEN: AISJB6.

COOPER, WILLIAM S.; HUIZINGA, P. 1982. The Maximum Entropy Principle and Its Application to the Design of Probabilistic Retrieval Systems. Information Technology: Research and Development. 1982; 1: 99-112. ISSN: 0144-817X.

CROFT, W. BRUCE. 1980. A Model of Cluster Searching Based on Classification. Information Systems. 1980; 5(3): 189-195. ISSN: 0306-4379; CODEN: INSYD6.

CROFT, W. BRUCE. 1982. An Overview of Information Systems. Information Technology: Research and Development. 1982 January; 1(1): 73-96. ISSN: 0144-817X.

CROFT, W. BRUCE. 1983a. Applications for Information Retrieval Techniques in the Office. In: Kuehn, Jennifer J., ed. Proceedings of the 6th Annual International Association for Computing Machinery Special Interest Group on Information Retrieval (ACM-SIGIR) Conference on Research and Development in Information Retrieval; 1983 June 6-8; Bethesda, MD. New York, NY: ACM, Inc.; 1983. 18-23. (A publication of SIGIR Forum; 1983 Summer; 17(4)). Available from: ACM Order Department, P.O. Box 64145, Baltimore, MD 21264. ISBN: 0-89791-107-5.

CROFT, W. BRUCE. 1983b. Experiments with Representation in a Document Retrieval System. Information Technology: Research and Development. 1983 January; 2(1): 1-21. ISSN: 0144-817X.

CROFT, W. BRUCE. 1984. A Comparison of Cosine Correlation. Information Technology: Research and Development. 1984; 3: 113-114. ISSN: 0144-817X.

CROFT, W. BRUCE. 1986. Boolean Queries and Term Dependencies in Probabilistic Retrieval Models. Journal of the American Society for Information Science. 1986 March; 37(2): 71-77. ISSN: 0002-8231; CODEN: AISJB6.

CROFT, W. BRUCE; HARPER, D. J. 1979. Using Probabilistic Models of Document Retrieval Without Relevance Information. Journal of Documentation. 1979 December; 35(4): 285-295. ISSN: 0022-0418.

CROFT, W. BRUCE; LEWIS, D. 1987. An Approach to Natural Language Processing for Document Retrieval. In: [Proceedings of the] Association for Computing Machinery Special Interest Group on Information Retrieval (ACM-SIGIR) 10th International Conference on Research and Development in Information Retrieval; 1987 June 3-5; New Orleans, LA. 26-32. ISBN: 0-89791-232-2. New York, NY: ACM, Inc., 1987.

CROFT, W. BRUCE; PARENTY, THOMAS J. 1985. A Comparison of a Network Structure and a Database System Used for Document Retrieval. Information Systems. 1985; 10(4): 377-390. ISSN: 0306-4379; CODEN: INSYD6.

CROFT, W. BRUCE; THOMPSON, ROGER H. 1984. The Use of Adaptive Mechanisms for Selection of Search Strategies in Document Retrieval Systems. See reference: VAN RIJSBERGEN, C. J., ed. 1984. 95-110.

CROFT, W. BRUCE; THOMPSON, ROGER H. 1985. An Expert Assistant for Document Retrieval. Amherst, MA: University of Massachusetts, Department of Computer and Information Science; 1985. 52p. (COINS Technical Report 85-05). Available from: W. Bruce Croft, Dept. of Computer and Information Science, University of Massachusetts, Amherst, MA 01003.

CROFT, W. BRUCE; THOMPSON, ROGER H. 1987. I^3R: A New Approach to the Design of Document Retrieval Systems. Journal of the American Society for Information Science. 1987. (In press). ISSN: 0002-8231; CODEN: AISJB6.

DEERWESTER, SCOTT; DUMAIS, SUSAN T.; FURNAS, GEORGE, W.; LANDAUER, THOMAS K.; HARSHMAN, RICHARD. 1987. Indexing by Latent Structure Analysis. 1987. To appear. 33p. Available from: Susan T. Dumais, Bell Communications Research, 435 South St., MRE 2L-371, Morristown, NJ 07960.

EL-HAMDOUCHI, A.; WILLETT, PETER. 1986. Hierarchic Document Clustering Using Ward's Method. See reference: RABITTI, FAUSTO, ed. 149-156.

ELKERTON, JAY; WILLIGES, ROBERT C. 1984. Information Retrieval Strategies in a File-Search Environment. Human Factors. 1984; 26(2): 171-184. ISSN: 0018-7208.

ENSER, P. G. B. 1985a. Automatic Classification of Book Material Represented by Back-of-the-Book Index. Journal of Documentation. 1985 September; 41(3): 135-155. ISSN: 0022-0418.

ENSER, P. G. B. 1985b. Experimenting with the Automatic Classification of Books. In: Informatics 8: Advances in Intelligent Retrieval: Proceedings of a Conference Jointly Sponsored by Aslib, the Aslib Informatics Group, and the Information Retrieval Specialist Group of the British Computer Society; 1985 April 16-17; Wadham College, Oxford, England. London, England: Aslib, The Association for Information Management, Information House; 1985. 68-83. Available from: Aslib, 26-27 Boswell Street, London WC1N3JZ, England. ISBN: 0-85142-195-4.

FINDLER, NICHOLAS V. 1980. Vers l'Optimisation Interactive de l'Organisation de la Base des Donnees [Toward Interactive Optimization of Database Organization]. In: Diday, E.; Lebart, L.; Pages, J. P.; Tomassone, R., eds. Data Analysis and Informatics: Proceedings of the 2nd International Symposium on Data Analysis and Informatics; 1979 October 17-19; Versailles, France. Amsterdam, The Netherlands: North-Holland Publishing Co.; 1980. 411-421. (In English). ISBN: 0-444-86005-3.

FOX, EDWARD A. 1983. Extending the Boolean and Vector Space Models of Information Retrieval with P-Norm Queries and Multiple Concept Types. Ithaca, NY: Cornell University; 1983. (Ph.D. dissertation). 386p. Available from: University Microfilms Int., Ann Arbor, MI.

FOX, EDWARD A. 1987a. Development of the CODER System: A Testbed for Artificial Intelligence Methods in Information Retrieval. Information Processing and Management. 1987; 23(4). (In press). ISSN: 0306-4573; CODEN: IPMADK.

FOX, EDWARD A. 1987b. Improved Retrieval Using a Relational Thesaurus Expansion of Boolean Logic Queries. In: Evens, Martha, ed. Proceedings of a Workshop on Relational Models of the Lexicon; 1984 July; Stanford, CA. (To appear). Available from: Edward A. Fox, Department of Computer Science, Virginia Tech, Blacksburg, VA 24061.

FREI, H. P.; JAUSLIN, J.-F. 1983. Graphical Presentation of Information and Services: A User-Oriented Interface. Information Technology: Research and Development. 1983 January; 2(1): 23-42. ISSN: 0144-817X.

FUHR, NORBERT. 1986. Two Models of Retrieval with Probabilistic Indexing. See reference: RABITTI, FAUSTO, ed. 249-257.

FUHR, NORBERT; KNORZ, G. E. 1984. Retrieval Text Evaluation of a Rule Based Automatic Indexing (AIR/PHYS). See reference: VAN RIJSBERGEN, C. J., ed. 1984. 391-408.

GOFFMAN, W. 1968. An Indirect Method of Information Retrieval. Information Storage and Retrieval. 1968; 4: 361-373. ISSN: 0020-0271.

GRIFFITHS, ALAN; LUCKHURST, H. CLAIRE; WILLETT, PETER. 1986. Using Interdocument Similarity Information in Document Retrieval Systems. Journal of the American Society for Information Science. 1986 January; 37(1): 3-11. ISSN: 0002-8231; CODEN: AISJB6.

GRIFFITHS, ALAN; ROBINSON, LESLEY A.; WILLETT, PETER. 1984. Hierarchic Agglomerative Clustering Methods for Automatic Document Classification. Journal of Documentation. 1984 September; 40(3): 175-205. ISSN: 0022-0418.

GUIDA, GIOVANNI; TASSO, CARLO. 1983. An Expert Intermediary System for Interactive Document Retrieval. Automatica (Great Britain). 1983; 19(6): 759-766. ISSN: 0005-1098.

HAHN, UDO. 1985. Expertensysteme als intelligente Informationsysteme: Konzepte für die funktionale Erweiterung des Information Retrieval [Expert Systems as Intelligent Information Systems: Perspectives for Functional Extension of Information Retrieval]. Nachrichten für Dokumentation (West Germany). 1985; 36(1): 2-12. ISSN: 0027-7436.

HARPER, D. J.; DUNNION, J.; SHERWOOD-SMITH, M.; VAN RIJSBERGEN, C. J. 1986. Minstrel-ODM: A Basic Office Data Model. Information Processing & Management. 1986; 22(2): 83-107. (Special Issue). ISSN: 0306-4573; CODEN: IPMADK.

HARPER, D. J.; VAN RIJSBERGEN, C. J. 1978. An Evaluation of Feedback in Document Retrieval Using Co-Occurrence Data. Journal of Documentation. 1978; 34(3): 189-216. ISSN: 0022-0418.

HOBBS, JERRY R.; WALKER, DONALD E.; AMSLER, ROBERT A. 1982. Natural Language Access to Structured Text. In: Horecky, Jan, ed. COLING 82: Proceedings of the 9th International Conference on Computational Linguistics; 1982 July 5-10; Prague, Czechoslovakia. Amsterdam, The Netherlands: North Holland Publishing Co.; 1982. 127-132. ISBN: 0-444-86393-1 (U.S.); LC: 82-7960.

HOEPPNER, WOLFGANG; MORIK, KATHARINA; MARBURGER, HEINZ. 1986. Talking It Over: The Natural Language Dialog System HAM-ANS. In: Bolc, Leonard; Jarke, Matthias, eds. Cooperative Interfaces to Information Systems. Berlin, West Germany: Springer-Verlag; 1986. 189-258. ISBN: 0-387-16599-1 (U.S.); LC: 86-13965.

HOLLAAR, LEE A. 1979. Unconventional Computer Architectures for Information Retrieval. In: Williams, Martha E., ed. Annual Review of

Information Science and Technology: Volume 14. White Plains, NY: Knowledge Industry Publications, Inc. for the American Society for Information Science; 1979. 129-151. ISSN: 0066-4200; ISBN: 0-914236-44-X.

HOLLAAR, LEE A. 1984. The Utah Text Retrieval Project A Status Report. See reference: VAN RIJSBERGEN, C. J., ed. 1984. 123-132.

HOLLAAR, LEE A. 1985a. A Testbed for Information Retrieval Research: The Utah Retrieval System Architecture. See reference: ASSOCIATION FOR COMPUTING MACHINERY. 1985. 227-232.

HOLLAAR, LEE A. 1985b. The Utah Search Engine. In: Parkhurst, Carol A., ed. ASIS '85: Proceedings of the American Society for Information Science (ASIS) 48th Annual Meeting: Volume 22; 1985 October 20-24; Las Vegas, NV. White Plains, NY: Knowledge Industry Publications, Inc. for ASIS; 1985. 369. ISSN: 0044-7870; ISBN: 0-86729-176-1; CODEN: PAISDQ.

INGWERSEN, PETER. 1984. A Cognitive View of Three Selected Online Search Facilities. Online Review. 1984; 8(5): 465-492. ISSN: 0309-314X.

JARDINE, N.; VAN RIJSBERGEN, C. J. 1971. The Use of Hierarchic Clustering in Information Retrieval. Information Storage and Retrieval. 1971; 7: 217-240. ISSN: 0020-0271.

KANTOR, PAUL B. 1983. Minimal Constraint Implementation of the Maximum Entropy Principle in the Design of Term-Weighting Systems. In: Vondran, Raymond, F.; Caputo, Anne; Wasserman, Carol; Diener, Richard A.V., eds. Productivity in the Information Age: Proceedings of the American Society for Information Science (ASIS) 46th Annual Meeting: Volume 20; 1983 October 2-6; Washington, DC. White Plains, NY: Knowledge Industry Publications, Inc. for ASIS; 1983. 28-31. ISSN: 0044-7870; ISBN: 0-86729-072-2; CODEN: PAISDQ.

KOLL, MATTHEW B.; NOREAULT, TERRY; MCGILL, MICHAEL J. 1984. Enhanced Retrieval Techniques on a Microcomputer. In: Williams, Martha E.; Hogan, Thomas H., comps. Proceedings of the National Online Meeting; 1984 April 10-12; New York, NY. Medford, NJ: Learned Information, Inc.; 1984. 165-170. ISBN: 0-938734-07-5.

KOLODNER, JANET L. 1983. Indexing and Retrieval Strategies in Natural Language Fact Retrieval. ACM Transactions on Database Systems. 1983 September; 8(3): 434-464. ISSN: 0362-5915.

KRAWCZAK, DEBORAH A.; SMITH, PHILIP J.; SHUTE, STEVEN J.; CHIGNELL, MARK. 1985. EP-X: A Knowledge-Based System to Aid in Searches of the Environmental Pollution Literature. In: The Engineering of Knowledge-Based Systems: [Proceedings of the] 2nd Conference on Artificial Intelligence Applications; 1985 December 11-13; Miami Beach, FL. Washington, DC: Institute of Electrical and Electronics Engineers (IEEE) Computer Society Press; 1985. 455-460. Available from: IEEE Service Center, 445 Hoes Lane, Piscataway, NJ 08904 (Catalog no. 85CH2215-2). ISBN: 0-8186-0688-6; LC: 85-62321.

KROPP, D.; WALCH, G. 1981. A Graph Structured Text Field Based on Word Fragments. Information Processing & Management. 1981; 17(6): 363-376. ISSN: 0306-4573; CODEN: IPMADK.

KWOK, K. L. 1985a. A Probabilistic Theory of Indexing and Similarity Measure Based on Cited and Citing Documents. Journal of the American Society for Information Science. 1985 September; 36(5): 342-351. ISSN: 0002-8231; CODEN: AISJB6.

KWOK, K. L. 1985b. A Probabilistic Theory of Indexing Using Author-Provided Relevance Information. In: Parkhurst, Carol A., ed. ASIS'85: Proceedings of the American Society for Information Science (ASIS) 48th Annual Meeting: Volume 22; 1985 October 20–24; Las Vegas, NV. White Plains, NY: Knowledge Industry Publications, Inc. for ASIS; 1985. 59–63. ISSN: 0044-7870; ISBN: 0-86729-176-1; CODEN: PAISDQ.

LEVINSON, ROBERT ARLEN. 1985. A Self-organizing Retrieval System for Graphs. Austin, TX: University of Texas at Austin; 1985. (Ph.D. Dissertation). 101p. Available from: University Microfilms Int., Ann Arbor, MI; No. 8527601.

MARCUS, RICHARD S. 1983. An Experimental Comparison of the Effectiveness of Computers and Humans as Search Intermediaries. Journal of the American Society for Information Science. 1983 November; 34(6): 381–404. ISSN: 0002-8231; CODEN: AISJB6.

MARON, M. E. 1982. Associative Search Techniques versus Probabilistic Retrieval Models. Journal of the American Society for Information Science. 1982 September; 33(5): 308–310. ISSN: 0002-8231; CODEN: AISJB6.

MARON, M. E.; CURRY, SEAN; THOMPSON, PAUL. 1986. An Inductive Search System: Theory, Design, and Implementation. Institute of Electrical and Electronics Engineers (IEEE) Transactions on Systems, Man, and Cybernetics. 1986 January/February; SMC-16(1): 21–28. ISSN: 0018-9472.

MCCALL, FIONA M.; WILLETT, PETER. 1986. Criteria for the Selection of Search Strategies in Best Match Document Retrieval Systems. International Journal of Man-Machine Studies. 1986 September; 25(3): 317–326. ISSN: 0020-7373.

MCCUNE, BRIAN P.; TONG, RICHARD M.; DEAN, JEFFREY S.; SHAPIRO, DANIEL G. 1983. RUBRIC: A System for Rule-Based Information Retrieval. In: COMPSAC 83: Proceedings of the Institute of Electrical and Electronics Engineers (IEEE) Computer Society's 7th International Computer Software and Applications Conference; 1983 November 7–11; Chicago, IL. Silver Spring, MD: IEEE Computer Society Press; 1983. 166–172. Available from: IEEE Service Center, 445 Hoes Lane, Piscataway, NJ 08854 (Catalog no. 83CH1940-6). ISSN: 0730-3157; ISBN: 0-8186-0509-X; LC: 83-640060.

MCCUNE, BRIAN P.; TONG, RICHARD M.; DEAN, JEFFREY S.; SHAPIRO, DANIEL G. 1985. RUBRIC: A System for Rule-Based Information Retrieval. Institute for Electrical and Electronics Engineers (IEEE) Transactions on Software Engineering. 1985 September; SE-11(9): 939–944. ISSN: 0098-5589.

MCGILL, MICHAEL J.; HUITFELDT, JENNIFER. 1979. Experimental Techniques of Information Retrieval. In: Williams, Martha E., ed. Annual Review of Information Science and Technology: Volume 14. White Plains, NY: Knowledge Industry Publications, Inc. for the American Society for Information Science; 1979. 93–127. ISSN: 0066-4200; ISBN: 0-914236-44-X; LC: 66-25096; CODEN: ARISBC.

MCGILL, MICHAEL J.; KOLL, MATTHEW B.; NOREAULT, TERRY. 1979. An Evaluation of Factors Affecting Document Ranking by Information Retrieval Systems. Syracuse, NY: Syracuse University, School of Information Studies; 1979. (Technical Report). Available from: Syracuse University, School of Information Studies, Huntington Hall, Syracuse, NY 13244.

METZLER, DOUGLAS P.; NOREAULT, TERRY; HEIDORN BRYAN. 1983. Syntactic Parsing for Information Retrieval. In: Vondran, Raymond F.; Caputo, Anne; Wasserman, Carol; Diener, Richard A. V., eds. Proceedings of the American Society for Information Science (ASIS) 46th Annual Meeting; 1983 October 2-6; Washington, DC. White Plains, NY: Knowledge Industry Publications, Inc.; 1983. 269-273. ISSN: 0044-7870; ISBN: 0-86729-072-2; CODEN: PAISDQ.

MIYAMOTO, SADAAKI; NAKAYAMA, K. 1986. Fuzzy Information Retrieval Based on a Fuzzy Pseudothesaurus. Institute of Electrical and Electronics Engineers (IEEE) Transactions on Systems, Man, and Cybernetics. 1986 March/April; SMC-16(2): 278-282. ISSN: 0018-9472.

MOHAN, KONDRAHALLI C.; WILLETT, PETER. 1985. Nearest Neighbor Searching in Serial Files Using Text Signatures. Journal of Information Science. 1985; 11(1): 31-39. ISSN: 0165-5515.

MUHLHAUSER, G. 1985. Dawn of Next Generation Information Retrieval. In: 9th International Online Information Meeting; 1985 December 3-5; London, England. Oxford, England: Learned Information, Ltd.; Medford, NJ: Learned Information, Inc.; 1985. 365-371. ISBN: 0-904933-50-4.

MUKHOPADHYAY, UTTAM; STEPHENS, LARRY M.; HUHNS, MICHAEL N.; BONNELL, RONALD D. 1986. An Intelligent System for Document Retrieval in Distributed Office Environments. Journal of the American Society for Information Science. 1986 May; 37(3): 123-135. ISSN: 0002-8231; CODEN: AISJB6.

MURTAGH, F. 1985. Clustering and Nearest Neighbor Searching. In: Informatics 8: Advances in Intelligent Retrieval: Proceedings of a Conference Jointly Sponsored by Aslib, the Aslib Informatics Group, and the Information Retrieval Specialist Group of the British Computer Society; 1985 April 16-17; Wadham College, Oxford, England. London, England: Aslib, The Association for Information Management, Information House; 1985. 54-65. Available from: Aslib, 26-27 Boswell Street, London WC1N3JZ, England. ISBN: 0-85142-195-4.

NOREAULT, TERRY; CHATHAM, R. 1982. A Procedure for the Estimation of Term Similarity Coefficients. Information Technology: Research and Development. 1982; 1(3): 189-196. ISSN: 0144-817X.

NOREAULT, TERRY; MCGILL, MICHAEL J.; KOLL, MATTHEW B. 1981. A Performance Evaluation of Similarity Measures, Document Term Weighting Schemes and Representations in a Boolean Environment. In: Oddy, R. N.; Robertson, S. E.; Van Rijsbergen, C. J.; Williams, P. W., eds. Information Retrieval Research: Papers given at the 1st Joint British Computer Society (BCS) and Association for Computing Machinery (ACM) Symposium: Research and Development in Information Retrieval; 1980 June; St. John's College, Cambridge, England. London: Butterworths; 1981. 57-76. ISBN: 0-408-10775-8.

ODDY, R. N. 1977. Information Retrieval through Man-Machine Dialogue. Journal of Documentation. 1977; 33: 1-14. ISSN: 0022-0418.

PANYR, JIRI. 1986a. Probabilistische Modelle in Information-Retrieval-Systemen [Probabilistic Models in Information-Retrieval-Systems]. Nachrichten für Dokumentation (West Germany). 1986 April; 37(2): 60-66. ISSN: 0027-7436.

PANYR, JIRI. 1986b. Die Theorie der Fuzzy Mengen und Information-Retrieval-Systeme [Fuzzy Set Theory and Information Retrieval Systems]. Nachrichten für Dokumentation (West Germany). 1986; 37(3): 163-167. ISSN: 0027-7436.

POGUE, CHRISTINE A.; WILLETT, PETER. 1984. An Evaluation of Document Retrieval from Serial Files Using the ICL Distributed Array Processor. Online Review. 1984; 8(6): 569-584. ISSN: 0309-314X.

POGUE, CHRISTINE A.; WILLETT, PETER. 1987. Use of Text Signatures for Document Retrieval in a Highly Parallel Environment. Parallel Computing. 1987. (In press). ISSN: 0167-8191.

POLLITT, A. S. 1983. End User Searching for Cancer Therapy Literature- A Rule Based Approach. In: Kuehn, Jennifer J., ed. Proceedings of the 6th Annual International Association for Computing Machinery Special Interest Group on Information Retrieval (ACM SIGIR) Conference on Research and Development in Information Retrieval; 1983 June 6-8; Bethesda, MD. New York, NY: ACM, Inc.; 1983. 136-145. (A publication of SIGIR Forum; 1983 Summer; 17(4)). Available from: ACM Order Department, P.O. Box 64145, Baltimore, MD 21264. ISBN: 0-89791-107-5.

POLLITT, A. S. 1985. End User Boolean Searching on Viewdata Using Numeric Keypads. In: 9th International Online Information Meeting; 1985 December 3-5; London, England. Oxford, England: Learned Information, Ltd.; Medford, NJ: Learned Information, Inc.; 1985. 373-379. ISBN: 0-904933-50-4.

POPOV, I. I.; KRAVCHENKO, A. E.; PAVLOV, A. N. 1985. Implementing Associative Search Strategies in a Documentary-Lexical Information Base. Automatic Documentation and Mathematical Linguistics. 1985; 19(1): 13-22. (Translated from: Nauchno-Tekhnicheskaya Informatsiya, Ser. 2. 1985; 19(1): 9-15). ISSN: 0005-1055.

RABITTI, FAUSTO, ed. 1986. [Proceedings of the] Association for Computing Machinery (ACM) Conference on Research and Development in Information Retrieval; 1986 September 8-10; Pisa, Italy. 283p. Available from: ACM Order Department, P.O. Box 64145, Baltimore, MD 21264 (Order no. 606860); or ACM-CNR CONFERENCE, IEI- Via S. Maria, 46, 56100 Pisa, Italy. ISBN: 0-89791-187-3.

RADECKI, TADEUSZ. 1982. Reducing the Perils of Merging Boolean and Weighted Retrieval Systems. Journal of Documentation. 1982 September; 38(3): 207-211. ISSN: 0022-0418.

RADECKI, TADEUSZ. 1983. Incorporation of Relevance Feedback into Boolean Retrieval Systems. In: Salton, Gerard; Schneider, Hans-Jochen, eds. Research and Development in Information Retrieval: Proceedings [of the 5th International Conference of the Association for Computing Machinery, Special Interest Group on Information Retrieval (ACM-SIGIR)]; 1982 May 18-20; Berlin, West Germany. Berlin: Springer-Verlag; 1983. 133-150. (Lecture Notes in Computer Science: 146). ISBN: 3-540-11978-7 (Berlin); ISBN: 0-387-11978-7 (New York).

RADECKI, TADEUSZ. 1985. A Theoretical Framework for Defining Similarity Measures for Boolean Search Request Formulations, Including Some Experimental Results. Information Processing & Management. 1985; 21(6): 501-524. ISSN: 0306-4573; CODEN: IPMADK.

RAU, LISA F. 1987. Knowledge Organization and Access in a Conceptual Information System. Information Processing & Management. 1987. (In press). ISSN: 0306-4573; CODEN: IPMADK.

ROBERTSON, S. E. 1977a. The Probability Ranking Principle in IR. Journal of Documentation. 1977 December; 33(4): 294-304. ISSN: 0022-0418.

ROBERTSON, S. E. 1977b. Theories and Models in Information Retrieval. Journal of Documenation. 1977 June; 33(2): 126-148. ISSN: 0022-0418.

ROBERTSON, S. E.; SPARCK JONES, KAREN. 1976. Relevance Weighting of Search Terms. Journal of the American Society for Information Science. 1976 May/June; 27(3): 129-146. ISSN: 0002-8231; CODEN: AISJB6.

ROBERTSON, S. E.; THOMPSON, C. L.; MACASKILL, M. J.; BOVEY, J. D. 1986. Weighting, Ranking and Relevance Feedback in a Front-End System. Journal of Information Science. 1986; 12(1-2): 71-75. ISSN: 0165-5515.

SACCO, GIOVANNI MARIA. 1984. OTTER: An Information Retrieval System for Office Automation. In: Ellis, Clarence A., ed. [Proceedings of the] 2nd Association for Computing Machinery, Special Interest Group on Office Automation (ACM-SIGOA) Conference on Office Information Systems; 1984 June 25-27; Toronto, Canada. New York, NY: ACM, Inc.; 1984. 104-112. (Published as: SIGOA Newsletter. 1984; 5(1-2)). Available from: ACM Order Department, P.O. Box 64145, Baltimore, MD 21264 (Order no. 611840). ISBN: 0-89791-140-7.

SALTON, GERARD. 1968. Automatic Information Organization and Retrieval. New York, NY: McGraw-Hill; 1968. 514p. LC: 68-25664.

SALTON, GERARD. 1979. Mathematics and Information Retrieval. Journal of Documentation. 1979; 35(1): 1-29. ISSN: 0022-0418.

SALTON, GERARD. 1981. A Blueprint for Automatic Indexing. SIGIR Forum. 1981; 16: 22-38. (A publication of the Association for Computing Machinery Special Interest Group on Information Retrieval). Available from: ACM SIGIR, 11 W. 42nd St., New York, NY 10036.

SALTON, GERARD. 1983. On the Representation of Query Term Relations by Soft Boolean Operators. In: Proceedings of the Association for Computational Linguistics (ACL) European Chapter 2nd Conference; 1985 March 27-29; Geneva, Switzerland. [n.p.]: ACL; 1985. 116-122. Available from: Donald E. Walker (ACL), Bell Communications Research, 445 South Street, MRE 2A379, Morristown, NJ 07960.

SALTON, GERARD. 1985. Some Characteristics of Future Information Systems. SIGIR Forum. 1985 Fall; 18(2-4): 28-39. (A publication of the Association for Computing Machinery Special Interest Group on Information Retrieval). Available from: ACM SIGIR, 11 W. 42nd St., New York, NY 10036.

SALTON, GERARD. 1986a. Another Look at Automatic Text-Retrieval Systems. Communications of the ACM. 1986 July; 29(7): 648-656. ISSN: 0001-0782.

SALTON, GERARD. 1986b. Recent Trends in Automatic Information Retrieval. See reference: RABITTI, FAUSTO, ed. 1-10.

SALTON, GERARD; BUCKLEY, CHRIS; YU, C. T. 1983a. In: Salton, Gerard; Schneider, Hans-Jochen, eds. Research and Development in Information Retrieval: Proceedings [of the 5th International Conference of the Association for Computing Machinery, Special Interest Group on Information Retrieval (ACM-SIGIR)]; 1982 May 18-20; Berlin, West Germany. Berlin: Springer-Verlag; 1983. 151-165. (Lecture Notes in Computer Science: 146). ISBN: 3-540-11978-7 (Berlin); ISBN: 0-387-11978-7 (New York).

SALTON, GERARD; FOX, EDWARD A.; VOORHEES, ELLEN M. 1985. Advanced Feedback Methods in Information Retrieval. Journal of the American Society for Information Science. 1985 May; 36(3): 200–210. ISSN: 0002-8231; CODEN: AISJB6.

SALTON, GERARD; FOX, EDWARD A.; WU, HARRY. 1983b. Extended Boolean Information Retrieval. Communications of the ACM. 1983 November; 26(11): 1022–1036. ISSN: 0001-0782.

SALTON, GERARD; MCGILL, MICHAEL J. 1983. Introduction to Modern Information Retrieval. New York, NY: McGraw-Hill; 1983. 400p. ISBN: 0-07-054484-0.

SALTON, GERARD; VOORHEES, ELLEN M. 1985. Automatic Assignment of Soft Boolean Operators. See reference: ASSOCIATION FOR COMPUTING MACHINERY. 54–69.

SARACEVIC, TEFKO. 1983. On a Method for Studying the Structure and Nature of Requests in Information Retrieval. In: Vondran, Raymond F.; Caputo, Anne; Wasserman, Carol; Diener, Richard A. V., eds. Productivity in the Information Age: Proceedings of the American Society for Information Science (ASIS) 46th Annual Meeting: Volume 20; 1983 October 2-6; Washington, DC. White Plains, NY: Knowledge Industry Publications, Inc.; 1983. 22-25. ISSN: 0044-7870; ISBN: 0-86729-072-2; CODEN: PAISDQ.

SCHANK, ROGER C. 1975. Conceptual Information Processing. Amsterdam, The Netherlands: North-Holland; New York, NY: American Elsevier; 1975. 374p. ISBN: 0-444-10773-8 (American Elsevier).

SCHANK, ROGER C.; ABELSON, ROBERT P. 1977. Scripts, Plans, Goals, and Understanding: An Inquiry into Human Knowledge Structures. Hillsdale, NJ: L. Erlbaum Associates; 1977. ISBN: 0-470-99033-3; LC: 76-051963.

SCHEK, HANS-JOERG. 1984. Nested Transactions in a Combined IRS-DBMS Architecture. See reference: VAN RIJSBERGEN, C. J., ed. 1984. 55-70.

SHAW, W. M., JR. 1986. An Investigation of Document Partitions. Information Processing & Management. 1986; 22(1): 19–28. ISSN: 0306-4573; CODEN: IPMADK.

SHOVAL, PERETZ. 1985. Principles, Procedures and Rules in an Expert System for Information Retrieval. Information Processing & Management. 1985; 21(6): 475–487. ISSN: 0306-4573; CODEN: IPMADK.

SIMMONS, ROBERT F. 1987. A Text Knowledge Base from the AI Handbook. Information Processing & Management. 1987. (In press). ISSN: 0306-4573; CODEN: IPMADK.

SMEATON, A. F.; VAN RIJSBERGEN, C. J. 1981. The Nearest Neighbor Problem in Information Retrieval: An Algorithm Using Upper Bounds. Proceedings of the Association for Computing Machinery Special Interest Group on Information Retrieval (ACM SIGIR) 4th International Conference on Information Storage and Retrieval. 1981 May 31-June 2; Oakland, CA. New York, NY: ACM, Inc., 1981. 83-87. ISBN: 0-89791052-4.

SMEATON, A. F.; VAN RIJSBERGEN, C. J. 1986. Information Retrieval in an Office Filing Facility and Future Work in Project Minstrel. Information Processing & Management. 1986; 22(2): 135-149. (Special issue). ISSN: 0306-4573; CODEN: IPMADK.

SNEATH, PETER H. A.; SOKAL, ROBERT R. 1973. Numerical Taxonomy: The Principles and Practice of Numerical Classification. San Francisco, CA: W. H. Freeman; 1973. 573p. ISBN: 0-7167-0697-0; LC: 72-001552.

SPARCK JONES, KAREN. 1974. Automatic Indexing. Journal of Documentation. 1974; 30(4): 393–432. ISSN: 0022-0418.
SPARCK JONES, KAREN; TAIT, J. I. 1984. Automatic Search Term Variant Generation. Journal of Documentation. 1984 March; 40(1): 50–66. ISSN: 0022-0418.
SPARCK JONES, KAREN; WEBSTER, C. A. 1980. Research on Relevance Weighting. Cambridge, England: University of Cambridge, Computer Laboratory; 1980. (British Library R&D Report 5553). Available from: Computer Laboratory, University of Cambridge, Corn Exchange Street, Cambridge CB2 3QG, England.
SUBRAMANIAN, VISWANATH; BISWAS, GAUTAM; BEZDEK, JAMES C. 1986. Document Retrieval Using a Fuzzy Knowledge-Based System. Optical Engineering. 1986 March; 25(3): 445–455. ISSN: 0091-3286.
TONG, RICHARD M.; ASKMAN, VICTOR N.; CUNNINGHAM, JAMES F.; TOLLANDER, CARL J. 1985. See reference: ASSOCIATION FOR COMPUTING MACHINERY. 243–251.
TONG, RICHARD M.; SHAPIRO, DANIEL G. 1985. Experimental Investigations of Uncertainty in a Rule-Based System for Information Retrieval. International Journal of Man-Machine Studies. 1985; 22(3): 265–282. ISSN: 0020-7373.
VAN RIJSBERGEN, C. J. 1977. A Theoretical Basis for the Use of Co-Occurrence Data in Information Retrieval. Journal of Documentation. 1977 June; 33(2): 106–119. ISSN: 0022-0418.
VAN RIJSBERGEN, C. J. 1979. Information Retrieval. 2nd ed. London, England: Butterworths; 1979. 208p. ISBN: 0-408-70929-4.
VAN RIJSBERGEN, C. J., ed. 1984. Research and Development in Information Retrieval: Proceedings of the 3rd Joint British Computer Society (BCS) and Association for Computing Machinery (ACM) Symposium; 1984 July 2-6; Cambridge, England. Cambridge, England: Cambridge University Press; 1984. 433p. (The British Computer Society Workshop Series). ISBN: 0-521-26865-6; LC: 84-45234.
VAN RIJSBERGEN, C. J. 1986a. A New Theoretical Framework for Information Retrieval. See reference: RABITTI, FAUSTO, ed. 194–200.
VAN RIJSBERGEN, C. J. 1986b. A Non-classical Logic for Information Retrieval. Computer Journal. 1986 December; 29(6): 481–485. ISSN: 0010-4620.
VAN RIJSBERGEN, C. J., ed. 1986c. Office Automation. Information Processing & Management. 1986; 22(2). (Special Issue). ISSN: 0306-4573; ISBN: 0-08-034217-5; CODEN: IPMADK.
VAN RIJSBERGEN, C. J.; ROBERTSON, S. E.; PORTER, M. F. 1980. New Models in Probabilistic Information Retrieval. Cambridge, England: University of Cambridge, Computer Laboratory; 1980. 123p. (British Library R&D Report 5587). Available from: Computer Laboratory, University of Cambridge, Corn Exchange Street, Cambridge CB2 3QG, England.
VLADUTZ, G.; COOK, J. 1984. Bibliographic Coupling and Subject Relatedness. In: Flood, Barbara; Witiak, Joanne; Hogan, Thomas H., comps. 1984: Challenges to an Information Society: Proceedings of the American Society for Information Science (ASIS) 47th Annual Meeting: Volume 21; 1984 October 21-25; Philadelphia, PA. White Plains, NY: Knowledge Industry Publications, Inc.; 1984. 204–207. ISSN: 0044-7870; ISBN: 0-86729-115-X; CODEN: PAISDQ.

VOORHEES, ELLEN M. 1986. The Efficiency of Inverted Index and Cluster Searches. See reference: RABITTI, FAUSTO, ed. 164-174.

WALKER, DONALD E.; HOBBS, JERRY R. 1981. Natural Language Access to Medical Text. Menlo Park, CA: SRI International, Artificial Intelligence Center, Computer Science and Technology Division; 1981 March. 20p. (Technical Note 240; Project 1944). Available from: SRI International, 333 Ravenswood Ave., Menlo Park, CA 94025.

WATTERS, C. R.; SHEPHERD, M. A.; GRUNDKE, E. W.; BODORIK, P. 1985. Integration of Menu Retrieval and Boolean Retrieval from a Full-Text Database. Online Review. 1985; 9(5): 391-402. ISSN: 0309-314X.

WILLETT, PETER. 1984a. A Nearest Neighbour Search Algorithm for Bibliographic Retrieval from Multilist Files. Information Technology: Research and Development. 1984 April; 3(2): 78-83. ISSN: 0144-817X.

WILLETT, PETER. 1984b. A Note on the Use of Nearest Neighbors for Implementing Single Linkage Document Classifications. Journal of the American Society for Information Science. 1984 May; 35(3): 149-152. ISSN: 0002-8231; CODEN: AISJB6.

WILLETT, PETER. 1985a. An Algorithm for the Calculation of Exact Term Discrimination Values. Information Processing & Management. 1985; 21(3): 225-232. ISSN: 0306-4753; CODEN: IPMADK.

WILLETT, PETER. 1985b. Query-Specific Automatic Document Classification. International Forum on Information and Documentation. 1985 April; 10(2): 28-32. ISSN: 0304-9701.

WILLETT, PETER. 1985c. Ranked Output Searching in Textual and Structural Data Bases. In: 9th International Online Information Meeting; 1985 December 3-5; London, England. Oxford, England: Learned Information, Ltd.; Medford, NJ: Learned Information, Inc.; 1985. 343-353. ISBN: 0-904933-50-4.

WILLETT, PETER; WINTERMAN, VIVIENNE; BAWDEN, DAVID. 1986. Implementation of Nearest-Neighbor Searching in an Online Chemical Structure Search System. Journal of Chemical Information and Computer Sciences. 1986; 26(1): 36-41. ISSN: 0095-2338; CODEN: JCISD8.

WINETT, SHEILA G.; FOX, EDWARD A. 1985. Using Information Retrieval Techniques in an Expert System. In: The Engineering of Knowledge-Based Systems: [Proceedings of the] 2nd Conference on Artificial Intelligence Applications; 1985 December 11-13; Miami Beach, FL. Washington, DC: Institute of Electrical and Electronics Engineers (IEEE) Computer Society Press; 1985. 230-235. Available from: IEEE Service Center, 445 Hoes Lane, Piscataway, NJ 08854 (Catalog no. 85CH2215-2). ISBN: 0-8186-0688-6; LC: 85-62321.

WONG, S. K. M.; RAGHAVAN, VIJAY V. 1984. Vector Space Model of Information Retrieval- A Reevaluation. See reference: VAN RIJSBERGEN, C. J., ed. 1984. 167-185.

WONG, S. K. M.; ZIARKO, WOJCIECH. 1985. On Generalized Vector Space Model in Information Retrieval. Annales Societatis Mathematicae Polonnae: Series IV: Fundamenta Informaticae. 1985; 8(2): 253-267. Available from: Fundamenta Informaticae, P. O. Box 22, Warsaw 00-901, Poland.

WONG, S. K. M.; ZIARKO, WOJCIECH; WONG, P. C. N. 1985. See reference: ASSOCIATION FOR COMPUTING MACHINERY. 18-25.

YU, C. T.; BUCKLEY, CHRIS; LAM, K.; SALTON, GERARD. 1983. A Generalized Term Dependence Model in Information Retrieval. Information Technology: Research and Development. 1983 October; 2(4): 129-154. ISSN: 0144-817X.

YU, C. T.; LEE, T. C. 1986. Non-binary Independence Model. See reference: RABITTI, FAUSTO, ed. 265-268.

YU, C. T.; LUK, W. S.; SIU, M. K. 1979. On Models of Information Retrieval Processes. Information Systems. 1979; 4(3): 205-218. ISSN: 0306-4379; CODEN: INSYD6.

ZARRI, GIAN PIERO. 1983. RESEDA, an Information Retrieval System Using Artificial Intelligence and Knowledge Representation Techniques. In: Kuehn, Jennifer J., ed. Proceedings of the 6th Annual International Association for Computing Machinery, Special Interest Group on Information Retrieval (ACM-SIGIR) Conference on Research and Development in Information Retrieval; 1983 June 6-8; Bethesda, MD. New York, NY: ACM, Inc.; 1983. 189-195. (A publication of SIGIR Forum, 1983 Summer; 17(4)). Available from: ACM Order Department, P. O. Box 64145, Baltimore, MD 21264. ISBN: 0-89791-107-5.

ZARRI, GIAN PIERO. 1984. Some Remarks about the Inference Techniques of RESEDA, an "Intelligent" Information Retrieval System. See reference: VAN RIJSBERGEN, C. J., ed. 1984. 281-300.

ZARRI, GIAN PIERO. 1985. Interactive Information Retrieval: An Artificial Intelligence Approach to Deal with Biographical Data. In: Informatics 8: Advances in Intelligent Retrieval: Proceedings of a Conference Jointly Sponsored by Aslib, the Aslib Informatics Group, and the Information Retrieval Specialist Group of the British Computer Society; 1985 April 16-17; Wadham College, Oxford, England. London, England: Aslib, The Association for Information Management, Information House; 1985. 101-119. Available from: Aslib, 26-27 Boswell Street, London WC1N3JZ, England. ISBN: 0-85142-195-4.

ZENNER, REMBRAND B. R. C.; DE CALUWE, RITA M. M.; KERRE, ETIENNE E. 1985. A New Approach to Information Retrieval Systems Using Fuzzy Expressions. Fuzzy Sets and Systems (The Netherlands). 1985; 17(1): 9-22. ISSN: 0165-0114.

5 Statistical Methods in Information Science Research

MARK T. KINNUCAN,
MICHAEL J. NELSON, and
BRYCE L. ALLEN
University of Western Ontario

INTRODUCTION

This chapter reviews how statistics, primarily inferential statistics, are used to analyze the data of research in information science. In terms of the kinds of statistics discussed, the coverage here is similar to that of a general intermediate textbook on statistics, such as that by BERENSON ET AL. The major topics include goodness-of-fit tests, analysis of variance (ANOVA) and similar tools, regression and related techniques, the analysis of contingency tables, and various other methods that we call dimensionality-reduction techniques (factor analysis, multidimensional scaling, and cluster analysis). Within some of the categories, especially ANOVA-like techniques and regression-like techniques, a distinction can be made between parametric and nonparametric approaches. The importance of this distinction is discussed, and examples of both kinds of approaches are presented. (The statistics used to analyze contingency tables are considered to be nonparametric statistics.)

The purpose of this chapter is to review the use of statistics for data analysis, not to review the application of statistics or probability theory to automatic indexing or information retrieval. Other authors have described or reviewed those applications, notably BOOKSTEIN, SCHWARTZ & EISENMANN, SALTON & MCGILL, and VAN RIJSBERGEN. See also the chapter

The authors are indebted to Patricia Burt and Margaret Ann Wilkinson for their helpful comments on an earlier draft of this chapter. The preparation of this review was supported in part by Grant No. A8088 to M. T. Kinnucan and Grant No. A2485 to M. J. Nelson, from the Natural Sciences and Engineering Research Council of Canada.

Annual Review of Information Science and Technology (ARIST), Volume 22, 1987
Martha E. Williams, Editor
Published for the American Society for Information Science (ASIS)
by Elsevier Science Publishers B.V.

by BELKIN & CROFT in this volume of *ARIST*. There is probably some overlap between data-analytic uses and automatic indexing/information retrieval uses of statistics, especially with respect to what we have called dimensionality-reduction techniques; however, the focus here is on how researchers use statistics to help them understand their data.

The presence of statistics in an article is rarely mentioned in the article's abstract. It is even more rare for articles to be indexed by the statistics used. Therefore, to find examples of the use of statistics in information science research during the past five or so years, the following journals and proceedings were scanned: *ACM Transactions on Office Information Systems, Canadian Journal of Information Science, Communications of the ACM, Information Processing and Management, Journal of Documentation, Journal of Information Science, Journal of the American Society for Information Science, Library and Information Science Research, Proceedings of the Annual Meeting of the American Society for Information Science, Scientometrics, Social Studies of Science,* and the various ACM (Association for Computing Machinery) SIGIR conference proceedings. This search turned up over 200 articles and papers that used statistical methods deemed to be within the scope of this review. The examples presented below were culled from this set. Some of them were chosen to illustrate correct application of the statistical tools, while others were chosen to illustrate common mistakes in the use of statistics.

Categories of Statistics

Statistics are commonly categorized as descriptive or inferential. Descriptive statistics depict the data at hand. Inferential statistics estimate characteristics of populations based on data from samples (OTT ET AL.). In information science, for example, it may be difficult to examine every article in a given field or to trace the progress of every author; samples must be taken in these cases. If the researcher wishes to generalize from his or her findings to the population as a whole, inferential statistics are necessary.

Another basic distinction in statistics is made among univariate, bivariate, and multivariate statistics. Univariate statistics describe characteristics of a single variable. Bivariate and multivariate statistics are used to describe the relationships between two (bivariate) or among more than two (multivariate) variables or to test hypotheses concerning the existence or nonexistence of such relationships.

Univariate descriptive statistics describe such simple characteristics as central tendency (e.g., mean and median), dispersion (e.g., range and standard deviation), and simple proportions. These measures are commonplace, and it would not be helpful, or even really possible, to review their use in the literature of information science. For the most part, this chapter is concerned instead with inferential statistics, although some of the dimensionality-reduction techniques are probably better thought of as descriptive methods.

Most inferential statistics are bivariate or multivariate because they are important in hypothesis testing, and most research hypotheses concern the presence or absence of relationships among two or more variables. However, there is one important univariate application of inferential techniques (and of hypothesis testing) that deserves mention and is reviewed in this chapter. Researchers can sometimes specify on theoretical grounds what they expect the shape of the frequency distribution of a variable to be. Lotka's law (NICHOLLS), which predicts that as one searches for increasingly prolific authors there is a specified decline in the number of such authors found, is a good example. Once an appropriate set of data is collected, a class of inferential techniques called goodness-of-fit tests allow the researcher to test the hypothesis that the theory does an adequate job of predicting the shape of the frequency distribution of those data.

The next simplest case involves two variables, and the appropriate statistical analysis depends on whether these variables are continuous or categorical. (Here "categorical" includes discrete variables that have only a few different values.) Basically there are three possibilities: 1) one of the variables is continuous and one is categorical, 2) both variables are continuous, and 3) both are categorical. Each possibility is discussed in turn below.

In a great many cases, one of the variables is considered the independent or predictor variable, and the other is then the dependent or criterion variable. ANOVA-like techniques (including t-tests and some nonparametric tests) are appropriate when the independent variable is categorical or has only a few discrete levels and the dependent variable is continuous. In such a situation one usually compares the means of the dependent measure for the different categories or levels of the independent variable. A classical example is an experiment in which the independent variable is an experimental manipulation and the dependent variable is some outcome measure. The role of the ANOVA technique is to say how likely it is that an observed difference in the sample means could have arisen by chance in the absence of any real effect of the experimental manipulation in the population as a whole. If the likelihood is small enough, the researcher concludes that the manipulation must have had an impact. Typically, researchers require the probability of such a random occurrence to be less than 0.05 before they are willing to believe in the effectiveness of the experimental manipulation. This reasoning is known as hypothesis testing.

Regression and related techniques are used when both variables are measured on a continuous scale. This typically happens when no experiment is involved. The purpose of regression and related techniques is to measure the strength of the relationship between the variables and to enable one to predict the value of the dependent variable for different values of the independent variable. Thus, regression is more appropriate for relationships, and ANOVA is more appropriate for comparisons (TAGUE, 1981a).

Hypothesis testing is possible in the regression context as well: how likely is it that the observed correlation between the variables could have arisen in the sample when there is actually no correlation between the variables in the population? However, hypothesis testing tends to be less important in regression analysis than in ANOVA.

The strength of a relationship can also be calculated in the ANOVA situation as it is in the regression situation, but this is rarely done. It should perhaps be done more often because it helps remind the researcher and the reader that statistical significance tells only part of the story. A result may be significant statistically but of little practical importance. This kind of situation would show up as a minute value of the strength of the relationship between the independent and dependent variables. Standard textbooks on experimental design and analysis, such as those of KEPPEL or MYERS, provide formulas for calculating the strength of the relationship as part of an ANOVA as well as guidelines for interpreting it.

Chi-square is appropriate when both variables are categorical or have only a few discrete levels. (The chi-square statistic also finds application in goodness-of-fit tests.) Discrete data may or may not arise from an experiment. When each variable has only a few possible values, the researcher can form a contingency table (also called a cross-classification) and can tally the number of occurrences of each particular combination of values of the two variables. By studying the pattern of the frequencies in such a table, one can attempt to discern if the variables are related. Like ANOVA, chi-square is used to test hypotheses. Usually, the chi-square test tells the researcher the probability that the pattern of association observed in the table could have occurred if the two variables were actually unrelated in the total population. The reasoning and decision rule are the same as for ANOVA. Measures of the strength of the relationship are available for contingency tables, too. Several such measures have been proposed, but none has gained universal acceptance.

To summarize, ANOVA, regression, and the chi-square analysis of contingency tables (and their cousins) are the primary statistical tools for the bivariate situation. Each has been generalized to the multivariate situation, where the pattern of relationships among three or more variables is explored. These generalizations are discussed in the appropriate sections below.

The testing of statistical hypotheses—whether univariate, bivariate, or multivariate—has traditionally involved the comparison of a null hypothesis (i.e., no relationship exists among the variables) with a research hypothesis that a relationship exists. The null hypothesis is the one that gets tested; it is either accepted or rejected. More recently, however, researchers and statisticians have begun to use a more flexible framework for thinking about the application of inferential statistics to data. It is now becoming customary to talk about statistical models—i.e., equations that predict the value of the dependent variable on the basis of the hypothesized nature of the relationship. In ANOVA and regression analysis these equations are known as linear models; in chi-square analysis they are known as log-linear models. Using statistical models in situations where traditional methods could be applied always leads to exactly the same results; the advantages of the models approach are that it allows more precise specification of hypotheses, especially in the multivariate situation, and that it provides a more comprehensive and unifying framework for thinking about all kinds of hypothesis testing situations.

Within the models view, an hypothesis test always consists of the comparison of two models, a full model and a restricted model. Each model tries to predict the values of the dependent variable, and each model has associated with it some free parameters that must be estimated from the data. The full model always contains all the parameters specified by the restricted model plus at least one more. Thus, the full model assumes that the relationship between the variables is more complex than does the restricted model. It also always predicts the dependent variable more accurately than the restricted model because it is drawing on more information. In comparing the two models the question is whether the increased accuracy of the full model is worth its increased complexity. If there is only a small gain in accuracy when the full model is used, the restricted model should be chosen as the best overall description of the relationship between the independent and the dependent variable. On the other hand, if the increase in accuracy outweighs the increase in complexity, the full model should be chosen.

Scope within Information Science

The use of statistics for data analysis implies that the researcher has collected observations in the real world—i.e., that the research is empirical. Empirical research and statistical evaluation are carried out in a wide array of social, natural, and engineering science fields, of which one example is, of course, information science. In information science there are three classes of measurable objects: artifacts, systems, and users. The artifacts of information science are documents or document surrogates: records of various kinds. They have attributes that can be studied quantitatively. This is essentially the task of bibliometrics and related fields such as scientometrics. Information systems also have attributes that can be measured, and statistical techniques can be used to analyze them. Finally, there is the broad class of user studies, which often survey the characteristics of people who use (or fail to use) information systems and how they use them. The results of user studies are well suited to statistical treatment.

Obviously a broad range of research in information science is concerned with countable or measurable objects and thus meets the primary requirement for the use of statistics. Of course, not all quantitative research in information science uses statistics. Some topics and approaches, while using mathematics, do not involve the empirical activities of counting or measuring. The presentation of a formal proof of an algorithm or the development of a theoretical model for system behavior are typical of quantitative, but not statistical, research in information science.

Some authors believe that the development of more and better empirical research is crucial to the development of information science as a science. For example, HAAS & KRAFT call for more extensive use of experimental methods in information science. They maintain that experiments are the best way to test theories and that theories should be given more attention in information science. However, not all questions are amenable to experimental methods, so the examples below include both experimental and nonexperimental research.

The main body of this review consists primarily of illustrations of the use of statistical techniques in reports of actual research in various areas of information science. Before moving on to these illustrations, mention should be made of those authors who have focused on statistical issues in information science.

WALLACE (1985) surveyed the literature of four applied social science fields: 1) library and information science, 2) business, 3) education, and 4) social work. He selected 25 journals in each field, and for each journal he counted the number of articles published in 1981 that contained either descriptive or inferential statistics. He found that library and information science had the lowest use of statistics generally, and of inferential statistics. It should be pointed out that Wallace included library science along with information science in his study, while no effort has been made to cover library science in this review.

The absence of inferential statistics from an article indicates either that the article does not report empirical research or that the research that was done did not use inferential techniques. Articles lacking empirical research and articles with research but no inferential statistics appear at different rates in the various areas of information science. Clearly, bibliometrics and scientometrics use empirical research, so they probably have the strongest tradition of inferential statistics. Many of the authors in bibliometrics and scientometrics use univar-iate or bivariate techniques. MORAVCSIK admonished workers in scientometrics to adopt a "multidimensional view" of their field. Commentators on Moravcsik's article noted that multivariate statistical techniques are designed precisely to allow a multidimensional view (FRAME; HAITUN, 1984).

While bibliometrics is empirical, it is definitely not experimental. The design of information retrieval systems lends itself naturally to comparisons of competing systems, so this area is more fertile ground for experimentation. Methods for conducting information retrieval experiments, as well as illustrative examples, are contained in a book edited by SPARCK JONES. The chapter by TAGUE (1981a), especially, describes the use of statistics to analyze the results of information retrieval experiments. Very few of the information retrieval experiments that are carried out are analyzed statistically, however. To give just one example, SALTON ET AL. could have used ANOVA to test the generality of the difference they found between two methods of relevance feedback in retrieval systems but did not choose to do so.

ROBERTSON also discussed statistical issues in his chapter of the Sparck Jones volume, pointing out some supposed problems that in his view prevent more extensive use of inferential statistics. Many of these apparent problems can however be dealt with by the choice of appropriate statistical models. For example, Robertson stated that most statistical methods are not valid when both the queries and the documents in an information retrieval experiment are sampled from larger populations. However, this situation is formally identical to research designs used in education and psychology. ANOVA techniques have been developed for those situations (KEPPEL).

Parametric and Nonparametric Statistics

Parametric approaches to statistical inference make reference to known theoretical distributions, such as the normal distribution, the t distribution, or the F distribution, in evaluating statistical hypotheses. Many authors have argued that there are situations in which reference to these theoretical distributions is inappropriate. In those situations, these authors argue, alternative approaches, which are known as nonparametric statistics, should be used. This view is by no means universally accepted, and the whole area is one of considerable controversy. Although the issues are subtle and interwoven, they essentially surround the assumptions underlying the use of parametric techniques and the importance of those assumptions.

Some of these assumptions concern the philosophical concept of levels of measurement. Four levels of measurement are usually identified: 1) nominal, 2) ordinal, 3) interval, and 4) ratio. A particular variable is said to be measured at one or another of these levels. The levels are defined by how freely one could change the scale on which a variable is measured without changing the information provided by the measurement. In statistics, the most crucial distinction is between the ordinal level and the interval level. When ordinal measurement is used, the only thing really being accomplished is the rank ordering of the observations. This means that any change at all can be made to the scale, as long as the observations retain the same order (from lowest to highest) as before. With interval measurement, on the other hand, the size of the difference between pairs of values is important, as well as the order of the observations. This means that for interval-level measurement, there is a greater restriction on the kinds of changes that one could make in the scale of measurement without affecting what information the measurement is providing than there is for ordinal-level measurement. For example, converting degrees Celsius to degrees Fahrenheit is a change in scale of measurement that is allowed for interval-level data. Telling a person the temperature in Fahrenheit gives that person the same information as relating the temperature in Celsius.

Proponents of nonparametrics claim that parametric methods should not be applied to ordinal data (TOWNSEND & ASHBY), whereas nonparametric methods, which typically are based on ranks, are well suited to ordinal data. Defenders of the parametric techniques maintain that no harm is caused by their use for ordinal data (GAITO; GREGOIRE & DRIVER). The issue is further muddied by the fact that it is not always completely obvious whether a particular set of data represents measurement at the ordinal or the interval level (CLIFF; MICHELL). MICHELL suggests that the issue will not be resolved until theoreticians reach a consensus on what is meant by saying that one is measuring something.

The second focus of the controversy concerns other assumptions underlying the use of parametric statistics. Typically, this includes such assumptions as the variables involved being normally distributed with equal variances. The question, put simply, is: "Does it matter if these assumptions are violated?". Proponents of nonparametric statistics say that the assumptions are crucial to the correct application of parametric techniques, while proponents of parametric methods say that the assumptions are often inconsequential. The correct answer is by no means clear. There are a variety of different criteria by which one could conclude whether the violation of an assumption were important in a particular instance or not, and there is a bewildering array of possible combinations of violations of the assumptions one might want to consider (see, for example, BRADLEY and EDGELL & NOON).

There are two reasons statisticians would like to know whether the assumptions of parametric techniques are important. First, it is generally believed that the parametric techniques are more powerful than the non-parametric techniques and thus should be used whenever possible (MYERS). However, even *that* belief has been challenged in some instances (BLAIR & HIGGINS). Second, the parametric techniques are more versatile than the nonparametric techniques in that specific models exist for a wider variety of research designs.

Within information science, the use of parametric methods seems to have been questioned most strongly in bibliometrics and scientometrics. This is not surprising since most bibliometric distributions are extremely skewed and, hence, nonnormal. HAITUN (1982) has concluded that there are two main categories of empirical frequency distributions, Gaussian (normal) and Zipfian (skewed). Parametric techniques are only appropriate, he says, for variables that have Gaussian distributions. (Haitun's arguments are reiterated by BROOKES (1984)). Haitun does not suggest the use of existing nonparametric techniques for Zipf-distributed data; rather he advocates the development of new techniques specifically for those data. HAITUN (1986) has more recently expanded on the reasons for his rejection of the use of Gaussian statistics to include measurement issues as well, although here again he is not simply advocating traditional nonparametric methods but proposing the adoption of totally new techniques. As mentioned above, the use of parametric statistics, despite violations of the various assumptions, does have its defenders. Apparently none of them has yet responded to Haitun's arguments.

GOODNESS-OF-FIT TESTS

If there is a theory that predicts a probability distribution for some variable, and a random sample from some population is measured on that variable, it is often useful to calculate the goodness of fit of the theoretical distribution to the observed distribution. The most common test of this sort is the chi-square goodness-of-fit test, which can be used when the theoretical and observed frequencies have been grouped into classes. For this reason, it is especially useful for testing discrete distributions, such as the distributions often encountered in bibliometrics. SICHEL, for example, used chi-square in tests of the fit of a theoretical distribution, known as the generalized inverse Gaussian-Poisson distribution, to several sample distributions such as journal productivity, journal use, and author productivity.

Even with discrete distributions it is sometimes necessary to group classes of the variable together into fewer, larger classes for the chi-square test to be valid. Many bibliometric distributions have a long, straggling tail. When SICHEL, HUBERT, BROOKES (1981) and others applied the chi-square test to such distributions, they grouped the tail into one class. This means that any detailed predictions their theories may have made about values of the distribution in the tail were not being tested.

A major assumption of any hypothesis test is that the observed distribution at hand represents a random sample from a population. Bibliometric distributions often arise from convenience samples or from measuring the entire population of interest. For this reason, some researchers, such as BAGUST, BROWNSEY & BURRELL and NELSON & TAGUE have preferred to use the chi-square value as an indicator of relative fit between various theoretical models rather than as an absolute test of the fit of a particular model.

Another test that has been used recently in information science is the Kolmogorov-Smirnov goodness–of–fit test. POTTER, PAO (1985; 1986), and NICHOLLS used the Kolmogorov-Smirnov test in studies of the validity of Lotka's law. There is also a two-sample Kolmogorov-Smirnov test for testing whether two independent samples come from the same unspecified theoretical distribution. BORGMAN used this version in testing models of online catalog use.

ANOVA AND RELATED TECHNIQUES

The techniques discussed in this section are used where the outcome measure is a continuous variable, and the researcher calculates the means or medians on this measure for various categories or treatment levels of the independent variables. Usually the researcher wants to be able to say that the mean or median value for some categories is higher than for others. In the simplest case, there is only a single independent variable with two categories, so the question becomes, "Which category, if either, produces higher scores on the outcome measure?". The role of inferential statistics in this case is to tell the researcher whether the observed difference between the categories is statistically reliable. Formally, the researcher tests a null hypothesis that the categories are equivalent against the alternative hypothesis that one of the categories produces higher scores than the other.

One Independent Variable with Two Categories

The t-test is probably the best-known statistical test for the two-category case. BLAIR & MARON used t-tests in their evaluation of a full-text retrieval system for lawyers. For example, they compared the recall rates of searches done by lawyers with the recall rates of searches done by paralegals and found no statistically significant differences. Blair and Maron correctly concluded that there is no evidence to support the hypothesis that lawyers can find more relevant documents than paralegals.

Sometimes, however, authors fail to heed the negative results of statistical tests. For example, HANSEN conducted a t-test, then stated that "precision increased from 26% to 38% when the electronic search followed the manual search. Even though this is not a statistically significant increase, it represents a 46% increase in precision based on search order" (p. 316). Later she ignored the negative results of the t-test in stating her conclusions. This practice defeats the purpose of the scientifically rigorous test of significance and should on principle be avoided.

Researchers sometimes apply statistics (such as the t-test) that are meant for the two-category case when they actually have more than two categories. They consider the categories a pair at a time and apply the test repeatedly. This approach to the use of statistics is known as making multiple comparisons. Specialized procedures have been developed for multiple comparisons, and those should be used instead of repeating the t-tests. Further, the researcher should make multiple comparisons only after first applying a test that considers all of the categories simultaneously. Two studies that inappropriately made multiple comparisons follow.

PERITZ (1983a) studied journal articles in sociology and looked at three categories (rather than two): methodological, theoretical, and empirical articles. Her research tried to determine which of those categories receives the most citations. She looked at the average number of times articles in the different categories were cited and made pairwise comparisons between methodological and theoretical articles and between methodological and empirical articles. A single significance test involving all three categories simultaneously would have been more appropriate (see "One Independent Variable with More Than Two Categories," below).

WANG ET AL. used the Wilcoxon signed ranks test (a nonparametric equivalent of the t-test) with eight categories. Their study measured precision in an information retrieval system. Rather than applying a single significance test to all eight categories simultaneously, they compared each of seven types of thesauri with a baseline condition of no thesaurus, applying a separate significance test each time. Their approach capitalized unfairly on chance, and masked possible differences among the different kinds of thesauri.

One Independent Variable with More than Two Categories

When the researcher does have more than two categories of the independent variable, the appropriate parametric statistical test is the one-way ANOVA. The Kruskal-Wallis test (CONOVER) is an appropriate nonparametric test. If the ANOVA or Kruskal-Wallis test shows that significant differences exist among the categories, multiple comparisons can be used as long as the multiple comparisons follow an established procedure for such comparisons.

WALLACE & BONZI and PONTIGO & LANCASTER both used one-way ANOVA to compare the quality of journals that occupy different zones in the Bradford distribution. The Bradford distribution describes the productivity of different journals in a particular discipline. Wallace and Bonzi used the number of citations that an article receives as a measure of the quality of the journal in which the article appeared. Their Bradford distribution had three zones, and the one-way ANOVA showed significant differences. They then used the Scheffe multiple-comparisons test, which allowed them to conclude that articles from journals in the two most-productive zones were cited more frequently than articles from journals in the least-productive zone.

Pontigo and Lancaster also tried to determine whether the quality of journals declines as one moves toward less-productive Bradford zones. Their sample of journals contained four Bradford zones rather than three. Like

Wallace and Bonzi, they sampled articles from journals in the different zones and counted the number of citations each article received. Unlike Wallace and Bonzi, however, they did not find support for their hypothesis. Pontigo and Lancaster did find a significant difference among the zones in terms of the number of citations. Inspection of the means for the different zones suggested that articles from journals in the middle two zones received more citations on the average than articles in either the most productive journals or the least productive zones. Pontigo and Lancaster could have conducted multiple comparisons to verify this difference between the zones. In their case, though, multiple comparisons were not really necessary. It doesn't really matter whether the middle zones received more citations than the most productive zone or the same number. In either case, their hypothesis would not have been confirmed.

Pontigo and Lancaster's study also illustrates one method of handling violations of ANOVA assumptions. These authors found that the variance of the citation scores was not the same for the different zones, so they applied a variance-stabilizing transformation before conducting the ANOVA test. (To be exact, they took the logarithm of each score.) BERENSON ET AL. describe different possible transformations one can use to bring data into conformity with ANOVA assumptions, and MYERS discusses some of the pros and cons of transforming data.

BROOKS (1985) used one-way ANOVA to analyze his data in a study concerning why authors cite particular documents. He sampled recently published authors and asked them to evaluate the references in their bibliographies. The study used a four-point rating scale to indicate degree of importance, and Brooks asked the authors to rate the references on seven such scales, one for each possible motive. He compared the mean ratings of the different motives to see which was most important. Like Wallace and Bonzi, Brooks first performed an overall ANOVA, then made multiple comparisons to pinpoint the significant differences among the means. A notable point of Brooks's data analysis is that, in addition to conducting tests of statistical significance, he reported an estimate of the strength of the relationship between the motives and the importance ratings. As mentioned above, these estimates are useful to have but are rarely encountered in reports of ANOVA.

Brooks (1985) actually performed the aforementioned analysis three times: once for each author, then for those authors in the sciences, then for those authors in the humanities. His three tests were partially redundant since the data on which they were calculated overlap. A more concise approach would have been to acknowledge at the outset that there are two different independent variables of interest (motives *and* disciplines). A form of ANOVA called two-way ANOVA is designed explicitly to address the situation of two independent variables.

Two Independent Variables

The main advantage offered by two-way ANOVA over repeated application of one-way ANOVA is that it allows the researcher to test explicitly for an interaction between the independent variables (or "factors," as they are commonly called). In Brooks's (1985) case a test of the interaction between motive and discipline would have told him whether the important motives are the same in the sciences and the humanities, or whether some motives are important in one but not the other.

Examples of the use of two-way ANOVA are readily found in the analysis of experiments. ROUSE ET AL., for instance, conducted an experiment in which they measured various aspects of the performance of six engineering students in using a simulated database. The ANOVA revealed that the students performed better when there were many links between the "documents" in the simulated database. The students also performed better when they were conducting a broad search. However, these two factors did not interact; that is, their effects were additive.

An experiment by WILLIAMS ET AL. on the use of information retrieval systems illustrates a significant interaction in two-way ANOVA. The two factors in their experiment were the type of interface and the academic discipline of the users (some were in library and information science, and some were in computer science). The users were asked to search a database using one of three interfaces. They evaluated the system they worked with by answering several questions about it. The questionnaire used seven-point rating scales. Two-way ANOVAs were conducted for each question, one of which yielded a significant interaction between the factors. The library and information science students found the use of one of the interfaces easier to remember than the others, while the computer science students thought that all of the interfaces were equally easy to use.

This discussion of ANOVA-related techniques has progressed thus far from the one-variable, two-category case (t-test) to the one-variable, many-category case (one-way ANOVA), then to the two-variable case wherein each variable can have many categories (two-way ANOVA). In both instances where a more comprehensive statistical technique was introduced, the simpler technique that had been under discussion could then be viewed as a specific application of the more comprehensive technique. In both instances where this happened, there were studies that had been analyzed using the simpler technique but that could have been analyzed by the more comprehensive technique.

It is possible to continue this process of generalization of ANOVA in at least three different directions. However, examples of these still more complex research designs are relatively rare in information science and with good reason. One of the main advantages of choosing the most complex statistical technique possible for a particular set of data is that it allows the data to be described in the most succinct way possible. However, in statistics, as in language, there is a tradeoff between succinctness and clarity of expression. The more complex techniques are probably, in some cases, too complex. It is sometimes very difficult to express in words exactly what the statistical test is telling the researcher. Nonetheless, the more complex tools are sometimes helpful, so brief mention is given to uses of these three extensions of ANOVA.

First, it is possible that one of the predictor variables may be a continuous variable rather than a categorical one. If the continuous predictor is not the primary variable of interest, then knowledge of its value can be used to reduce the error of prediction made by both the full and the restricted model in a test of the effect of the categorical independent variable. This procedure is called analysis of covariance (ANCOVA), and it has been used in information science by COOPER in a study of system response time in an online bibliographic search system.

Second, it is possible that the research design consists of three independent variables rather than two. In this case statistical tests are made of the three possible two-way interactions of the independent variables, as well as of the three-way interaction. A three-way interaction exists when the nature of a two-way interaction is different when one considers the two-way interaction within each of the different categories of the third independent variable. HAUPTMANN & GREEN used three-way ANOVA in a comparison of different interfaces to a graphics package.

Third, the statistical analysis may take more than one dependent variable into consideration in a single test. This procedure is known as multivariate analysis of variance (MANOVA). For each main effect or interaction term, MANOVA finds the linear combination of the dependent variables that maximizes the size of that effect or term, then tests to see whether or not the effect is statistically significant. However, with MANOVA it may be difficult to interpret how the test has combined the dependent variables. EISENBERG & BARRY used MANOVA in their study of presentation-order effects on judgments of document relevance.

MANOVA is very closely related to another statistical technique called discriminant analysis. In discriminant analysis the several continuous variables are considered the predictors and category membership is considered the outcome variable, while in MANOVA the reverse is the case. AVERSA used discriminant analysis in a study of citation patterns over time.

REGRESSION AND RELATED TECHNIQUES

Correlation

Regression and correlation are closely related techniques. Correlation is used to measure the relationship between variables in a study. It measures mutual association but does not measure the response of one variable to the influence of another, as does regression. For example, WEINBERG & CUNNINGHAM investigated the relationship between the hierarchical level of a term in a thesaurus and the number of postings that term received in an online database by calculating Pearson's r correlation coefficient. Pearson's r measures the linear relationship between two continuous variables. For some databases, Weinberg and Cunningham found a small but statistically significant negative correlation (i.e., the more specific the term, the fewer the postings). In most cases, however, the correlation was not significantly different from zero, which is to say, the variables were not linearly related.

If the variables are not measured at an interval or ratio level, or if they are not sampled from a bivariate normal distribution, then some statisticians recommend that a nonparametric correlation measure such as Spearman's rho, Kendall's tau, or the gamma correlation coefficient be used instead of

Pearson's r. These measures are based on the ranks of observations. Thus, they are appropriate with ordinal-level data. HE & PAO used Spearman's rho to measure the correlation between a discipline influence score they devised for journals and influence as perceived by professionals in veterinary medicine. GORDON used Spearman's rho to show that measures derived from citation counts correlate rather highly with users' rankings of journals in the social sciences. An example of the use of Kendall's tau can be found in BARENDREGT ET AL.

In some cases, where several variables are being studied, investigators use a correlation matrix of all possible combinations of variables. More general multivariate techniques will often provide more information in these cases. For example, in a study of the operating costs of information retrieval systems, KOWALSKI & ZGRZYWA computed the pairwise correlations of several variables, such as length of query, number of modifications to the query, and number of answers. By performing tests of significance of the correlation coefficients they showed which variables were related to each other. However, multiple regression (see the next section) would have enabled them to build a model of the combined effects of the different variables on costs.

It is not always necessary to follow a regression approach when dealing with several correlation coefficients. For example, in a study of different ways of measuring the similarity between requests and documents, KOLL derived some specific hypotheses. For each hypothesis two correlation coefficients were compared to see which one was higher, and an appropriate statistical test was used to guide Koll's decisions about whether one of the correlation coefficients was in fact higher than the other.

When three or more variables are intercorrelated, it is sometimes helpful to calculate the correlation between two of them while holding the other(s) constant. This technique enables the researcher to see how strongly the two variables of interest are related over and above their shared association with the other variables. In a correlational study it is often impossible to control the extraneous variables physically, so methods to control them statistically have been devised. Correlations calculated in this manner are called partial correlation coefficients. PERITZ (1983b) found a moderate positive correlation between the number of citations a paper contains and the number of times that paper is cited, even after she controlled for journal, year, subject category, number of authors, and seniority. Thus, she was able to conclude that this relationship is not spurious.

Regression

Regression uses a linear model to predict the value of a dependent variable from the value of one or more independent variables. If one independent variable is involved, the procedure is called bivariate, or simple, linear regression. With two or more independent variables, it is called multiple regression.

SHEPHERD & PHILLIPS used simple regression in a study of the similarity between user profiles that are gathered for selective-dissemination-of-information service and queries that are formulated by the same users for retrospective searches. In their experiment, they investigated the relationship between two different ways of measuring the similarity between simulated profiles and simulated queries. The regression analysis enabled them to determine how accurately one similarity measure could be predicted from another.

According to GUNST & MASON, most uses of multiple regression can be classified into three categories: 1) for prediction, 2) for model specification, and 3) for parameter estimation. In prediction the emphasis is on estimating accurate values of the dependent variable for any combination of the independent variables. In model specification, the emphasis shifts to finding the best combination of predictor variables and assessing their relative importance in prediction. (Relative importance is expressed as a weight, called the "beta weight," associated with each predictor in the model.) Finally, in parameter estimation, the regression equation is usually derived from a theory. The question of which independent variables are to be included is not at issue, so regression analysis is used to provide accurate estimates of the weights (parameters) associated with the predictor variables.

As an example of the first type of use, BENNION & KARSCHAMROON developed a model for predicting the perceived usefulness of physics journals (as measured by a survey) from various citation and journal circulation statistics. In this case, the goal was to predict usefulness from numbers readily available in *Journal Citation Reports.*

In model specification, there is an interest not only in predicting, but also in finding the correct combination of predictor variables and the relationship of those predictors to the dependent variable. FEDOROWICZ developed a multiple regression model of access times in the MEDLINE retrieval system. Her model used several characteristics of inverted files as predictors. She concluded with a plea for more use of regression modeling techniques in information science.

As examples of the third type of application, DROTT & GRIFFITH used regression to estimate the parameters of the Bradford distribution, and NICHOLLS explored the usefulness of regression for estimating the parameters of Lotka's law.

One problem that can affect multiple regression models is multicollinearity, a condition that occurs when strong interrelationships exist among the predictor variables. Most textbooks on regression describe how to test for multicollinearity. It is most disastrous when multiple regression is being used for model specification. When multicollinearity is present, the estimates of the beta weights become unreliable. RUSHINEK & RUSHINEK (1986a) failed to detect multicollinearity when they used multiple regression to predict user satisfaction with communication monitors for data communication. Their regression model included ten independent variables, some of which were completely redundant. Their own table 2 showed that only five of the predictors were statistically significant. (Incidentally, table 4 in their paper shows only asterisks where values should be printed. It was probably a computer printout that should not have been printed without editorial augmentation.)

Although the basic correlation and regression measures do not allow a researcher to determine cause and effect, an extension of regression, called path analysis, can test models of causality through correlations. For example BAROUDI ET AL. showed that user involvement in the development of a system enhances both system use and the user's satisfaction and that an increase in user satisfaction leads to greater system use.

Time Series

Observations taken in a regular sequence over time are called time–series data—e.g., daily circulations in a library. Time-series data usually fail to meet one of the assumptions of the basic regression model: that the observations are independent of one another. What happens in one time period usually affects what happens in the next time period. Specialized variations of regression analysis have been developed in recognition of this fact.

Sometimes with time-series data, only a single time series is studied and the researcher attempts to predict future values of that variable. In other cases, the regression of one time-series variable on others is studied.

One of the most successful methods for dealing with the first situation is the ARIMA (autoregressive integrated moving average) approach developed by BOX & JENKINS. ARIMA is used extensively in economic forecasting. Within information science, BROOKS (1984) and BROOKS & FORYS compared several forecasting models for time series, including an ARIMA model, for data on monthly circulation statistics. They found no significant differences among the different methods, although the ARIMA model had the smallest error. TAGUE ET AL. fit an ARIMA model to daily circulation statistics and to the number of papers published on particular topics over a number of years. Not all forecasters use approaches designed especially for time series, however. For example, in a study of the informational value of titles of articles, DIENER used a simple regression model with time as the independent variable.

Econometric models have been developed to study the regression of time-series variables on one another (the second situation mentioned above). MCALLISTER & CONDON used such a model to relate levels of federal funding to numbers of articles published in biomedical journals.

CONTINGENCY TABLES

Chi-Square and Related Measures

Two-way contingency tables are customarily analyzed with the help of the chi-square statistic. A contingency table is formed when two categorical or discrete variables are cross-classified, forming a matrix in which the column totals represent the number of observations in the categories of one variable and the row totals represent the number of observations in the categories of the other variable. The cell entries are the number of observations for that particular combination. Contingency-table analysis tries to determine whether the two variables are associated. Chi-square serves as a test statistic for the null hypothesis that the two variables are independent. If the two variables are associated, subsequent analysis and conclusions should focus on the nature of the association.

WALLACE (1985) found that chi-square analysis was the most frequently used method of analysis in the library and information science journals he scanned. This frequency may be associated with the common use of questionnaires in the research reported in those journals. Such surveys tend to produce categorical data, which require Chi-square or similar methods to test for the presence of an association. A number of instances in which questionnaire results were analyzed using Chi-square were found in the literature scanned for this review. Some examples are presented below, along with other applications of contingency-table analysis.

The simplest form of contingency-table analysis is the situation in which the two variables being analyzed can each assume two values, resulting in a 2 x 2 contingency table. MONDSCHEIN analyzed data of this type. He found that 83% of online search results that were requested orally were judged by the requester to be satisfactory, while only 11% of the online search results that were requested in writing were judged to be satisfactory. Chi-square analysis indicated that this difference in percentages was statistically significant.

PADDOCK & SCAMELL produced 2 x 2 contingency tables from the results of a questionnaire they administered to business managers. The managers were divided into those who were involved in office automation and those who were not. Four of the managers involved in office automation said they had experience with steering committees, while nine did not. None of the 12 managers not involved in office automation had steering committee experience. Since the actual numbers involved were so small, an ordinary chi-square analysis was not possible. Instead, the authors used an alternative, Fisher's exact test, which showed that the relationship between involvement in office automation and steering committee experience was statistically significant.

Analyzing a 2 x 2 table is in many respects like conducting a t-test. As discussed in the ANOVA section, t-tests are sometimes conducted repeatedly when ANOVA is more appropriate. Likewise, researchers sometimes analyze a contingency table with more than two rows or two columns as though it were a series of 2 x 2 tables. For example, MAHAPATRA & BISWAS studied the co-occurrence of five PRECIS role operators in an indexing study. Rather than presenting comprehensive tables of their data, they treated the data in a piecemeal fashion, using 2 x 2 tables. This approach makes it difficult for the reader to get an overall picture of the relationships among the operators. A further problem is that the data themselves appear nowhere in the article. The tables contain chi-square values, but the frequencies themselves are not presented.

When a contingency table contains more than two rows or more than two columns (or both), the appropriate statistical test is still a chi-square test, but in this case the chi-square test is more analogous to one-way ANOVA than to a t-test. For example, KAHN reported a survey of data administrators in three environments. Respondents were asked to comment on the importance of data management standards, among other things. Four categories of importance were considered. Kahn discovered that administrators in companies with information resources management were more likely to consider standards important than administrators in companies with other data administration environments.

The chi-square method provides a way to determine whether the association between two categorical variables is statistically significant. It does not indicate the degree to which the two variables are associated. There are numerous measures of association that can be calculated from a contingency table, some of which are based on the chi-square statistic. It is helpful to have a measure of association reported along with the results of a chi-square test. A good survey of some measures of association can be found in OTT ET AL.

SALASIN & CEDAR used both Chi-square and a measure of association in a study of the information needs of a nationwide sample of mental health workers. One of their results was that the source for information chosen by the respondents depended on the topic of the inquiry. For example, periodicals were more likely to be consulted for research issues, while state staff were more likely to be consulted for government statutes. Although chi-square testing showed that the association between source and topic was statistically significant, the relationship was not particularly strong. The phi coefficient measured the strength at 0.09 on a scale from 0 to 1.

It is important to keep in mind that tests of significance and measures of association are not equivalent. HURT (1983), for example, inappropriately attempted to treat the gamma measure of association as a test statistic. His decision rule required near-perfect association before it would be considered statistically significant. This led him to conclude that two variables were independent when they actually seem to be related.

Sometimes researchers wish to demonstrate that two variables are independent of one another rather than associated. Usually this occurs when one of the variables is considered a nuisance variable. For example, FAIRHALL used chi-square analysis to demonstrate that focusing skills of online searchers are unrelated to the searchers' knowledge of the content area of the search. In a similar vein, TAPSCOTT ET AL. used chi-square tests to show that their experimental and control groups were equivalent to each other before the beginning of the experiment.

When more than two variables are being analyzed, one can use multiple chi-square tests, but this approach has drawbacks; it is like performing several one-way ANOVAs when a two-way ANOVA is indicated. For example, WELBORN ET AL. contrasted users of an online catalog with users of a card catalog with respect to various demographic variables. Each demographic variable was associated with the use variable separately. In this situation, chi-square analysis cannot determine how these different demographic variables interact because it considers the variables two at a time.

Log-Linear Models

There is, however, an approach to contingency-table analysis that can analyze more than two variables simultaneously and thus (for example) build a model of the "typical" online catalog user. This approach uses what are called log-linear models. Log-linear analysis is actually a generalization of chi-square analysis, just as two-way ANOVA is a generalization of one-way ANOVA. Log-linear models are relatively new and have not gained wide use in information science research. FIENBERG gives an excellent introduction to log-linear analysis.

RICE used log-linear analysis to study communications networks between individuals in work groups. This method allowed the characterization of eight different roles of individuals within information networks.

NOMA (1982a) provides an example of a specialized application of log-linear models. He constructed a matrix in which he counted the number of times each journal in a set of eight journals cited each of the other journals in the set including itself. He points out that the main diagonal in the matrix, which contains the frequency of self-citation, must be treated differently from the other cells, and he introduced a specific log-linear model that enables this to be done.

DIMENSIONALITY-REDUCTION TECHNIQUES

The data-analysis methods reviewed in this section are: 1) factor analysis, 2) multidimensional scaling (MDS), and 3) cluster analysis. (Cluster analysis is also known as numerical taxonomy, a reflection of the fact that it was developed in large part for use in biological classification.) These methods are referred to as dimensionality-reduction methods because their function is to simplify what might at first appear to be a complex pattern of associations among many entities. In geometry, this process of simplification is viewed as the projection of a larger number of dimensions onto a smaller number of dimensions (as when a three-dimensional globe is projected onto a two-dimensional map in the well-known Mercator projection).

All three methods of data analysis share a common feature—i.e., they operate on a square, symmetric matrix. The rows and columns of the matrix refer to the same entities, and the contents of the matrix consist of a measure of association or proximity between pairs of these entities. A familiar example of such a matrix is the table of distances between pairs of cities that accompanies many road maps.

As far as input is concerned, MDS is more like cluster analysis; with respect to output, it is more like factor analysis. That is, in factor analysis the entities are usually variables and the entries in the matrix are the covariances or correlations between those variables; in MDS and cluster analysis, the entities are objects (sometimes referred to generically as "stimuli") and the entries in the matrix are usually similarities or dissimilarities between the objects. As for output, the result of a cluster analysis is either a single partition of the objects into mutually exclusive sets or a nested hierarchy of such partitions that can be represented as a tree diagram. With factor analysis and MDS, the result of the analysis is a representation of the variables (factor analysis) or objects (MDS) as points in a space of low dimensionality. Choosing an appropriate number of dimensions (factors) to represent the data is one of the tasks facing the data analyst using MDS or factor analysis.

Examples of the use of the three techniques in information science follow. Each has been extended and elaborated on here in various ways. An introduction to factor analysis can be found in HARMAN, to MDS in KRUSKAL & WISH, and to cluster analysis in HARTIGAN.

Factor Analysis

DZIDA ET AL. and CULNAN were interested in characterizing users' perceptions of the quality of online computer systems. Both studies involved ratings by users of a large number of attributes of such systems. The authors then used factor analysis to find out which attributes are viewed by the users essentially as alternative ways of expressing the same aspect of system quality. Dzida et al., for example, decided that users think of quality for the most part as consisting of seven different aspects (or dimensions).

In a very different application of factor analysis, ERES studied aggregate measures of economic and information activity in 87 countries. She identified three different dimensions of information activity from her data and went on to perform multiple regression analyses to determine how well each of these dimensions of information activity could be predicted from various economic indicators.

Multidimensional Scaling

Most of the applications of MDS in information science have concerned analyses of the similarities between authors in a discipline in order to produce maps of that discipline. Typically, similarity is measured through co-citation analysis, although occasionally ratings of experts are collected to determine the validity of the co-citation approach (MCCAIN). Recent examples of the use of MDS for "discipline mapping" can be found in the publications of LEYDESDORFF, MIYAMOTO & NAKAYAMA, SMALL & GREENLEE, and WHITE & GRIFFITH, as well as in the work of MCCAIN. It is interesting that Leydesdorff and White and Griffith conducted factor analyses of their data as well as MDS analyses. Although the techniques are very similar, they tend to produce somewhat different "pictures" of a given set of data. Also, researchers need to be careful in analyzing correlation matrices with MDS since the technique involves the assumption that a correlation can be thought of as a measure of similarity, which creates problems when negative correlations are encountered.

NOMA (1984) questioned the assumptions underlying the use of co-citations as a basis for calculating similarities for input to MDS. He argued that this could be done only if all of the citing authors share a common point of view, which is unlikely for real data. As a way around this problem, Noma suggested the use of an extension of MDS, centroid scaling (NOMA, 1982a; 1982b). Essentially, centroid scaling allows scattering of both the citing authors and the cited authors in the multidimensional space. This approach contrasts with the traditional use of MDS for co-citation analysis, in which the similarities among the citing authors are not represented. A different extension of MDS, which could be applied if different categories of citing authors could readily be identified, is three-way MDS (KRUSKAL & WISH).

Cluster Analysis

The techniques of cluster analysis as a data-analytic tool share many commonalities with the statistical approaches to the clustering of documents or of terms for information retrieval. Sometimes the same algorithms are used in both situations (MURTAGH). A major difference is that document or term clustering often involves the clustering of hundreds or thousands of items, while applications for data analysis typically involve fewer than 50 objects.

Since cluster analysis and MDS operate on the same kind of data, it is not surprising that cluster analysis has been applied extensively to analyze co-citations (SHAW, 1986; SMALL & SWEENEY). In many instances, the same data are analyzed by both MDS and cluster analysis (MCCAIN; SMALL ET AL.), resulting in some redundancy but also revealing complementary aspects of the structure of the data. MDS highlights gross similarities and differences between the objects, while cluster analysis focuses attention on the details.

Some authors are concerned that MDS and cluster analysis are commonly used as descriptive techniques (the term "exploratory" is often used). Very little hypothesis testing is conducted in most applications of MDS or cluster analysis (BOLL & ZWEIZIG; MCGRATH). Although SHAW (1985) has worked on a statistical test for one kind of cluster analysis, much work remains to be done in this area.

CONCLUSIONS

This review has attempted to convey some idea of how widely statistics are used in information science research. One can find examples of the entire spectrum of statistical tools. This is perhaps a reflection of the interdisciplinary nature of information science.

We do not entirely share the views of WALLACE (1985), who was dismayed by the infrequent use of statistics in library and information science. It is less important that many or most articles use statistics than that those articles that do use statistics use them well. Many examples of well-conducted statistical tests and procedures in information science research are cited in this chapter.

The situation is not entirely satisfactory, however. This review also uncovered examples of the misuse of statistics, examples that should have been caught by the peer-review process of the professional journals in which they were published. A common problem was the repeated use of a simple statistical test when a single application (or at least fewer applications) of a more comprehensive test should have been carried out. Any time that one observes a single statistical test being conducted repeatedly, it is worth inquiring whether there might not be a better way to accomplish the same thing.

Other problems are harder to spot. Fortunately, they are also rarer. Nonetheless, articles in which the logical errors are made in the choice or calculation of statistics should be revised before being published. To accomplish this editors in the field may well need to enlist the help of more statisticians in reviewing articles that employ sophisticated statistical methods.

BIBLIOGRAPHY

AUSTER, ETHEL; LAWTON, STEPHEN B. 1983. Improving Performance: The Relationship between Negotiation Behaviors of Search Analysts and User Satisfaction with Online Bibliographic Retrieval. In: Vondran, Raymond F.; Caputo, Anne; Wasserman, Carol; Diener, Richard A.V., eds. Productivity in the Information Age: Proceedings of the American Society for Information Science (ASIS) 46th Annual Meeting: Volume 20; 1983 October 2-6; Washington, DC. White Plains, NY: Knowledge Industry Publications, Inc. for ASIS; 1983. 125-127. (Anova). ISSN: 0044-7870; ISBN: 0-86729-115-X.

AVERSA, ELIZABETH SMITH. 1985. Citation Patterns of Highly Cited Papers and Their Relationship to Literature Aging: A Study of the Working Literature. Scientometrics. 1985; 7(3-6): 383-389. (Discriminant). ISSN: 0138-9130.

BAGUST, A. 1983. A Circulation Model for Busy Public Libraries. Journal of Documentation. 1983 March; 39(1): 24-37. (Chi-square). ISSN: 0022-0418.

BARENDREGT, L. G.; BENSCHOP, C. A.; DE HEER, T. 1985. Subjective Trial of the Performance of the Information Trace Method. Information Processing & Management. 1985; 21(2): 103-111. (Rank Correlation). ISSN: 0306-4573.

BAROUDI, JACK J.; OLSON, MARGRETHE H.; IVES, BLAKE. 1986. An Empirical Study of the Impact of User Involvement on System Usage and Information Satisfaction. Communications of the ACM. 1986 March; 29(3): 193-196. (Correlation). ISSN: 0001-0782.

BEHESHTI, JAMSHID; TAGUE, JEAN M. 1984. Morse's Markov Model of Book Use Revisited. Journal of the American Society for Information Science. 1984; 35(5): 259-267. (Regression). ISSN: 0002-8231.

BELKIN, NICHOLAS J.; CROFT, W. BRUCE. 1987. Retrieval Techniques. In: Williams, Martha E., ed. Annual Review of Information Science and Technology: Volume 22. Amsterdam, The Netherlands: Elsevier Science Publishers B.V. for ASIS; 1987. 109-145. ISSN: 0066-4200; ISBN: 0-444-70302-0.

BENNION, BRUCE C.; KARSCHAMROON, SUNEE. 1984. Multivariate Regression Models for Estimating Journal Usefulness in Physics. Journal of Documentation. 1984 September; 40(3): 217-227. (Regression). ISSN: 0022-0418.

BERENSON, MARK L.; LEVINE, DAVID M.; GOLDSTEIN, M. 1983. Intermediate Statistical Methods and Applications: A Computer Package Approach. Englewood Cliffs, NJ: Prentice-Hall, Inc.; 1983. 579p. (General). ISBN: 0-13-470781-8.

BLAIR, DAVID C.; MARON, M. E. 1985. An Evaluation of Retrieval Effectiveness for a Full-Text Document-Retrieval System. Communications of the ACM. 1985 March; 28(3): 289-294. (T-test). ISSN: 0001-0782.

BLAIR, R. C.; HIGGINS, J. J. 1985. Comparison of the Power of the Paired Samples T Test to that of Wilcoxon's Signed-Ranks Test Under Various Population Shapes. Psychological Bulletin. 1985; 97: 119-128. (General). ISSN: 0033-2909.

BOLL, JOHN J.; ZWEIZIG, DOUGLAS. 1985. Mapping a Curriculum by Computer. Journal of the American Society for Information Science. 1985; 36(5): 252-253. (Letter to the editor). ISSN: 0002-8231.

BOOKSTEIN, ABRAHAM. 1985. Probability and Fuzzy-Set Applications to Information Retrieval. In: Williams, Martha E., ed. Annual Review of Information Science and Technology: Volume 20. Washington, DC: American Society for Information Science; 1985. 117-151. (General). ISSN: 0066-4200; ISBN: 0-86729-175-3.

BORGMAN, CHRISTINE L. 1983. End User Behavior on an Online Information Retrieval System: A Computer Monitoring Study. In: Kuehn, Jennifer J. ed. Research and Development in Information Retrieval: 6th Annual International ACM SIGIR (Association for Computing Machinery Special Interest Group on Information Retrieval) Conference. New York, NY: ACM; 1983. 162-176. (Kolmogorov-Smirnov). ISBN: 0-89791-107-5.

BOX, GEORGE E. P.; JENKINS, GWILYM M. 1970. Time Series Analysis: Forecasting and Control. San Francisco, CA: Holden-Day; 1970. 553p. LC: 77-79534.

BRADLEY, J. V. 1978. Robustness?. British Journal of Mathematical and Statistical Psychology. 1978; 31: 144-152. (General). ISSN: 0007-1102.

BROOKES, B. C. 1981. A Critical Commentary on Leimkuhler's 'Exact' Formulation of the Bradford Law. Journal of Documentation. 1981 June; 37(2): 77-88. (Chi-square). ISSN: 0022-0418.

BROOKES, B. C. 1984. Towards Informetrics: Haitun, Laplace, Zipf, Bradford and the Alvey Programme. Journal of Documentation. 1984 June; 40(2): 120-143. (Global). ISSN: 0022-0418.

BROOKS, TERRENCE A. 1984. Naive vs. Sophisticated Methods of Forecasting Public Library Circulations. Library and Information Science Research. 1984; 6: 205-214. (Time series). ISSN: 0740-8188.

BROOKS, TERRENCE A. 1985. Private Acts and Public Objects: An Investigation of Citer Motivations. Journal of the American Society for Information Science. 1985; 36(4): 223-229. (Anova). ISSN: 0002-8231.

BROOKS, TERRENCE A. 1986. Evidence of Complex Citer Motivations. Journal of the American Society for Information Science. 1986; 37(1): 34-36. (Factor Analysis). ISSN: 0002-8231.

BROOKS, TERRENCE A.; FORYS, JOHN W., JR. 1986. Smoothing Forecasting Methods for Academic Library Circulations: An Evaluation and Recommendation. Library and Information Science Research. 1986; 8: 29-39. (Regression; Time series). ISSN: 0740-8188.

BROWNSEY, K. W. R.; BURRELL, Q. L. 1986. Library Circulation Distributions: Some Observations on the PLR Sample. Journal of Documentation. 1986 March; 42(1): 22-45. (Chi-square goodness of fit). ISSN: 0022-0418.

BRUER, JOHN T. 1985. Methodological Quality and Citation Frequency of the Continuing Medical Education Literature. Journal of Documentation. 1985 September; 41(3): 165-172. (Kruskal-Wallis). ISSN: 0022-0418.

CHUDAMANI, K. S.; SHALINI, R. 1983. Journal Acquisition—Cost Effectiveness of Models. Information Processing & Management. 1983; 19(5): 307-311. (Rank Correlation). ISSN: 0306-4573.

CLIFF, NORMAN. 1982. What Is and Isn't Measurement. In: Keren, Gideon, ed. Statistical and Methodological Issues in Psychology and Social Sciences Research. Hillsdale, NJ: Lawrence Erlbaum Assoc.; 1982. 1-40. (General). ISBN: 0-89859-062-0.

COADY, REGINALD P. 1983. Testing for Markov-Chain Properties in the Circulation of Science Monographs. Information Processing & Management. 1983; 19(5): 279–284. (Markov). ISSN: 0306-4573.

CONOVER, W. J. 1980. Practical Nonparametric Statistics. 2nd edition. New York, NY: John Wiley; 1980. 493p. (General). ISBN: 0-471-02867-3.

COOPER, MICHAEL D. 1983. Response Time Variations in an Online Search System. Journal of the American Society for Information Science. 1983; 34(6): 374–380. (Anova). ISSN: 0002-8231.

CULNAN, MARY J. 1984. The Dimensions of Accessibility to Online Information: Implications for Implementing Office Information Systems. ACM Transactions on Office Information Systems. 1984 April; 2(2): 141–150. (Factor Analysis). ISSN: 0734-2047.

DANILOWICZ, CZESLAW; SZARSKI, HENRYK. 1981. Selection of Scientific Journals Based on the Data Obtained from an Information Service System. Information Processing & Management. 1981; 17: 13–19. (Rank Correlation). ISSN: 0306-4573.

DAVID, H. G.; PIIP, L.; HALY, A. R. 1981. The Examination of Research Trends By Analysis of Publication Numbers. Journal of Information Science. 1981; 3: 283–288. (Correlation). ISSN: 0165-5515.

DAVIES, ROY. 1985. Q-Analysis: A Methodology for Librarianship and Information Science. Journal of Documentation. 1985 December; 41(4): 221–246. (Cluster Analysis). ISSN: 0022-0418.

DIENER, RICHARD A. V. 1984. Informational Dynamics of Journal Article Titles. Journal of the American Society for Information Science. 1984 July; 35(4): 222–227. (Regression). ISSN: 0002-8231.

DILLON, MARTIN; FEDERHART, PEGGY. 1984. Statistical Recognition of Content Terms in General Text. Journal of the American Society for Information Science. 1984; 35(1): 3–10. (Discriminant analysis). ISSN: 0002-8231.

DROTT, M. CARL; GRIFFITH, BELVER C. 1978. An Empirical Examination of Bradford's Law and the Scattering of Scientific Literature. Journal of the American Society for Information Science. 1978 September; 29(5): 238–246. (Regression). ISSN: 0002-8231.

DZIDA, W.; HERDA, S.; ITZFELDT, W. D. 1978. User-Perceived Quality of Interactive Systems. IEEE Transactions on Software Engineering. 1978; 4(4): 270–276. (Factor Analysis). ISSN: 0098-5589.

EDGELL, S. E.; NOON, S. M. 1984. Effect of Violation of Normality on the T Test of the Correlation Coefficient. Psychological Bulletin. 1984; 95: 576–583. (General). ISSN: 0033-2909.

EISENBERG, MICHAEL; BARRY, CAROL. 1986. Order Effects: A Preliminary Study of the Possible Influence of Presentation Order on User Judgments of Document Relevance. In: Hurd, Julie M., ed. ASIS '86: Proceedings of the American Society for Information Science (ASIS) 49th Annual Meeting: Volume 23; 1986 September 28–October 2; Chicago, IL. Medford, NJ: Learned Information, Inc. for ASIS; 1986. 80–86. (Anova). ISSN: 0044-7870; ISBN: 0-938734-14-8.

ERES, BETH KREVITT. 1985. Socioeconomic Conditions Related to Information Activity in Less Developed Countries. Journal of the American Society for Information Science. 1985 May; 36(3): 213–219. (Factor Analysis). ISSN: 0002-8231.

FAIRHALL, DONALD. 1985. In Search of Searching Skills. Journal of Information Science. 1985; 10: 111–123. (Chi-square). ISSN: 0165-5515.

FEDOROWICZ, JANE. 1984. Database Evaluation Using Multiple Regression Techniques. SIGMOD Record. 1984 June; 14(2): 70–76. (Regression).
FIENBERG, STEPHEN. 1980. Analysis of Cross-Classified Categorical Data. 2nd edition. Cambridge, MA: MIT Press; 1980. 198p. (General). ISBN: 0-262-06071-X.
FRAME, J. D. 1984. Multidimensionality is Alive and Well in Applied Statistics. Scientometrics. 1984; 6(2): 97–102. (General). ISSN: 0138-9130.
GAITO, J. 1980. Measurement Scales and Statistics: Resurgence of an Old Misconception. Psychological Bulletin. 1980; 87: 564–567. (General). ISSN: 0033-2909.
GELLER, V. J.; LESK, M. E. 1983. User Interfaces to Information Systems: Choices vs. Commands. In: Kuehn, Jennifer J., ed. Research and Development in Information Retrieval: 6th Annual International ACM SIGIR (Association for Computing Machinery Special Interest Group on Information Retrieval) Conference; 1983 June 6–8; Bethesda, MD. New York, NY: ACM; 1983. 130–135. (Anova). ISBN: 0-89791-107-5.
GORDON, MICHAEL D. 1982. Citation Ranking versus Subjective Evaluation in the Determination of Journal Hierarchies in the Social Sciences. Journal of the American Society for Information Science. 1982 January; 33(1): 55–57. (Rank Correlation). ISSN: 0002-8231.
GREGOIRE, T. G.; DRIVER, B. L. 1987. Analysis of Ordinal Data to Detect Population Differences. Psychological Bulletin. 1987; 101(1): 159–165. (General). ISSN: 0033-2909.
GRIFFITHS, ALAN; LUCKHURST, H. CLAIRE; WILLETT, PETER. 1986. Using Interdocument Similarity Information in Document Retrieval Systems. Journal of the American Society for Information Science. 1986; 37(1): 3–11. (Clustering). ISSN: 0002-8231.
GUNST, RICHARD F.; MASON, ROBERT L. 1980. Regression Analysis and Its Application: A Data-Oriented Approach. New York, NY: Marcel Dekker; 1980. 402p. ISBN: 0-8247-6993-7.
HAAS, DAVID F.; KRAFT, DONALD H. 1984. Experimental and Quasi-Experimental Designs for Research in Information Science. Information Processing & Management. 1984; 20(1–2): 229–237. (General). ISSN: 0306-4573.
HAITUN, S. D. 1982. Stationary Scientometric Distributions: Part II. Non-Gaussian Nature of Scientific Activities. Scientometrics. 1982; 4(2): 89–104. (General). ISSN: 0138-9130.
HAITUN, S. D. 1984. Life in a Multidimensional World. Scientometrics. 1984; 6(2): 93–96. (General). ISSN: 0138-9130.
HAITUN, S. D. 1986. Problems of Quantitative Analysis of Scientific Activities: The Non-Additivity of Data. Part I: Statement and Solution. Scientometrics. 1986; 10(1–2): 3–16. (General). ISSN: 0138-9130.
HANSEN, KATHLEEN A. 1986. The Effect of Presearch Experience on the Success of Naive (End-User) Searches. Journal of the American Society for Information Science. 1986; 37(5): 315–318. (T-test). ISSN: 0002-8231.
HARMAN, HARRY H. 1976. Modern Factor Analysis. 3rd edition. Chicago, IL: University of Chicago Press; 1976. 487p. (General). ISBN: 0-226-31652-1.
HARTIGAN, J. A. 1975. Clustering Algorithms. New York, NY: Wiley Interscience; 1975. 368p. (General). ISBN: 0-471-36545-X.

HAUPTMAN, ALEXANDER C.; GREEN, BERT F. 1983. A Comparison of Command, Menu-Selection and Natural-Language Computer Programs. Behaviour and Information Technology. 1983; 2(2): 163–178. (Anova). ISSN: 0144-929X.

HAYES, ROBERT M.; POLLACK, ANN M.; NORDHAUS, SHIRLEY. 1983. An Application of the Cobb-Douglas Model to the Association of Research Libraries. Library and Information Science Research. 1983; 5: 291–325. (Regression). ISSN: 0740-8188.

HE, CHUNPEI; PAO, MIRANDA LEE. 1986. A Discipline-Specific Journal Selection Algorithm. Information Processing & Management. 1986; 22(5): 405–416. (Rank Correlation). ISSN: 0306-4573.

HUBERT, J. J. 1981. A Rank-Frequency Model for Scientific Productivity. Scientometrics. 1981; 3(3): 191–202. (Chi-square goodness of fit). ISSN: 0138-9130.

HURT, C. D. 1983. A Comparison of a Bibliometric Approach and an Historical Approach to the Identification of Important Literature. Information Processing & Management. 1983; 19(3): 151–157. (Chi-square). ISSN: 0306-4573.

HURT, C. D. 1985. Library School Faculty Strengths in Data Processing Canadian-U.S. Differences. Information Processing & Management. 1985; 21(2): 157–163. (MultiComp Z-test). ISSN: 0306-4573.

KAHN, BEVERLY K. 1983. Some Realities of Data Administration. Communications of the ACM. 1983 October; 26(10): 794–799. (Chi-square). ISSN: 0001-0782.

KEPPEL, GEOFFREY. 1982. Design and Analysis: A Researcher's Handbook. Englewood Cliffs, NJ: Prentice-Hall; 1982. 624p. (General). ISBN: 0-13-200048-2.

KOLL, MATTHEW B. 1981. Information Retrieval Theory and Design Based On a Model of the User's Concept Relations. In: Oddy, R. N.; Robertson, S. G.; Van Rijsbergen, C. J.; Williams, P. W., eds. Information Retrieval Research. London, England: Butterworths; 1981. 77–93. (Correlation). ISBN: 0-408-10775-8.

KOWALSKI, KAZIMIERZ; ZGRZYWA, ALEKSANDER. 1984. Evaluation of Bibliographic Data Base Operation in an SDI System. Journal of Information Science. 1984; 8: 57–61. (Correlation). ISSN: 0165-5515.

KRUSKAL, JOSEPH B.; WISH, MYRON. 1978. Multidimensional Scaling. Beverly Hills, CA: Sage Publications; 1978. 93p. (MDS). ISBN: 0-8039-0940-3.

LACY, WILLIAM B.; BUSCH, LAWRENCE. 1983. Informal Scientific Communication in the Agricultural Sciences. Information Processing & Management. 1983; 19(4): 193–202. (Correlation). ISSN: 0306-4573.

LAM, K.; YU, C. T. 1982. A Clustered Search Algorithm Incorporating Arbitrary Term Dependencies. ACM Transactions on Database Systems. 1982 September; 7(3): 500–508. ISSN: 0362-5951.

LEYDESDORFF, L. 1986. The Development of Frames of Reference. Scientometrics. 1986; 9(3-4): 103–125. (MDS). ISSN: 0138-9130.

LOGAN, E. L.; WOELFL, NANCY N. 1986. Individual Differences in On-line Searching Behavior of Novice Searchers. In: Hurd, Julie M., ed. ASIS '86: Proceedings of the American Society for Information Science (ASIS) 49th Annual Meeting: Volume 23; 1986 September 28–October 2; Chicago, IL. Medford, NJ: Learned Information, Inc. for ASIS; 1986. 163–166. (Correlation). ISSN: 0044-7870; ISBN: 0-938734-14-8.

MAHAPATRA, MANORANJAN; BISWAS, SUBAL CHANDRA. 1986. Interdependence of PRECIS Role Operators: A Quantitative Analysis of

Their Associations. Journal of the American Society for Information Science. 1986; 37(1): 20–25. (Chi-square). ISSN: 0002-8231.

MCALLISTER, P. R.; CONDON, T. 1985. Econometric Analysis of Biomedical Research Publishing Patterns. Scientometrics. 1985; 7(1/2): 55–75. (Regression; Time Series). ISSN: 0138-9130.

MCCAIN, KATHERINE W. 1986. Cocited Author Mapping As a Valid Representation of Intellectual Structure. Journal of the American Society for Information Science. 1986; 37(3): 111–122. (Multidimensional scaling). ISSN: 0002-8231.

MCGILL, MICHAEL J.; HUITFELDT, JENNIFER. 1979. Experimental Techniques of Information Retrieval. In: Williams, Martha E., ed. Annual Review of Information Science and Technology: Volume 14. White Plains, NY: Knowledge Industry Publications, Inc. for ASIS; 1979. 93–127. (General). ISSN: 0066-4200; ISBN: 0-914236-44-X.

MCGRATH, WILLIAM E. 1986. Theory Building and Hypothesis Testing in Multidimensional Scaling Research. Journal of the American Society for Information Science. 1986 September; 37(5): 355. (Letter to the editor). ISSN: 0002-8231.

MICHELL, JOEL. 1986. Measurement Scales and Statistics: A Clash of Paradigms. Psychological Bulletin. 1986; 100: 398–407. (General). ISSN: 0033-2909.

MIYAMOTO, S.; NAKAYAMA, K. 1983. A Technique of Two-Stage Clustering Applied to Environmental and Civil Engineering and Related Methods of Scientific Analysis. Journal of the American Society for Information Science. 1983; 34(3): 192–201. (MDS). ISSN: 0002-8231.

MONDSCHEIN, LAWRENCE G. 1983. Factors Involved in the Enhancement of the SDI. In: Vondran, Raymond F.; Caputo, Anne; Wasserman, Carol; Diener, Richard A. V., eds. Productivity in the Information Age: Proceedings of the American Society for Information Science (ASIS) 46th Annual Meeting: Volume 20; 1983 October 2–6; Washington, DC. White Plains, NY: Knowledge Industry Publications, Inc. for ASIS; 1983. 92–98. (Chi-square). ISSN: 0044-7870; ISBN: 0-86729-072-2.

MORAVCSIK, M. J. 1984. Life in a Multidimensional World. Scientometrics. 1984; 6(2): 75–86. (General). ISSN: 0138-9130.

MOREHEAD, DAVID R.; PEJTERSEN, ANNELISE M.; ROUSE, WILLIAM B. 1984. The Value of Information and Computer-Aided Information Seeking: Problem Formulation and Application to Fiction Retrieval. Information Processing & Management. 1984; 20(5/6): 583–601. (Correlation). ISSN: 0306-4573.

MOREHEAD, DAVID R.; ROUSE, WILLIAM B. 1985. Online Assessment of the Value of Information for Searchers of a Bibliographic Data Base. Information Processing & Management. 1985; 21(2): 83–101. (Anova). ISSN: 0306-4573.

MURTAGH, F. 1984. Structure of Hierarchic Clusterings: Implications for Information Retrieval and for Multivariate Data Analysis. Information Processing & Management. 1984; 20(5/6): 611–617. (Cluster Analysis). ISSN: 0306-4573.

MYERS, J. L. 1979. Fundamentals of Experimental Design. Boston, MA: Allyn and Bacon; 1979. 524p. (General). ISBN: 0-205-06615-1.

NADEL, EDWARD. 1983. Commitment and Co-Citation: An Indicator of Incommensurability in Patterns of Formal Communication. Social Studies of Science. 1983; 13: 255–282. (Factor Analysis). ISSN: 0306-3127.

NELSON, MICHAEL J.; TAGUE, JEAN M. 1985. Split Size-Rank Models for the Distribution of Index Terms. Journal of the American Society for Information Science. 1985 September; 36(5): 283–296. (Chi-square goodness of fit). ISSN: 0002-8231.

NICHOLLS, PAUL TRAVIS. 1986. Empirical Validation of Lotka's Law. Information Processing & Management. 1986; 22(5): 417–419. (Kolmogorov-Smirnov). ISSN: 0306-4573.

NOMA, ELLIOT. 1982a. An Improved Method for Analyzing Square Scientometric Transaction Matrices. Scientometrics. 1982; 4(4): 297–316. (Log Linear Models). ISSN: 0138-9130.

NOMA, ELLIOT. 1982b. Untangling Citation Networks. Information Processing & Management. 1982; 18(2): 43–53. (Dual Scaling). ISSN: 0306-4573.

NOMA, ELLIOT. 1984. Co-Citation Analysis and the Invisible College. Journal of the American Society for Information Science. 1984; 35(1): 29–33. (Multidimensional scaling). ISSN: 0002-8231.

ODDY, ROBERT N.; PALMQUIST, RUTH A.; CRAWFORD, MARGARET A. 1986. Representation of Anomalous States of Knowledge in Information Retrieval. In: Hurd, Julie M., ed. ASIS '86: Proceedings of the American Society for Information Science (ASIS) 49th Annual Meeting: Volume 23; 1986 September 28–October 2; Chicago, IL. Medford, NJ: Learned Information, Inc. for ASIS; 1986. 248–254. (Correlation). ISSN: 0044-7870; ISBN: 0-938734-14-8.

OTT, L.; LARSON, R. F.; MENDENHALL, W. 1983. Statistics: A Tool for the Social Sciences. Boston, MA: Duxbury Press; 1983. 494p. (General). ISBN: 0-87150-400-6.

PADDOCK, CHARLES E.; SCAMELL, RICHARD W. 1984. Office Automation Projects and Their Impact on Organization, Planning and Control. ACM Transactions on Office Information Systems. 1984 October; 2(4): 289–302. (Chi-square). ISSN: 0734-2047.

PAO, MIRANDA LEE. 1985. Lotka's Law: A Testing Procedure. Information Processing & Management. 1985; 21(4): 305–320. (Kolmogorov-Smirnov). ISSN: 0306-4573.

PAO, MIRANDA LEE. 1986. An Empirical Examination of Lotka's Law. Journal of the American Society for Information Science. 1986 January; 37(1): 26–33. (Kolmogrov-Smirnov). ISSN: 0002-8231.

PAO, MIRANDA LEE; MCCREERY, LAURIE. 1986. Bibliometric Application of Markov Chains. Information Processing & Management. 1986; 22(1): 7–17. (Markov). ISSN: 0306-4573.

PERITZ, BLUMA C. 1983a. Are Methodological Papers More Cited than Theoretical Ones? The Case of Sociology. Scientometrics. 1983; 5: 211–218. (Anova). ISSN: 0138-9130.

PERITZ, BLUMA C. 1983b. A Note on "Scholarliness" and "Impact". Journal of the American Society for Information Science. 1983; 34(5): 360–362. (Brief communication; Correlation). ISSN: 0002-8231.

PONTIGO, J.; LANCASTER, F. W. 1986. Qualitative Aspects of the Bradford Distribution. Scientometrics. 1986; 9(1–2): 59–70. (Anova). ISSN: 0138-9130.

POTTER, WILLIAM GRAY. 1981. Lotka's Law Revisited. Library Trends. 1981 Summer; 30(1): 21–39. (Kolmogorov-Smirnov). ISSN: 0024-2594.

RADECKI, TADEUSZ. 1985. A Theoretical Framework for Defining Similarity Measures for Boolean Search Request Formulations, Including

Some Experimental Results. Information Processing & Management. 1985; 21(6): 501-524. (Correlation). ISSN: 0306-4573.

RICE, RONALD E. 1982. Group Interaction via Information Systems: Structural Attributes and Role Development. In: Petrarca, Anthony E.; Taylor, Celianna I.; Kohn, Robert S., eds. Information Interaction: Proceedings of the American Society for Information Science (ASIS) 45th Annual Meeting: Volume 19; 1982 October 17-21; Columbus, OH. White Plains, NY: Knowledge Industry Publications, Inc. for ASIS; 1982. 239-242. (Chi-square). ISSN: 0044-7870; ISBN: 0-86729-038-2.

RICE, RONALD E.; TOROBIN, JACK. 1986. Expectations about the Impacts of Electronic Messaging. In: Hurd, Julie M., ed. ASIS '86: Proceedings of the American Society for Information Science (ASIS) 49th Annual Meeting: Volume 23; 1986 September 28-October 2; Chicago, IL. Medford, NJ: Learned Information, Inc. for ASIS; 1986. (Multiple Regression). ISSN: 0044-7870; ISBN: 0-938734-14-8.

ROBERTSON, STEPHEN E. 1981. The Methodology of the Information Retrieval Experiment. In: Sparck Jones, Karen, ed. Information Retrieval Experiment. London, England: Butterworths; 1981. 9-31. ISBN: 0-408-10648-4.

ROUSE, WILLIAM B., ROUSE, SANDRA H.; MOREHEAD, DAVID R. 1982. Human Information Seeking: Online Searching of Bibliographic Citation Networks. Information Processing & Management. 1982; 18(3): 141-149. (Anova). ISSN: 0306-4573.

RUSHINEK, AVI; RUSHINEK, SARA F. 1986a. The Effects of Communication Monitors on User Satisfaction. Information Processing & Management. 1986; 22(4): 345-351. (Multiple Regression). ISSN: 0306-4573.

RUSHINEK, AVI; RUSHINEK, SARA F. 1986b. What Makes Users Happy? Communications of the ACM. 1986 July; 29(7): 594-598. (Multiple Regression). ISSN: 0001-0782.

SALASIN, J.; CEDAR, T. 1985. Information-Seeking Behavior in an Applied Research/Service Delivery Setting. Journal of the American Society for Information Science. 1985; 36(2): 94-102. (Chi-square). ISSN: 0002-8231.

SALTON, GERARD; MCGILL, MICHAEL J. 1983. Introduction to Modern Information Retrieval. New York, NY: McGraw-Hill; 1983. 400p. (General). ISBN: 0-07-054484-0.

SALTON, GERARD; VOORHEES, E.; FOX, E. A. 1984. A Comparison of Two Methods for Boolean Query Relevancy Feedback. Information Processing & Management. 1984; 20(5/6): 637-651. (General). ISSN: 0306-4573.

SANDISON, A. 1983. Follow-Up and Basic Searches at the Lane Medical Library, Stanford University. Library Research. 1983; 5: 101-108. (Regression density). ISSN: 0164-0763.

SCHWARTZ, CANDY; EISENMANN, LAURA MALIN. 1986. Subject Analysis. In: Williams, Martha E., ed. Annual Review of Information Science and Technology: Volume 21. White Plains, NY: Knowledge Industry Publications, Inc. for ASIS; 1986. 37-61. (General). ISSN: 0066-4200.

SHAW, W. M., JR. 1985. Critical Thresholds in Co-Citation Graphs. Journal of the American Society for Information Science. 1985; 36(1): 38-43. (Clustering). ISSN: 0002-8231.

SHAW, W. M., JR. 1986. An Investigation of Document Partitions. Information Processing & Management. 1986; 22(1): 19-28. (Clustering; Information theory). ISSN: 0306-4573.

SHEPHERD, MICHAEL A.; PHILLIPS, W. J. 1986. The Profile-Query Relationship. Journal of the American Society for Information Science. 1986; 37(3): 146-152. (Regression). ISSN: 0002-8231.

SICHEL, H. S. 1985. A Bibliometric Distribution which Really Works. Journal of the American Society for Information Science. 1985 September; 36(5): 314-321. (Chi-square). ISSN: 0002-8231.

SIMONTON, DEAN KEITH. 1984. Is the Marginality Effect All That Marginal? Social Studies of Science. 1984; 14: 621-622. (Multiple Regression). ISSN: 0306-3127.

SMALL, H.; GREENLEE, E. 1986. Collagen Research in the 1970s. Scientometrics. 1986; 10(1-2): 95-117. (MDS). ISSN: 0138-9130.

SMALL, H.; SWEENEY, E. 1985. Clustering the Science Citation Index Using Co-Citations, I: A Comparison of Methods. Scientometrics. 1985; 7(3-6): 391-409. (Clustering, MDS). ISSN: 0138-9130.

SMALL, H.; SWEENEY, E.; GREENLEE, E. 1985. Clustering the Science Citation Index Using Co-Citations, II: Mapping Science. Scientometrics. 1985; 8(5-6): 321-340. (Clustering, MDS). ISSN: 0138-9130.

SNIZEK, W. E. 1986. A Re-Examination of the Ortega Hypothesis: The Dutch Case. Scientometrics. 1986; 9(1-2): 3-11. (Multiple Regression). ISSN: 0138-9130.

SOLAK, JERZY J. 1983. Comparative Document Evaluation as a Feedback in On-Line Information Retrieval. Information Processing & Management. 1983; 19(3): 141-149. (General). ISSN: 0306-4573.

SPARCK JONES, KAREN, ed. 1981. Information Retrieval Experiment. London, England: Butterworths; 1981. 360p. (General). ISBN: 0-408-10648-4.

STEPHENSON, MARY SUE. 1985. The Research Method Used in Subfields and the Growth of Published Literature in Those Subfields: Vertebrate Paleontology and Geochemistry. Journal of the American Society for Information Science. 1985; 36(2): 130-133. (Brief communication; Correlation). ISSN: 0002-8231.

SUBRAMANYAM, K. 1984. Research Productivity and Breadth of Interest of Computer Scientists. Journal of the American Society for Information Science. 1984; 35(6): 369-371. (Brief communication; Correlation). ISSN: 0002-8231.

TAGUE, JEAN M. 1981a. The Pragmatics of Information Retrieval Experimentation. In: Sparck Jones, Karen, ed. Information Retrieval Experiment. London, England: Butterworths; 1981. 59-104. ISBN: 0-408-10648-4.

TAGUE, JEAN M. 1981b. User-Responsive Subject Control in Bibliographic Retrieval Systems. Information Processing & Management. 1981; 17: 149-159. (Anova). ISSN: 0306-4573.

TAGUE, JEAN M.; NELSON, MICHAEL J.; MURPHY, LARRY J. 1975. ARIMA Forecasts in Information Studies. In: Husbands, Charles W.; Tighe, Ruth L., eds. Information Revolution: Proceedings of the American Society for Information Science (ASIS) 38th Annual Meeting: Volume 12; 1975 October 26-30; Boston, MA. Washington, DC: American Society for Information Science; 1975. 47-48. (Time series). ISBN: 0-87715-412-0; LC: 64-8303.

TAPSCOTT, DON; GREENBERG, MORLEY; MACFARLANE, DAVID. 1980. Researching Office Information Communication Systems. Canadian Journal of Information Science. 1980; 5: 61–71. (Chi-square). ISSN: 0380-9218.

TOMER, CHRISTENGER. 1986. A Statistical Assessment of Two Measures of Citation: The Impact Factor and the Immediacy Index. Information Processing & Management. 1986; 22(3): 251–258. (Correlation). ISSN: 0306-4573.

TOWNSEND, J.; ASHBY, F. G. 1984. Measurement Scales and Statistics: The Misconception Misconceived. Psychological Bulletin. 1984; 96: 394–401. (General). ISSN: 0033-2909.

TURNER, STEPHEN J.; O'BRIEN, GREGORY. 1984. A Fuzzy Set Theory Approach to Periodical Binding Decisions. Journal of the American Society for Information Science. 1984; 35(4): 228–234. (Discriminant analysis). ISSN: 0002-8231.

VAN RIJSBERGEN, C. J. 1979. Information Retrieval. 2nd edition. London, England: Butterworths; 1979. 218p. (General). ISBN: 0-408-70929-4.

VIGIL, PETER J. 1982. Utilization of Boolean Not to Facilitate Online Searching Effectiveness and Comprehension. In: Petrarca, Anthony E.; Taylor, Celianna I.; Kohn, Robert S., eds. Information Interaction: Proceedings of the American Society for Information Science (ASIS) 45th Annual Meeting: Volume 19; 1982 October 17–21; Columbus, OH. White Plains, NY: Knowledge Industry Publications, Inc. for ASIS; 1982. 316–319. (T-test). ISSN: 0044-7870; ISBN: 0-86729-038-2.

WALLACE, DANNY P. 1985. The Use of Statistical Methods in Library and Information Science. Journal of the American Society for Information Science. 1985; 36(6): 402–410. (General). ISSN: 0002-8231.

WALLACE, DANNY P.; BONZI, SUSAN. 1985. The Relationship Between Journal Productivity and Quality. In: ASIS '85: Proceedings of the American Society for Information Science (ASIS) 48th Annual Meeting; 1985 October 20–24; Las Vegas, NV. White Plains, NY: Knowledge Industry Publications, Inc. for ASIS; 1985. 193–196. ISSN: 0044-7870; ISBN: 0-86729-176-1.

WANG, YIH-CHEN; VANDENDORPE, JAMES; EVENS, MARTHA. 1985. Relational Thesauri in Information Retrieval. Journal of the American Society for Information Science. 1985; 36(1): 15–27. (Anova). ISSN: 0002-8231.

WEINBERG, BELLA HASS; CUNNINGHAM, JULIE A. 1984. Term Specificity and Online Postings: Inverse Relationship? In: Flood, Barbara; Witiak, Joanne; Hogan, Thomas H., comps. 1984: Challenges to an Information Society: Proceedings of the American Society for Information Science (ASIS) 47th Annual Meeting: Volume 21; 1984 October 21–25; Philadelphia, PA. White Plains, NY: Knowledge Industry Publications, Inc. for ASIS; 1984. 144–147. (Correlation). ISSN: 0044-7870; ISBN: 0-86729-115-X.

WEINER, JOHN M.; STOWE, STEPHEN M.; FULLER, SHERRILYNNE S.; GILMAN, NELSON J. 1982. The Size of the Document Set and Conceptual Structure Identification. In: Petrarca, Anthony E.; Taylor, Celianna I.; Kohn, Robert S., eds. Information Interaction: Proceedings of the American Society for Information Science (ASIS) 45th Annual Meeting: Volume 19; 1982 October 17–21; Columbus, OH. White Plains, NY: Knowledge Industry Publications, Inc. for ASIS; 1982. 327–329. (T-test). ISSN: 0044-7870; ISBN: 0-86729-038-2.

WELBORN, VICTORIA; DAVIS, PHYLLIS B.; LYNCH, SARAGAIL
RUNYON. 1982. Card Catalog and LCS Users: A Comparison. In:
Petrarca, Anthony E.; Taylor, Celianna I.; Kohn, Robert S., eds. Infor-
mation Interaction: Proceedings of the American Society for Informa-
tion Science (ASIS) 45th Annual Meeting: Volume 19; 1982 October
17–21; Columbus, OH. White Plains, NY: Knowledge Industry Publica-
tions, Inc. for ASIS; 1982. 330–334. (Chi-square). ISSN: 0044-7870;
ISBN: 0-86729-038-2.
WHITE, HOWARD D.; GRIFFITH, BELVER C. 1982. Authors As Markers
of Intellectual Space: Co-Citation in Studies of Science, Technology and
Society. Journal of Documentation. 1982 December; 38(4): 255–272.
(Factor Analysis, MDS). ISSN: 0022-0418.
WILLIAMS, MARTHA E.; KINNUCAN, MARK; SMITH, LINDA C.;
LANNOM, LAURENCE; CHO, DONGSUNG. 1986. Comparative
Analysis of Online Retrieval Interfaces. In: Hurd, Julie M., ed. ASIS
'86: Proceedings of the American Society for Information Science
(ASIS) 49th Annual Meeting: Volume 23; 1986 September 28–October
2; Chicago, IL. Medford, NJ: Learned Information, Inc. for ASIS; 1986.
365–370. (Anova). ISSN: 0044-7870; ISBN: 0-938734-14-8.
YABLONSKY, A. I. 1985. Stable Non-Gaussian Distributions in Scientometrics. Scientometrics. 1985; 7(3-6): 459–470. (General). ISSN:
0138-9130.
YERKEY, A. NEIL. 1983. A Cluster Analysis of Retrieval Patterns Among
Bibliographic Databases. Journal of the American Society for Information Science. 1983; 34(5): 350–355. (Clustering). ISSN: 0002-8231.

6 Electronic Image Information

LOIS F. LUNIN
Herner and Company and
Cornell University Medical College

INTRODUCTION

History

In the world of computing, the digital image has been a concept since the mid-1950s, when digital image processing technology for the enhancement, restoration, coding, and transmission of images began to appear, especially in intelligence and space applications (BROWN; MARCEAU). The lay public became aware of the usefulness and potential of digital information processing when NASA moon images were processed in real time before our eyes on television.

Digital image processing became an active discipline about the mid-1960s when the third-generation digital computers began to offer the speed and storage capabilities required for practical implementations of image processing algorithms (GONZALEZ ET AL.). Electronic imaging is now fairly common and is found in the industrial, scientific, medical, military, and law enforcement sectors, among others. Since the mid-1970s, array processors have played a significant role in processing the enormous amounts of data necessary in image and signal processing applications. Array processing can be found in geographic image processing, medical imaging, satellite mapping, and image enhancement of photographic data from space (BOERHOUT).

Image processing is now a large and sophisticated field that incorporates many recent computer vision techniques. Today, image processing is esti-

The author is grateful to members of the Samuel J. Wood Library, The New York Hospital-Cornell Medical Center, New York City, and to Nancy Wright, Corporate Librarian and Vice President, Herner and Company, for their assistance with library research.

Annual Review of Information Science and Technology (ARIST), Volume 22, 1987
Martha E. Williams, Editor
Published for the American Society for Information Science (ASIS)
by Elsevier Science Publishers B.V.

mated to be a $400 million market, with a growth potential of 30-50% over the next five years (GONZALEZ ET AL.).

The recent surge of activity in digital image processing is due to several factors: the importance of access to large and growing numbers of digitized images, advances in microelectronic technology, and the development of more sophisticated algorithms. Recently, much attention has been directed to image databases for other reasons. First, the database concept has become popular in conventional data management and, together with progress in database management system (DBMS) research, the idea of centralized common data has been broadened to include pictorial information. Second, a body of standard methodologies in image processing has been established in the past 20 years. Third, recent progress in memory and device technologies can now cope with the vast storage requirements (TAMURA & YOKOYA).

This chapter is confined to two-dimensional black-and-white representations of images such as photographs, drawings, text, radiology films, biological cells, and maps. Not included are computer graphics, pattern recognition, image understanding, scene analyses, computer vision, and copyright. Because this is the first *ARIST* chapter on image information, some basic background information on image processing is included. This review is designed to present image information of interest to the information science community. During its preparation, computer searches of several databases and manual searches of many indexing and abstracting publications were conducted. Trade magazines, journals, and conference proceedings were particularly rewarding. While the review is intended to be broad and comprehensive, it is by no means exhaustive.

Definitions

WEBSTER'S SEVENTH NEW COLLEGIATE DICTIONARY offers several meanings for image, among which is the following: "An image is a representation, likeness, or imitation of an object or thing, a vivid or graphic description, something introduced to represent something else." According to an ancient etymology, BARTHES says the word image should be linked to the root *imitari*. Further, he raises the question of whether it is possible to conceive of an analogical "code" as opposed to a digital one. While the question of how to represent image information in code form is examined, the chapter deals more with the conversion of an image to digital form. CASTLEMAN explains that images occur in various forms, some visible, some not; some abstract, some real; some suitable for computer analysis and others not. He warns that the lack of an awareness of these different types can lead to considerable confusion among people communicating ideas about images. Since images form an overwhelming part of our experience from birth, we tend to take them for granted.

"Digital image processing, like other 'glamour' fields, suffers from myths, misconceptions, misunderstandings, and misinformation. It is a vast umbrella under which fall diverse aspects of optics, electronics, mathematics, photography, and computer technology. It is a truly multidisciplinary endeavor plagued with imprecise jargon" (CASTLEMAN, p. 3). To aid the reader, a glossary of electronic image-related terms used in this review appears at the end of this chapter.

ELEMENTS OF ELECTRONIC IMAGES

A picture is a visual representation drawn, photographed, printed, or otherwise presented in two dimensions. In this review, an image refers to the representation of a picture in terms of the binary 1s and 0s upon which digital computing is based. Several authors explain the basic principles and procedures (CASTLEMAN; GRIGSBY, 1986; HELMS; IASA (Insurance Accounting System Association); MARCEAU; F. MOORE, 1986; NUGENT, 1986; STAR; THIELEN).

A photograph or printed page becomes a digital image for computer manipulation in the following way. Imagine that a fine grid or mesh is placed over ‧ a printed page or a photograph, thus dividing it into many small regions called picture elements or pixels. At each pixel location the brightness is sampled and quantized (assigned a numerical value). This process generates a number representing the brightness or darkness of the image at that point. When this process is done for each pixel, the image is represented by a rectangular array of integers. Thus, each pixel has a location or address (its line or row number and its sample or column number) and an integer value called the gray level. This array of digital data (the digital image), produced by the digitizing process, is now stored for ready access by a suitable device and is ready for computer processing. The computer reads the input image into the computer. During output, the computer operates on one or several lines and generates the output image pixel-by-pixel. As the output image is created, it is written on the output storage device line-by-line. During the processing, the pixels can be modified by the operator as desired. After processing, the final product is displayed by a process that is the reverse of digitization (CASTLEMAN).

Computers treat images as arrays or series of elements. The number of elements in an array—i.e., the number of pixels used horizontally and vertically—determines the resolution level of the image. If only a few pixels are used, the image tends to look granular and area outlines have a jagged look. If many pixels are used, area outlines appear as smooth curves and gradations in shading seem continuous.

The number of shades of gray that can be numerically represented by the pixels determines the gray scale of the image. If only two shades are used- black and white—the result is a high-contrast image with poor definition. If many shades of gray are used, there can be smooth transitions from light to dark. Popular choices for the number of pixels in an image are based either on powers of 2 (256 x 256, 512 x 512, or 1,024 x 1,024) or on hardware standards such as the 525-line commercial television system. The gray levels are encoded in binary form. The number of bits for a given pixel determines the number of unique gray values or colors available. Eight-bit pixels provide 256 different gray values in black and white or 256 unique colors.

The number of bits per inch determines the resolution of the stored image and the fineness of detail that can be recreated. This number is also called lines per inch because scanning generally takes place one scan line at a time down the page. Current scanning devices generally operate between 200 and 400 lines per inch. A scan of 200 lines per inch is satisfactory for office correspondence and the body text of most printed matter (text as small as 4-point type).

The major difference between digital images and the character information with which data processing and bibliographic systems normally deal is the sheer quantity of information. Usually a single character is encoded in one byte; the average typewritten page contains less than 4,000 characters. If the typical 6 x 9 inch monograph page is digitized, that page image takes 25 times more storage than a corresponding character-encoded page (NUGENT, 1986).

IMAGE PROCESSING

The origin of digital image processing can be traced to the early 1920s when digitized pictures of world news events were first transmitted between London and New York via the Barlane submarine transmission system. Coded in digital form for cable transmission, the pictures were then reconstructed by specialized printing equipment. With this innovation, the time to transport a picture across the Atlantic dropped from more than a week to less than three hours (GONZALEZ ET AL.).

Traditionally, image processing has been defined as a function pertaining to pictures captured from the physical world as opposed to those created electronically using a computer graphics system. Moreover, image processing as a term has always encompassed specific applications, such as remote sensing and electronic publishing. New developments are changing those definitions, and, as a result, many systems vendors are avoiding the term "image processing" and naming their systems according to the industries or applications they serve, e.g., electronic publishing, remote sensing.

Image processing is a special form of two-dimensional (and sometimes three-dimensional) signal processing. A generalized image processing system has five main components: 1) digitization or input, 2) storage, 3) processing, 4) communications (transmission), and 5) output.

Input (Digitization)

Scanners are the primary input devices for an image system. During the scanning process each line on the document page is divided into thousands of pixels. If a page of print is being scanned, each pixel is interpreted as either a black or a white point on the page. The scanned image becomes a bit stream of 1s and 0s (hence, bit-mapped images) that represent black or white dots on the original document.

The process of digitizing or converting a two-dimensional analog signal (image) into a digital image has two separate parts. In the first part, called sampling, a number of discrete points are selected from the continuous image and measured. In the second, called quantization, values are assigned to the intensities at each of these points. The total number of quantization levels defines the gray scale—i.e., the number of shades that can be represented. Several authors have described the process in detail: BOYNE & OTANO, HELMS, MAURO, C. MOORE (1986a), PRESTON & MOLINARI, and THIELEN.

Scanning densities are usually measured by the number of pixels per inch on the horizontal and vertical axes. Most organizations require 200 x 200 pixels per inch, although some, such as engineering departments with blueprints and drawings, require a much higher scanning density.

Input systems also include the indexing process. Creating an identifying entry for each document in the image system usually is done at this stage, and more complex index information is added later.

Storage

Alphanumeric characters are stored differently from image information. HOOTON (1985) emphasizes the two main differences between optical character recognition (OCR) and digital raster image storage. First, in OCR the image is digitized and then run through a recognition algorithm to enable the computer to recognize the information. In image storage, the image is digitized and stored in a form that, when reconstituted, can be recognized only by a human. Second, with OCR, it takes only 8 bits to represent the character "A." In contrast, it might take 8,000 bits of information to represent the image of an "A" by analyzing it into its light and dark picture elements.

Processing

Digital image processing comprises a range of operations that manipulate frame data to alter an image or to extract and refine meaningful information. Those processing techniques permit the enhancement of contrast and detail, the combining of two images, and various forms of filtering (HALL; PRESTON & MOLINARI).

Five processes are normally associated with digital image processing: 1) enhancement, 2) restoration, 3) compression, 4) segmentation, and 5) description (GONZALEZ ET AL.).

Enhancement. Image enhancement deals with improving pictorial information for human interpretation and machine perception.

Restoration. Restoration is concerned with improving a given image by reconstructing or recovering an image that has been degraded.

Compression. Image compression techniques are useful whenever it is important to reduce storage or transmission requirements. Compression algorithms can be categorized as reversible or irreversible. Reversible techniques, also known as error-free methods, preserve the information contained in the original image, making perfect reconstruction possible. Irreversible methods degrade the image because some information is lost.

Image compression is a particularly important aspect of image processing. As an application of information theory it continues to be an active research area with hundreds of image compression algorithms and variants being proposed. While advances in image compression technology are leading to integration of digital imaging into general-purpose computer systems, image compression still has a way to go.

Segmentation. Segmentation subdivides an image into objects or regions of interest. Segmentation algorithms are generally based on one of two basic principles: discontinuity and similarity. Edge detection is the principal approach in discontinuity; thresholding and region growing are the principal approaches in similarity.

Description. Description extracts features from an object for recognition purposes. Ideally, as GONZALEZ ET AL. write, these features should be independent of the object's location and orientation, and they should contain enough information so that one object can be discriminated from another. Descriptors are usually based on shape and intensity. Shape descriptors attempt to capture an object's fixed properties.

Output

Image processing data are bit oriented rather than character oriented, and, thus, the resolution of both hard-copy and terminal display devices must be very high. The higher the resolution, the more the device costs. Generally, a resolution of 200 x 200 lines per square inch is appropriate for printing, while one of 100 x 100 lines per square inch is acceptable for workstation viewing. Integrating a high-resolution terminal with data processing systems is an important issue. Most applications will require communication with both the data processing computer as well as the image controller. Currently, many image processing systems use two terminals. Newer image processing systems use a single terminal that connects to two separate networks (IASA).

Communications

Designing a communications network for an image processing system is challenging because there is more information in an image than traditional data networks are used to handling. For example, an 8½ x 11-inch page contains nearly 4 million bits of information (at 200 x 200 pixels per inch) compared with 1,920 bytes (characters) on a typical online screen. The network usually also manages different types of information (e.g., text) with very different size and response requirements. If the requester is remote from the storage, the communication network requires the use of telephone lines. At line speeds of 9,600 characters per second, transmission times of 10 seconds per document may be unacceptable. Finally, overall standards for linking components in an image processing system do not exist, although work is under way.

Communications network designers have met these challenges by using several approaches. One is compression, which can reduce the volume of information on the network by 80-95%, depending on the density of information in the original document. Compression and decompression must take place at the terminals and printers; thus, the networks handle only compressed documents. Another approach is to use a separate network for the image traffic because document images contain much more information than data packets (IASA).

Image Processing Issues

Image processing is being called the next major breakthrough in advanced office technology. Yet even at this stage, IASA warns, there are a number of issues and concerns that should be addressed by any organization interested in considering an image system. It groups these issues into administrative and technical. The issues involve such diverse matters as cost, security, technology, immaturity, legal admissibility, standards, major vendor commitment, installed base, retention of documents, conversion, and organizational control.

Since image processing has given computers the ability to monitor and control visual information, PRESTON & MOLINARI conclude that applications have moved from the realm of science fiction to the commonplace. They account for the dramatic increase in real-world uses by the introduction of image processing hardware that offers greater processing power, lower cost, and growing compatibility with personal computers such as the IBM PC. The authors examine basic imaging concepts and identify some design pitfalls.

HARDWARE

The primary components used in a typical digital image system include document scanning devices, image system controllers and processors, optical disk storage devices, printers, high-resolution workstations, and system software (C. MOORE, 1986a).

Scanners

A scanner is a device that examines printed characters or graphics and represents them as electronic signals. Scanners are the primary input devices for an image system and include conventional cameras, optical digitizers, satellite scanners, and facsimile devices. High-speed scanners can process up to one page per second, whereas slower desktop scanners might scan only 10 to 12 pages per minute. Scanners can be combined with bar code readers and OCR devices so that the scanning process captures both raster and alphanumeric information as the documents are scanned.

Image Controllers

Image controllers can range from microprocessors to minicomputers. Although vendors group image components differently, in most configurations the scanners, terminals, and storage devices are all linked to the image controller. This device provides image enhancement, compression and decompression of scanned images, conversion from analog signals to digital bit streams, and the physical routing of images throughout the system.

Storage

Recently, the optical disk has captured the interest and imagination of potential users because of its extraordinary potential. The write-once nature of the medium is a primary advantage; this feature protects images that might have to serve as original documents. Another advantage is that a prodigious amount of data can be stored on its small surface—up to 40,000 images on one 12-inch disk, using both sides. Large storage capacity is essential for digital image systems since scanned documents require much more direct-access storage than typed documents. A typed page created through word processing requires 2.5 kB of storage, in contrast to the 50 kB of storage (after compression) required by an 8½ x 11-inch page scanned at 200 pixels per inch of storage.

Workstations, Printers, and Output Devices

Output devices are usually modified laser printers that range from slow, conventional desktop units to high-speed printers. Other devices may include computer-output-to-microfiche (COM) for micrographics applications.

The image display terminal is one of the more expensive components of an image system. The terminals are often intelligent high-resolution devices that display from 200 to 400 pixels per inch and zoom, pan, rotate, and window the image. Sometimes the terminals are linked to the mainframe as well as to the image system.

New Hardware Devices

Computing architectures are being designed and built for image processing and management. These and new technologies such as the very large-scale integrated (VLSI) circuits can be used to implement more directly the locally connected, parallel computations that are common in intrinsic image computation (BROWN).

WILSON (1986a) defines the problems faced by designers of image processors: fast computational elements, large memory requirements, and heavy I/O (input/output) loading. In the quest for "real time" image processing, many attempts have been made to relieve the host CPU (central processing unit) of the intensive and often iterative imaging functions characteristic of image processing. Array processors were among the first types of machine to address the problems of quickly processing images and have been used for several years to increase system performance.

Recent research efforts and advances in image information systems include: 1) high-resolution scanners, 2) stand-alone image processors incorporating microprocessors and array processors that can be interfaced with most major mini- and microcomputers, and 3) high-resolution color graphics raster displays. There is also renewed interest in using microfiche for storage and retrieval of large quantities of images and videodisk technology (CHANG & YANG). A minicomputer that stores the microfilm address and descriptors and controls the retrieval of specific document images can be used to create and maintain an image database.

In many image processing applications a new breed of graphics controllers will perform functions associated with primitive image manipulation and processing to offload primitive image processing functions from the host CPU. In others, they will be microcoded to perform a greater percentage of image processing functions.

In addition to these hardware components and systems, the recent advances in data communications, especially broadband local area networks (LANs), provide the means to interconnect many workstations for multimedia communication, which includes images. CRISMAN offers a summary of research efforts and directions in computer vision research that is applicable to hardware.

While imaging systems are already on the market, some lingering issues must be addressed concerning the associated technology. According to Insurance Accounting Systems Association (IASA), these include: untested components, indexing documents, backup and recovery, optical storage reliability, document scanning speed, mainframe compatibility, performance, response time, limited software, lack of integration, and unclear networking requirements.

SOFTWARE

For image systems the software consists of system and applications packages. System software includes image enhancement, compression and decompression, physical routing of images, network protocols, and access methods. Applications software can include: image work flow software; management statistics; audit trail software; electronic mail (EM) and text overlay on images plus keyword search for ASCII (American Standard Code for Information Interchange) text; windowing to images, text, and data; and screen design utilities.

IASA emphasizes that the image system index is the most important component in the host system software; it is the fundamental interface between users and the image display hardware. It also advises that there is little generic software written for image processing systems. Image technology depends strongly on software, yet software development remains one of the major challenges in bringing more image systems to the market. IASA describes the software for each of the principal components of an image processing system illustrated in its article (IASA).

WILSON (1986a) observes that third-party software application packages and device drivers are beginning to have an impact on the imaging market. Vendors will find that fast hardware is not enough and that products must be closely tied to well-known applications software to succeed. He, too, advises that such software is hard to find for image processing.

IMAGE DATABASE: CRITERIA, CONCEPTS, AND CHALLENGES

Image databases differ significantly from conventional databases in content, organization, storage, coding and abstraction, query languages, search strategies, and output, to name a few parameters. Some of the problems posed by image information and image databases may be helped somewhat by thinking two-dimensionally. *Flatland* (E. A. ABBOTT) is a delightful description of a two-dimensional world inhabited by straight lines, triangles, squares, pentagons, hexagons, and circles. When these inhabitants are introduced to the concept of a three-dimensional world, they have great difficulty in comprehending it. Turning the concept around, a look at life in a two-dimensional world can stir the imagination and help us think structurally and organizationally about spatial data.

Conceptual bases for thinking about image databases are offered in two excellent review articles on image databases and image database systems by TAMURA & YOKOYA and NAGY. The articles are discussed here at some length because the authors present overviews that clarify current image database problems and because they suggest directions for further work, especially Nagy.

Tamura and Yokoya found that much attention has been paid to the design of image database systems in recent years. However, they also found that the term "image database" has been used in different ways in the literature—e.g., as a new compression method for efficient storage of images, as a hierarchical data structure scheme for storing complex pictures of great complexity, as an evaluation process involving various similarity measures for image database retrieval, and so on. They observed that all of these techniques had been investigated independently before the concept of image database became a recognized topic.

They point out the essential problems in image database design, discuss approaches to image databases and the elements of image database systems, and present several representative image database systems. They define an image database as a system in which a large amount of image data and their related information are stored in an integrated manner. Without a management system, the collection of images should not be called an image database. They add, however, that a collection of image data itself is sometimes called an image database. They note two extremes in image databases. At one end, all images are collected and stored without any compression, and only the pointers to those images are managed by the DBMS; at the other end, all image features are completely symbolized and managed uniformly with secondary information.

In Nagy's review, much of the work appearing under the heading of "image database" describes either image nondatabase systems (i.e., software designed to manipulate sizeable collections of image data without recourse to database concepts) or nonimage database systems (wherein the DBMS contains image descriptors or attributes rather than the images themselves). Also, much of the work done under the heading "image database" deals with somewhat peripheral aspects.

All together the development of these image database systems represents a huge amount of thought and effort, but few, if any of these systems approach the critical size necessary for operational application. Even though some of them may, in principle, be portable, Nagy has not seen any that are used at several locations. Finally, he adds that none is based on existing commercial systems. Among the outstanding exceptions, he says, is computer-assisted tomography (CAT, CT). He offers characteristics of successful applications and suggests the types of applications that should become feasible.

Tamura and Yokoya review recent work according to three classes of criteria: 1) function, 2) extent of data abstraction, and 3) database models. They conclude that although the significant concepts of image database have been established, few systems can be called true image databases. They point out areas of concern in image database design. One is how to represent structural information in a database mode. Another concerns just what information is to be retrieved. Is the desired output a physical image, descriptive information, or just relevant information? What is a unit of an image entity? When are derivable image features computed? They also point out that coding techniques peculiar to image databases call for further study.

From their review, Tamura and Yokoya conclude that increasing interest is being directed to the development of image databases, and they suggest that with progress in optical disk technology this trend will become more obvious. They point out, however, that there is no theoretical background in image database design as yet, and they suggest that several large image databases in various fields will have been developed and used before a general methodology is established.

Nagy lists several features of current database systems that would be useful in reducing the time necessary to develop new image processing applications. There appear to be two conceptual barriers: 1) the discrete representation of continuous entities in conventional database systems and 2) the absence of total ordering in two dimensions. He is examining ways of overcoming some of these difficulties.

Table 1 compares various features of conventional databases with image databases to illustrate their many differences.

Research Directions

Coding techniques. Several investigators have been working on coding structures; these include hybrid quadtrees. (RAMAN & IYENGAR); partitioning, then organizing into a hierarchical structure (YANG & CHANG); pyramidal and predictive coding and contour texture models (KUNT ET AL.); recursive hierarchical decomposition of the image (COHEN ET AL.); and encoding, segmentation, and boundary formation (YANG ET AL.).

Syntactics and abstractions. Syntactic and semantic abstraction rules (CHANG & LIU), syntactic pattern grammar (DON & FU, 1985), and stochastic tree grammar (DON & FU, 1986) have been formulated and applied to image analysis and extraction. To construct picture indexes, Chang and Liu formulated abstraction operations for performing picture object

TABLE 1

Conventional vs. Image Databases:
Similarities and Differences[1]

Aspect	Conventional Database	Image Database
Characteristics of the data	Average record contains relatively small amounts of data	Each image contains vast amounts of data
	No dimension	Spatiality; two-dimensional data storage
	Entities represented and representation itself are discrete	Continuous entities in extent and intensity, both in independent variables (coordinate) and in dependent variables (reflectance, optical density, etc.)
	NA	Concerned with brightness, "color" (gray level of each pixel)
	Principal entities are alphanumeric	Principal entities are images, regions, curves, points (form, line, structure, texture)
	Absolute representation	Relative representation with spatial transformations, amplitude scaling, arbitrary sampling, and quantization
Acquisition	Distributed, aperiodic	Centralized, periodic
Verification	Data are routinely validated, especially in financial applications	Verification and control of image data are sparse
Data models	Hierarchical, network, and relational	Relational
Types of operations performed	Delete and update occur in some databases	Delete and update rarely occur
	No manipulation	Manipulation to enhance, segment, change colors
Input, processing, output	Input by keyboarding, OCR, etc.	Input by scanner
	Digital to digital	Analog to digital
	Moderate I/O loading	Heavy I/O loading
	No compression generally	Compression frequently

TABLE 1. Continued

Conventional vs. Image Databases:
Similarities and Differences[1]

Aspect	Conventional Database	Image Database
Ordering	Most primary keys may be rank ordered by sorting (according to numeric or collation order). Allows application of effective search algorithm.	Absence of total ordering in two dimensions. Partial ordering: nearest-neighbor searches, polygon intersection.
Encoding	No coding for compression; no decoding and reconstruction required.	Coding. Gray-level pictures are compressed by frame-coding techniques; line drawings are broken down into chain codes or series of coordinates of points. Original image must be decoded and reconstructed.
Storage; hierarchy of terminology, encoding	Generic-specific relationships of indexing vocabulary	Pyramid structures, quad trees, etc.
Queries; retrieval	Retrieval by identifier, conditional statement	Retrieval by identifier, conditional statement: structural information (e.g., skeleton, shape); similarity to a given sample. Some question about what is retrieved: physical image? descriptive information? relevant information?
	NA	Size, scale, orientation, sampling, quantization are all aspects; thus, any spatial location or intensity value may be specified in a query.
	Query language provides interface between user and database	Query language should be designed to manipulate both symbolic and pictorial data.
	NA	Sketch database: abstract images that contain outline information on shapes and locations (e.g., rib cage, heart). Can be used as pictorial keys to retrieve desired images.
	NA	Logical pictures: collections of picture objects considered as masks for extracting meaningful parts from an entire image.

TABLE 1. Continued

Conventional vs. Image Databases:
Similarities and Differences[1]

Aspect	Conventional Database	Image Database
Queries; retrieval (Cont.)	Queries are straightforward.	Query translation poses problems: queries must be translated in reverse order of execution; the desired view must be specified.
Hardware; equipment required	Keyboard; OCR, etc.	Scanner Image controller
	Computer	Computer, probably with special architecture
	Optical disk, tape, disk, etc.	Medium with vast storage capability, (e.g., optical disk)
	CRT	High-resolution terminal
	Printer	Laser printer; high-speed printer
	Modem	?
	I/O devices common, widespread	I/O devices far more complex and specialized
Software	Much software available, general and special-purpose	Little generic software available
Communications: Networks	Conventional networks	Generally separate networks because of huge size of data; thus greater transmission time.
Current storage and transmission facilities	Adequate	Inadequate for some applications; for example, a single frame of 4-band Landsat image contains upwards of 100 MB: at 9,600 baud, transmission would require about 4 hours.
Quantity Location	Thousands operational, many distributed; hundreds available on commercial systems.	Few of operational size; exception: CT. Few used at several locations; none on existing commercial systems.
Legality of data	NA	Questions re legal admissibility (e.g., signatures, compressed radiographic information)
Users	Large user community	Relatively small user community
User friendliness	Often transparent	Not transparent
Application areas	Many in private sector	Many in public domain with government rather than private funding Seem to be more heterogeneous

TABLE 1. Continued

Conventional vs. Image Databases:
Similarities and Differences[1]

Aspect	Conventional Database	Image Database
Design complexity	Relatively simple	Projects that combine image processing with database; technological ideas suffer excessive complexity.
Database design tools	Notion of schemes	Apparently none
	Normalization	Apparently none
	Data description language	Very little
	Data dictionaries	Apparently none
	Offline and online documentation	Very little
	Specialized query languages	In research
	Query optimization interfaces	In research

[1] Based principally on information in NAGY and in TAMURA & YOKOYA.

clustering and classification. To substantiate the abstraction operations, they also formulated syntactic and semantic abstraction rules. DON & FU (1985) use a pattern grammar to describe a class of patterns rather than just a single pattern. In a subsequent paper, DON & FU (1986) introduce a parallel algorithm for syntactic image segmentation using stochastic tree grammar as a context-generating model. One of the main concepts of the syntactic approach to image analysis and pattern recognition is the idea of decomposing an image into subimages that are simpler to analyze than the original image.

Data models and DBMS. Relational models for describing properties and relationships of objects (YAMAGUCHI ET AL.) and data models for an integrated pictorial DBMS (ASTHANA & CHING; GROSKY) have been formulated. CHOCK ET AL. offer the database structure and data manipulation capabilities of a generalized picture DBMS. CHANG & YANG describe an integrated DBMS that combines a relational DBMS with an image manipulation system.

Query language and search. Several investigators have been working on pictorial query languages (CHANG & FU; LIN & CHANG); others have developed techniques for indexing spatial objects and spatial data (IYENGAR & MILLER; ROUSSOPOULOS & LEIFKER). Various techniques to allow the user to retrieve data by manipulating images are described by FIELDS and NETRAVALI & BOWEN.

Comment

As Nagy discerned, the field appears ready for the application of concepts that will allow a unified structure to emerge from the details. One major reason for the lack of fusion appears to be the difficulty with the kind of information being dealt with. Yet, technology will undoubtedly impel the development of image databases before a general methodology has been developed.

The next section, "Applications," contains descriptions of ongoing efforts in pilot or operational image-handling efforts in several fields. In some of these databases, the database approach is relatively simple: assign an identifier to each image, then treat the identifier as one of the data types in a symbolic database system. A few others are using techniques applied in image processing analysis. Increasingly, as computer architecture is designed to handle image databases and as more elegant systems are planned, the theory and design of image databases will become increasingly important and will present new challenges.

APPLICATIONS

Offices and Archives

Digital image processing will eventually change the way information is stored, accessed, transmitted, and analyzed in organizations that handle large numbers of documents—e.g., insurance, financial, and engineering companies, libraries, museums, hospitals, universities, and federal, state, and local governments.

Business. C. MOORE (1986a) lists five areas in which business needs digital image systems: transactions processing, records management, office automation, storage of manufacturing and engineering drawings, and information distribution. C. MOORE (1986b) also addresses the developing use and application of digital image processing by private industry, ways to implement image systems, and key success factors.

Microcomputer-based imaging systems are being considered as a key element in the office of the future, particularly in organizations that handle much handwritten or signed paperwork and forms (MOSKOWITZ). However, a key concern of most businesses is that image systems should use the microcomputers already present in the office; in the long term, image systems probably will become integrated with in-place office automation systems. GRIGSBY (1985) compares micrographic and optical disk document image processing in the automated office; one large office image information system is described by MATULLO.

Government. The U.S. government has been largely responsible for the initial interest and growth of image systems (C. MOORE, 1986a), and most of the information available about image systems has been provided by federal agencies that have pioneered such systems (Library of Congress, Internal Revenue Service, U.S. Patent and Trademark Office, U.S. Army, U.S. Air

Force, and others). In contrast, publicly available information on the use of digital image systems in the private sector is limited (C. MOORE, 1986b). The National Archives and Records Administration (NARA) initiated a research test system, ODISS (Optical Digital Image Storage System) in 1984. Approximately 1.5 million page images will be stored on optical disks (HOOTON, 1985; 1986). Also in 1984 the Internal Revenue Service (IRS) awarded a contract for the Files Archival Image Storage and Retrieval System (FAISR). As of 1985, the FAISR effort is a research test conducted to determine the desirability of implementing such technology nationally. In describing FAISR, F. MOORE (1985) offers observations and recommendations on the procurement and implementation of digital optical disk imaging systems.

Multimedia viewing stations. CERVA & STRONG describe MITRE Corp.'s development of a prototype multimedia viewing station that allows a user access to images stored in various formats. Among the capabilities the station supports are image digitization and acquisition, image processing, digital image transmission, and analog image archiving. Reminiscent of Vannevar Bush's Memex is another multimedia system developed for demonstration purposes; using personal computers and a LAN, the system is linked with image retrieval software, micrographic storage devices, and management graphics hardware and software, and communicates with a host mainframe.

Art Libraries, Museums, and Architecture

State of the art. Collections of pictures are used in various ways, such as for public use and for research. Authors of articles often state that written language cannot substitute for the visual appearance of the work of art. Unlike text or data, pictorial material often derives its value from the object itself with much of the message conveyed through format, design, paper, and texture as well as artistic subtleties, strength of lines, and inscriptions (PARKER, 1985a).

Libraries of visual materials have problems that are similar to as well as different from libraries of books and periodicals. Among the differences, visual libraries lack the shared standards enjoyed by book libraries and they lack title pages on which cataloging entries are based; works from unfamiliar cultures on which little research or analysis has been done present serious cataloging problems; the current location of a work of art may not always be known because museum art may be deaccessioned, transferred to other collections, sold, lent, lost, or stolen and it is difficult to differentiate one work of art from all others by written description alone (ROBERTS, 1983). No matter how many data elements are used, subject access to visual materials ultimately relies on visual rather than textual distinctions (ESKIND).

FAWCETT (1982) traces the development of art history in relation to its attendant documentation, both textual and visual. He discusses problems of control and organization, with emphasis on subject analysis. He reviews past difficulties and offers some pointers for those involved in current projects. DE LAURIER and SCHREIBER also discuss technology, systems, and art.

OSBORN looks at video archiving: new problems will be created when image archiving can be distributed in the way that conventional databases are shared, via telecommunications and microcomputers, principally because indexing schemas for categorizing visuals are not standardized. GREENHALGH also discusses new technologies for data and image storage in relation to the history of art. He cites three main problems—all interrelated—in producing images on video display units: quality, storage, and speed of transmission from storage to screen.

Museums. Despite all the attendant difficulties of using surrogates, images can be useful in providing an illustrated catalog or directory of visual collections. NYERGES discusses the tremendous potential of videodisk technology for use in museums and lists several such videodisk projects, which include collections management, research, fund raising, visitor interpretation and education, and exhibition design and programming. One of the earliest museum ventures was undertaken by the Museum of Fine Arts in Boston in January 1980 (SORKOW). Another early venture was the National Air and Space Museum's information system, in which their entire photographic collection was put on videodisks, 50,000 images to a side, 100,000 to a disk. BOYNE & OTANO report on the system assembly. One advantage in using this technique is that the images can be viewed without having to take the actual material from the shelf, thus helping to preserve the original and save time of library personnel.

Other collections. ARTSearch microcomputer-based videodisk system contains 21,000 full-color images of the 12,000 individual objects in the Helen L. Allen Textile Collection (*OPTICAL INFORMATION SYSTEMS*, 1986b). There are also interior design databases that show house furnishings ranging from furniture, fabrics, and lighting to accessories, and a system that allows art dealers to view images of artworks for sale on a computer screen.

Architecture. PURCELL & OKUN present two case studies on information technology and architectural images. KAMISHER describes the Images System that integrates 30,000 images of Islamic architecture on a videodisk with a database of information that covers a subset of 4,000 images. A second database contains statistical and financial data on contemporary architectural projects in Islamic countries. NURCOMBE identifies some problems and gaps in architecture databases and predicts some likely developments.

Rock art. DICKMAN describes a system of recording and storing images of rock art on optical disk. One major advantage of the digitized pictorial information is that it can be modulated and enhanced, thus allowing better visibility and study of the art.

Maps, Cartography, and Geography

Because image acquisition and map production are now moving from film to digital form, there is a need to structure digital image and map data for cartographic applications (MCKEOWN, 1983). Primary constituents of map data are points, line segments, and polygons. The data structure has so far been based on computer graphics techniques. Consequently, the capabilities of cartographic systems are emphasized in the generation of maps rather than in the flexible manipulation of spatial data. Nowadays, image processing techniques tend to be incorporated in geographical information systems, also called cartographic database systems. The Image Based Information System (IBIS) developed at the Jet Propulsion Lab (JPL) shows a new style of geographic information system. Here all data manipulations are based on image domain, although three data types (tabular, graphical, and image) exist. Also, several attempts at image database design are directed to geographical data management (TAMURA & YOKOYA).

Research. CHANG ET AL. describe a generalized zooming technique that can be used for flexible information retrieval and manipulation for a pictorial database system in a distributed database environment. As digital picture processing applications have developed—some in handling geographic data—various pictorial database systems have also developed (CHANG & FU).

A study by DAVE ET AL. suggests that the extent of reduction in the dimensionality of multilayer large images depends on several factors, such as the size of the image, number of classes in the image, relative importance of the classes, and the nature of the information to be extracted. DAVE describes DIMAPS (Digital Image Manipulation, Analysis and Processing Systems). MEIER discusses a graph grammar approach to geographical databases, and FISCHLER describes an integrated testbed system.

Applications. MCKEOWN (1982; 1983; 1984) presents an overview of database aspects in digital cartography, addresses the problems of cartographic databases, and describes MAPS (Map Assisted Photo Interpretation System). He discusses methods of data acquisition, query specification, and geometric operations on map data. The image/map database approach allows the system to perform a great variety of tasks, such as image selection, spatial computation, semantic computation, and image synthesis.

EROS (Earth Resources Observation System) is a database of computerized images from Landsat satellite remote sensing activities along with images from various other federal agencies. INORAC (Inquiry, Ordering, Accounting) is the retrieval system used in this case (JACK).

With greater availability of digital data for large areas of the earth's surface, oceans, and atmosphere at reasonable degrees of resolution, the environmental base for a global environmental assessment system is beginning to take shape (FREEMAN). Included in the system are the components of a global information system for marine environmental assessments and its expert support system and image processing. FREEMAN & SMITH predict that information systems based on or incorporating remotely sensed data, image processors, and geographic system software promise to revolutionize the availability of environmental information.

Moving to outer space, RYLAND discusses storing "outer space" data, both textual data and images, such as pictures of the surface of Mars and digital terrain maps of the United States, on laser disk.

Commercial uses. Many attempts have been made to relate maps to information contained in data files and conversely to relate information from these files to maps. POIZNER describes the type of maps—statistical or data oriented—that have become useful to marketing and sales executives.

Libraries

Electronic imaging technologies are now considered feasible for library applications (LYNCH & BROWNRIGG, 1986a). Two national U.S. libraries— the Library of Congress (LC) and the National Library of Medicine (NLM)— have been exploring the application of these technologies to longstanding management problems.

Library of Congress. LC entered the world of electronic images in 1979 to develop a system to scan catalog cards, store them at high density on an optical digital disk, and permit their retrieval and printing on demand using high-quality laser printing devices (PRICE, 1984). Digitization of the scanned cards allowed computer-based image enhancement techniques to remove the effect of soiling and water damage and to print crisp new cards as needed.

LC's current Optical Disk Pilot Program has two facets. The Print Project involves the design and construction of a digital imaging system to permit rapid, high-resolution scanning, storage, and retrieval of images in black and white and as halftones. The Non-Print Project includes almost 50,000 images from LC's Prints and Photographs Division to represent many typical preservation and access problems. Users will be able to access the system by scanning through the images and then calling up the descriptive data or by conducting searches against the descriptive data and then calling up the associated images. Several papers describe various aspects of the project (KRAYESKI & SWORA; MANNS & SWORA; NUGENT, 1983; PARKER, 1985b; PRICE, 1984; 1985).

In contrast to other experiments or projects involving large optical disk systems in which uniform material is handled, LC has highly varied input and retrieval requirements for the storage of and access to a wide variety of materials commonly held by libraries (MANNS & SWORA). LC has six reasons for using optical disk storage: 1) preservation, 2) service, 3) access, 4) space, 5) compaction, and 6) image enhancement. There are five elements in the program: 1) systems for input, 2) storage, 3) processing, 4) output, and 5) the high-speed printer.

KRAYESKI (1986) lists the operations that have been accomplished to date: scanning documents at 300 x 300 pixels per inch; enhancing the images to produce "halftones" and gray scale for greater clarity of graphic materials; linking bibliographic citations with the images; recording the images on optical digital disk; retrieving the index and images on the video display terminal using the linked systems approach; and printing the character and image information at 300 x 300 pixels per inch on remote laser printers.

National Library of Medicine. A prototype system is under development at the Lister Hill National Center for Biomedical Communications (LHNCBC) for the electronic storage and retrieval of document images. The system has been implemented for electronic scanning, digitization, storage, retrieval, and display of images of biomedical documents; when coupled to a bibliographic database, it will demonstrate the potential utility of an image database system for library applications. Future developments include the incorporation of techniques for document image compression and gray rendition (THOMA ET AL., 1985). The storage of document images in digital form on optical digital disks is promising for several reasons: high storage density, rapid random accessibility, and compactness. Issues that need to be investigated are throughput rate for different types of documents, quality control procedures, speed of image access and retrieval, image quality, and the need for image enhancement techniques.

National Library of Canada. This library produced a videodisk in 1981–1982 to demonstrate the videodisk's ability to store and present library materials in a variety of formats (SONNEMANN).

Subject access. Developments in current online systems are having a tremendous impact on what information is cataloged or indexed, how it is cataloged or indexed, how one searches for information, and what information is searched for. These topics are equally relevant for image information systems. DOSZKOCS and NUGENT (1984) discuss modern subject access in relation to image handling. Nugent emphasizes that image automation presents a new need in subject access, the need for languages to describe pictures. By "subject access" he means the ability to find a particular picture or set of pictures within a large collection on the basis of the subject content of the image rather than on its physical description, its creator, or its membership in a larger collection. This same need was also expressed by authors in the section on "Art Libraries, Museums, and Architecture." Nugent suggests that this approach be called the "content based" or "intelligence analyst" approach to pictures as opposed to the traditional bibliographic approach. He calls for the development of a formalized means of describing the content of pictures, advanced query languages, new database structures (most likely relational), and new computer architectures designed initially for picture retrieval and eventually for picture processing.

Image recognition methods based on artificial intelligence (AI) and scene analysis may become necessary to distinguish the items in a two-dimensional image and their salient characteristics. NUGENT & HARDING say this is a broad and pressing problem not at all peculiar to libraries. They believe that the automated processing of image information is the greatest challenge of the next decade. They also believe that the integrated combination of full text and nontext information will be the central feature of what they have called "The Image Automation Revolution," and they discuss various roadblocks to be overcome.

Communication. Facsimile transmission machines that can function as high-speed high-resolution document digitizers (and also as printers) are well within reach and will become standard with the adoption of the Group IV specification of standards for facsimile machines. These machines will be able to operate as computer network peripherals (LYNCH & BROWNRIGG, 1985). Group IV machines will provide remote document delivery and act as network input and output devices for digital image processing.

The APOLLO (Article Procurement with OnLine Local Ordering) project is a digital facsimile system that digitizes text, graphic, and pictorial information contained in the same document. APOLLO is aimed at exploring advanced electronic delivery of documents and data using high-speed digital satellite links to allow information to be easily and quickly accessible throughout Europe. RAITT expects service to begin in 1987 and then soon become commercial.

Predictions. OSBORN predicts that a new problem will be created when image archiving can be distributed the way that standard databases have been distributed—i.e., by telecommunications and microcomputers. There will be a real need for a standardized indexing scheme that can do for image classification what the Dewey Decimal System did for knowledge classification. A system of this sort will undoubtedly be developed as video archiving becomes more widespread or as an auto-indexing, auto-abstracting image analysis scheme is developed based on AI.

As microcomputers and storage technology become more powerful and less expensive, it becomes feasible to have an inexpensive machine to provide a spatial data management "front end" for a distributed video archive. A natural language front end to an image database will allow requests in natural language and spatial techniques that permit browsing the database for its offerings (OSBORN).

Medicine

Much of the information base in medicine is visual. Images are generated, used, and stored for patient care (diagnosis), research, education, and administration. In an average hospital, up to one million images are generated every year, and archiving has become a major problem. Medical images are diverse and come from procedures used in diagnostic radiology, pathology, cardiology, urology, and radiation oncology. Ideally all of them should be integrated in a total medical information system. Most, if not all, of the existing image database systems are experimental or prototype; their cost performance and user interface must be improved before they can be used in clinical medicine (TAMURA & YOKOYA).

Patient information systems. Defined as complete patient databases, patient information systems are still in their infancy, although there are some complete systems operating within departments of hospitals. Most of these systems do not include image databases. One new system under development is the laser LifeCard of Blue Cross/Blue Shield, which is said to store up to 300 pages of medical information. The card is planned to contain the patient's digitized photograph and signature for positive personal identification, as well as medical history.

Education. Several educational image projects are available on videodisk, and the pharmaceutical industry is developing a variety of these in the health care area. Many of the disks are being developed in cooperation with hospitals or universities (HOFFOS). LAMIELLE & DE HEAULME describe VIDIX, a project being developed in France to couple a bank of referential images to a textual base. The images will be reached by menus or by retrieval using free rather than formal language "understanding" of the legends accompanying the images.

Cytology and pathology. Several universities have concentrated on biomedical image processing with a practical application in tissue-section analysis.

For example, SHERMAN ET AL. describe bladder cancer diagnosis by image analysis of cells in voided urine through the use of a small computer.

Dentistry. SOUTHARD found that optical disk technology provides low cost per megabyte and rapid access for archiving and searching dental images; the disk displays dental images of at least equal and sometimes better diagnostic quality than the conventional film method. The system's image resolution requirements depend on the level of anatomic or pathologic detail to be retained.

Computed/noncomputed images. Medical imaging systems tend to fall into two classes: systems that produce noncomputed images and those that produce computed images. For example, film-based radiographs involve no computation. Even when the film is digitized into pixels, the image is still noncomputed since the resulting digital image is only a numerical representation of the original image. However, a tomographic image such as CT is clearly computed (HANEY ET AL.).

Radiology. Much of the experience in image databases has and is occurring in radiology departments of hospitals, and much developmental work is also taking place in such companies as Philips, AT&T, IBM, and General Electric. LUNIN ET AL. note that an academic radiology department servicing a 1,200-bed hospital generates about 2,000 megabytes of images a day. The proportion of digital-to-analog diagnostic exams has been increasing. These images and their accompanying clinical text and data result from examinations in nuclear medicine, CT, digital fluorography, roentgenology, and recently, magnetic resonance imaging. As more and more images are generated, many of them in digital form, they are beginning to exceed the capacity of traditional forms of interpretation, interim storage, retrieval, review, and archiving.

Medical image database. Image formats reflect the multidimensional nature of the image information. A header record precedes the data records and specifies the number of views, number of lines, and number of samples per line along with several other descriptive parameters. An image is represented by filling each data record with one raster line of values and incrementing the line index for the next record. If several images are to be stored in one file, they can be assigned to different view angles (HANEY ET AL.).

Several investigators have looked at database structure for storage and retrieval. CAHILL ET AL. describe an intelligent database for image analysis, storage, and retrieval in radiology. The database was required to perform nonvisual searches of radiologic images vs. analog modalities of image storage, retrieval, archiving, and a comparison of the image quality produced by one modality with another. STEFANEK & CHANG report a method for incorporating numerical data derived from images into a relational pictorial database that is virtual. ACKERMAN instituted the use of AI to image understanding and attempts to use knowledge about the domain of the image to analyze it. He advised that the practicality of this approach is a long way off.

VAUGHAN ET AL. report that an image database can be implemented with a general-purpose commercially available database. The database should contain image, demographic, and quantitative data. They adopted a common file format for images called Medical Image Format (MIF), derived from

Landsat image processing systems. The image database implementation described proved viable for retrieving images in both the clinical and research environment.

DEVALK ET AL. view the problem of medical image storage and propose the use of optical disks in the prototype Image Information System intended as an adjunct to the Dutch Hospital Information Systems (Dutch Images project). Some major problems concern volume of data, storage structure and strategy, lack of image format standardization, image processing speed and methods, network capacity, and cost/benefit analysis.

Not all images are equivalent in terms of their storage and retrieval requirements; some contain more diagnostically significant information than others. There is a temptation to archive any image that may be important. As JAMES ET AL. note, questions concerning the design and management of the system arise: how are the choices made, who makes them, and what are the legal ramifications? The design of hardware for the storage system and of software for data compression also becomes an important factor in implementing any digital system. Proponents of digital image storage and retrieval systems emphasize the ability of digital hardware to handle a large volume of image information. Yet quite often, as the authors point out, a promising technology fails because the system designer is naive about the specialized environment in the health care community. For example, DBMS designed for businesses that operate during normal hours are rarely successful in handling patient care data. As JAMES ET AL. emphasize, "Medical, legal, and ethical aspects of the use of digital technology in the storage, manipulation, and retrieval of medical image data are important new problems facing hospitals and radiologists" (p. 851).

Image management systems. A picture archiving and communication system (PACS) is an image management system rather than an image processing system (MARCEAU). ARENSON suggests that MIMS—medical image management system—is probably a better term. Image management refers to the acquisition, storage, retrieval, communication, and display of digital images. In a PACS (or MIMS), an image can be archived within seconds after its creation. Because the archived image is never moved, it will never be lost. Thus, a major effect of the use of digital images is to eliminate the problems introduced by human error in keeping track of pictures that must be moved in order to be used.

The requirements of a PACS are flexibility, compatibility, and speed. Picture storage is best implemented with virtual addressing for picture storage hierarchy. Picture coding must also be efficient, flexible, and consistent. Conflicts will arise between user and designer as the system is developed, and the human interface must be considered (MEYER–EBRECHT & WENDLER). Several investigators have studied aspects of image management systems, among them COX ET AL. who did work on clinical requirements for such systems and TORIWAKE & HASEGAWA who described a DBMS for an x-ray database. ROBINSON recommended that the desired diagnostic information management system should have five features. He added that at least two major technology issues must be faced before implementation of a hospital-wide diagnostic imaging management system: 1) efficient digital archiving devices capable of storing and retrieving large amounts of digital diagnostic information; 2) hospital-wide area networks capable of transferring large amounts of diagnostic imaging data among clinical departments. Although a PACS or image management system can be interfaced with a radiology or hospital information system, it will be a challenge to successfully integrate an image management system with hospital information systems since many very different information systems exist, each with its own kinds of information needs and communication protocols.

Communication systems. Picture communication systems in medicine have been described by LINDBERG, SCHEIBE, and others. These systems require the ability to transfer relatively large blocks of data as a unit. Design factors of such systems range from the choice of transmission media through details such as modulation and encoding. Lindberg describes the state of the art of computer networks within health care. Central to the many designs and experiments in this field is the concept of a distributed diagnostic image management system. Plausible systems have been worked out using a wide range of architectures and LANs including star systems, hierarchical schemes, and ring and double ring designs. Still to be determined is the cost effectiveness of such imaging networks in small and large hospitals and in behavioral aspects of the expected diffusion of this technology.

SIEBERT ET AL. described a broadband image communication network linking a central image display and interpretation center to three CT scanners located in separate community hospitals. They review channel characteristics and experience to date.

Comment

Applied research and development of image information systems is occurring in a number of applications in many different fields, principally where many images must be accessed quickly and easily. The applications range from scholarly archival studies to life-saving events in hospitals. Common to all are the need to organize, manage, and communicate the image information. The major impetus for the increase in image databases is the vast storage capability of the optical disk. Other motivations are provided by image preservation and conservation issues. Lagging far behind the technology is the development of new subject access methods and languages specifically developed to retrieve images.

SYSTEMS DESIGN

Designing an image processing application should follow the general fundamental procedures advocated for any system development: planning, design, installation, and production support. O'CONNOR outlines specific critical decisions that must be made.

CHANG & YANG describe the two basic tasks facing designers of image information systems. The first is the storage, retrieval, and processing of many images; the second is the storage, retrieval, and processing of very large or very complex images. Also to be considered is the intended use of an image information system—i.e., whether it is intended mainly to retrieve images or to process and manipulate pictures. The authors explain that these considerations can lead to the design of entirely different systems to support the intended application. Common to these design problems are two central issues: 1) how to provide a unified approach to retrieve and manipulate image information; and 2) how to use data structures to improve and develop algorithms for image information retrieval and manipulation. TAMURA & YOKOYA advise that if a system includes processed images in the database, the history of the processing operations should be included with the resulting images.

STANDARDS

A continuing barrier to rapid progress in image processing and pattern recognition research is the lack of a universal method of transferring image data between different facilities. The proliferation of different image data formats has compounded the problem. PREWITT ET AL. recommend the establishment of logical formats and standards for images and image data headers as the first step.

BAXTER ET AL. recommend a standard data format that is simple to use and expandable enough to meet future needs, one through which data can be located easily, and one for which the media are compatible with the widest possible collection of equipment. This format would go a long way toward eliminating many of the current difficulties in exchanging digital image data regardless of the medium. These authors also discuss requirements for a protocol for exchanging image information between users of possibly dissimilar equipment.

HANEY ET AL. call for a standard that is sufficiently flexible to allow for the encoding of the raw data that were used to form the image.

In office settings, the lack of compatibility standards means that disks cannot be used in another vendor's system and that for each system a different searching technique must be learned (GRIGSBY, 1986). Several authors have stressed the need for standards in art, among them DUFF, FAWCETT (1979), KIRKPATRICK, OHLGREN (1982), RINEHART, and SUNDER-LAND.

Standards for image transmission and typesetting are still needed in libraries and publishing and must be integrated with the work of the Association of American Publishers, such as their Electronic Manuscript Project, and work within ISO according to LYNCH & BROWNRIGG (1986b; 1986c). These authors say that existing de facto standards can form a basis for work in this area. In recent years, digital image representations, compression, and processing have begun to come under consideration by standards organizations. The International Telegraph and Telephone Consultative Committee (CCITT) has sanctioned a few compression schemes for use in the facsimile industry. Certain industry standards organizations are considering sanctioning specific representation and compression schemes for their applications areas.

Image compression as an application of information theory continues to be an active research area with hundreds of image-compression algorithms and variants being proposed. While researchers are primarily concerned with optimizing performance, standards organizations are also concerned with compatibility and practicality of implementation (MAURO). The industry will need standards to allow the simple and easy interface of components from various vendors into any system configuration that meets the needs of a specific organization. As IASA points out, this is a big challenge. Three different groups are working on standards for data formats and communication protocols in image systems: users, professional societies, and official standards organizations (SCHNEIDER).

TRENDS, MARKETS, PREDICTIONS

Historically, euphoria, then tempered optimism, then realism accompany the introduction of new information technologies and systems. At this time, a blend of tempered optimism and realism about image information systems pervades the journal literature, conference proceedings, and business and trade press.

Trends

FELDMAN expects a boom on the image–systems front. While the lack of interface, resolution, and data-compression standards in the image arena will continue to pose problems, he reports that technology keeps forging ahead. The personal computer (PC) has created a gateway to popular image processing, which is seen by experts as the second fastest-growing application in the microcomputer marketplace, right behind desktop publishing (HICKS). In effect, one industry is driving the other. By incorporating image processing into publishing systems, users can combine images with text in one package.

The merging of software with both traditional and nontraditional image storage hardware was called one of the most significant developments and trends of 1985 and 1986 (*MODERN OFFICE TECHNOLOGY*). This merging was considered significant because it forces vendors and customers to think in terms of applications and usefulness rather than storage media or technology.

In 1986 many image processor manufacturers were looking to introduce "imaging workstations" in the hope that third-party software vendors would come up with application-specific software (WILSON, 1986c). Still, much imaging code is yet to be written, and the next stage of image processing is expected to be based on software advances. Wilson notes that manufacturers building imaging systems for specific markets have a more difficult task than their graphics counterparts since imaging workstations will need to be supported with higher bandwidth networks, more sophisticated processors, and large memory subsystems.

In the hardware arena, HELGERSON reports recent significant improvements in scanning products, and it appears that the market—rather than a standards-setting organization—will endorse and demand levels of resolution, throughput, protocols, and data editing and manipulation standards. She expects scanning products to fill a large gap in the development of information systems; she adds that no other method of converting existing documents and visual materials to digital form is as fast or as cheap as scanners. With the development of software that can move hard-copy information—whether pictorial or text—into the digital world, OWEN expects that facsimile may become one of the most powerful productivity tools around.

"Windowing" and providing simultaneous access to document images, word processing, and mainframe data, is already a reality (MILLER). Yet integration of all of these technologies (AI, display, image manipulation, cut-and-paste capabilities, etc.) remains to be achieved.

On a less optimistic and more sobering note, F. MOORE (1986) says that digital image technology is in its infancy. Many of the available products are not ideally suited for all business needs and applications or, indeed, for those in other areas. There are many image system vendors, and too many of their products are designed to be all things to all users. Although the use of digital image technology to archive and retrieve documents is slowly gaining acceptance as a practical alternative to paper files and conventional micrographics, the implementation of such systems has been slower than originally forecast, according to HAMILTON. He says this is due to various factors, among them a lack of media standards, questionable legal stature of digital images, and courtroom evidence defining acceptable image quality. Various commercial component suppliers can provide the pieces of a document imaging system. At the low end of the market, complete systems with packaged software and hardware are rapidly becoming available from a number of suppliers. However, for the large users in government, defense, and industry, Hamilton believes that no standard commercial product will satisfy their complex and unique requirements without extensive customization.

Further notes of caution, BETTS refers to Hooton's comments that although most federal agencies are enthusiastic about an optical technology for image storage, they are taking the cautious approach of conducting pilot projects and studies to evaluate all of the technical, legal, budgetary, and managerial issues surrounding such a major change in government operations. Hospitals, too, are investigating cautiously.

Markets

Yet optimism persists concerning market potential. As reported in the *S. KLEIN COMPUTER GRAPHICS REVIEW*, a research study by Frost & Sullivan, Inc. (F&S), entitled "The U.S. Commercial Market for Image Processing Systems" identifies several markets that are ripe for imaging systems. One potential market niche ready for commercial exploitation is the creation of image libraries, such as the Landsat picture database. Opportunities also exist in all applications that can provide faster computing power at lower cost and an inexpensive means for generating photographic quality hard copy directly from computer memory. Still other opportunities relate to color displays that can reproduce the subtle shadings that so far are limited to black and white. Another is the use of standardized low-cost communication networks and inexpensive remote display devices to transmit images inexpensively over distance (*S. KLEIN COMPUTER GRAPHICS REVIEW*).

Some Predictions

The F&S study foresees a booming market for image processing systems and components over the next five years, with annual revenues projected to quadruple from $415 million in 1986 to $1.6 billion in 1990 at current dollars (*S. KLEIN COMPUTER GRAPHICS REVIEW*). The report projects growth according to seven major application categories: 1) medical, 2) remote sensing, 3) geophysical, 4) artificial vision, 5) printing and publishing, 6) graphic arts, and 7) other. STEWART predicts that the medical videodisk area alone will represent an annual software dollar volume of over $94 million in 1990.

Systems integrators are warned of a "shift in identity" as vendors in the field increasingly refer to themselves as companies that offer products in remote sensing, machine vision, electronic publishing, and so forth rather than as image processing companies per se. (The only identifiable imaging sector, the F&S study says, may be the vendors of components that are built into the imaging systems, such as video cameras and frame buffers.) Compounding the identity crisis is the growing merger between image processing and computer graphics. Market fragmentation and product specialization have dissuaded general-purpose computer companies from entering the imaging arena. Only about 15 vendors currently offer imaging system components.

Image compression will continue to be an active research area and, in the short term, will play a significant role in the development of commercially viable computer-based digital imaging systems. In the long term, however, its role is less clear; the need for compression technology may diminish with higher bandwidth networks and mass storage devices (*S. KLEIN COMPUTER GRAPHICS REVIEW*).

CONCLUSIONS

The number of scholarly and trade publications with articles on image information appears to be growing rapidly, and the number of conferences and workshops on the topic is also increasing. The progress in storage technologies, the increasing abilities of PCs to handle images, improved communication systems—as well as the increasing sales by vendors—are all growing evidence of heightened interest in and acceptance of image systems and image databases.

The need has been proclaimed. The hardware and software are being developed. Standards are being discussed. The field is ready for unifying concepts. Yet one of the main areas of concern is how to represent structural information in a database mode. This calls for thinking in terms of spatial, continuous, gray-scale information. The designs and techniques for image databases still seem to have a long way to go before users can retrieve by "language of the image" and browse among these databases easily and efficiently.

A few last words to the potential users of image databases:

> An ancient Japanese court painter skillfully painted images of fish on a wooden panel in the Imperial Palace in Kyoto. The fish were so lifelike that in the ensuing century rumor spread that they escaped from the panel at night to swim in the pond. Such behavior necessitated a later court painter to draw a net on top of the fish to prevent such behavior. This example should serve as a figurative reminder to physicians [and others] who occasionally treat images as though they had a life and a reality of their own, independent of us, the observer (JAFFE, p. 14).

> Serious consideration of optical illusions should convince the reader that objective reality is never fully accessible to the human mind and that all imaging processes, including medical, possess inherent structured or unstructured artifacts that impair the ability to access "objective" (scientific) truth in the signal (JAFFE, p. 44).

Glossary of Electronic Image Terminology

Bandwidth—the range of frequencies that a communication line can transmit.

Bit Stream—during transmission, the string in which the bits composing the characters follow each other without separation.

Digitize—the process of converting an image from its original form into binary representation of numerical quantities. In the process, the image is divided into small sections called picture elements or pixels or pels.

Encode—to convert information to machine- or computer-readable form representing individual characters in a message. A step in the process of converting an analog signal into a digital signal.

Enhancement—a technique of data manipulation that improves details or contrast of an electronic image by sharpening edges, removing blurs, increasing brightness, and so forth.

Frame Buffer—a memory device that stores the image data pixel by pixel. Frame buffers are used to refresh a grid (raster) image. See also Raster.

Frame Grabber—a device that stores one screenful of information; also called frame storer.

Grammar—a way to describe how to generate patterns by applying rules to a few symbols.

Gray Level—the numeric description of a pixel; used to determine the brightness of the corresponding point on a display screen; also spelled "grey."

Gray Scale—the number of shades of gray in an image that can be represented numerically. The number of bits available to any pixel of the series of elements (arrays) determines the number of "colors" or gray scale values that element can have.

Icon—a symbolic, pictorial representation of any function— e.g., signature, logo, and so forth.

Image—a representation, likeness, or imitation of an object.

Image Algebra—a technique used to avoid the need for the computer to compare an object's dimensions with measurements stored in memory.

Image Information—data (point, line, density, etc.) comprising the image. The word "picture" or "pictorial" is sometimes used as the equivalent of the word "image."

Look-Up Table (LUT)—a table of stored data for reference. The table can be used to transform any particular value in image memory to any arbitrarily displayed value.

PACS—Picture Archiving and Communication System. An image management system rather than an image processing system. Term used primarily in radiology.

Pixel—an abbreviation of "picture element." Also called pel. The minimum display element; represented on the screen as a point with a specified color or intensity level.

Primitive—a basic display element: point, segment, alphanumeric character, or marker.

Pyramid—a generalized image data structure that consists of the same image at successively increasing (or decreasing) levels of resolution.

Quantize—to represent a measured value of a number (integer). A step in the process of converting an analog signal into a digital signal. This step measures a sample to determine a representative numerical value that is then encoded.

Raster—a grid. Best imagined as a sheet of graph paper with each "square" or element representing a unit of information. A commonly used raster contains 512 x 512 elements or 282, 144 squares that need to be described by units of information. A raster display both stores and shows data as horizontal rows of uniform grid or pixels. See also Vector Graphics.

Resolution—the number of pixels per unit of area. A display with a fine grid contains more pixels than one with a coarse grid and thus has a higher resolution and can reproduce more detail in an image.

Sampling—measuring the gray level of an image at a pixel location. The process of taking measurable samples of an analog signal at periodic intervals. A step in the process of converting an analog signal into a digital signal. The three steps in analog-to-digital conversion are sampling, quantizing, and encoding.

Scan—the process by which a document or hard-copy image is examined in the process of conversion to machine-storable image format. See also Digitize.

Scanner—a device for examining printed characters or graphics and representing them as electronic signals.

Scanning—the process of reading data by regular incremental horizontal sweeps to cover the entire image or screen.

Segmentation—the process that divides an image into objects or regions of interest. Segmentation algorithms are usually based on either discontinuity or similarity.

Semantic Network—represents objects and relationships between objects as a graph structure of nodes and arcs. The arcs usually represent relations between nodes.

Spatial Data—Often refers to the distribution of a variable or the relationships among variables in a geographic region (e.g., demographic features, marketing distribution). Also called locational data.

Thresholding—a process in which a particular shade of gray is selected, manually or automatically, and all pixels lighter than that are converted to white and all darker ones are converted to black.

Vector Graphics—line drawing. A vector display device stores and displays data as line segments identified by the x,y coordinates of their end points. See also Raster.

Zoom—to scale a display so that it is magnified or reduced on the screen.

BIBLIOGRAPHY

ABBOTT, EDWIN A. 1952. Flatland: A Romance of Many Dimensions. 6th ed., rev. New York, NY: Dover Publications, Inc.; 1952. 103p. ISBN: 0-88307-571-7.

ABBOTT, G. L. 1985. Video-Based Information Systems in Academic Library Media Centers. Library Trends. 1985 Summer; 34: 151-159. ISSN: 0024-2594.

ACKERMAN, LAWRENCE. 1986. Toward Automated Image Analysis: Future Possibilities in Historical Perspective. Radiologic Clinics of North America. 1986 March; 24(1): 79-85. ISSN: 0033-8389.

ADELSON, EDWARD H.; ANDERSON, C. H.; BERGEN, J. R.; BURT, P.J.; OGDEN, J. M. 1984. Pyramid Methods in Image Processing. RCA Engineer. 1984 November-December; 29(6): 33-41.

ALEKSANDER, I. 1982. Modern Pattern Recognition and the Classification of Works of Art. Art Libraries Journal. 1982 Summer; 7(2): 61-66. ISSN: 0307-4722.

ARENSON, R. L. 1986. Teaching with Computers. Radiologic Clinics of North America. 1986 March; 24(1): 97-103. ISSN: 0033-8389.

ASTHANA, R. G. S.; CHING, J. 1984. A Campus Query System: Image to Database Correspondence. In: Proceedings of the International Conference on Systems, Man and Cybernetics: Volume 2; 1983 December 29: 1984 January 7; Bombay and New Delhi, India. New York, NY: IEEE; 1984. 995-999.

BABAOGLU, GOKALP. 1987. Image Processing Enters the PC Arena. Washington Technology. 1987 January 8; 1(21): 10.

BALLARD, DANA H.; BROWN, CHRISTOPHER M. 1982. Computer Vision. Englewood Cliffs, NJ: Prentice-Hall; 1982. ISBN: 0-13-165316-4.

BARKER, P. G.; NAJAH, M. 1985. Pictorial Interfaces to Data Bases. International Journal of Man-Machine Studies. 1985 October; 23(4): 423-442. ISSN: 0020-7373.

BARTHES, ROLAND. 1983. Image, Music, Text. Translated by Stephen Heath. New York, NY: Hill and Wang; 1983. 220p. ISBN: 0-8090-1387-8.

BATTY, D.; STEVENS, P. 1982. Automated Retrieval Systems for Photo-Image Collections: Problems and a Solution. In: Petrarca, Anthony; Taylor, Celianna I.; Kohn, Robert S., eds. Information Interaction: Proceedings of the American Society for Information Science (ASIS) 45th

Annual Meeting; 1982 October 17-21; Columbus, OH. White Plains, NY: Knowledge Industry Publications, for ASIS; 1982. 23-25. ISSN: 0044-7870; ISBN: 0-86729-038-2.

BAXTER, BRENT; HITCHNER, LEWIS; MAGUIRE, GERALD, JR. 1982. Characteristics of a Protocol for Exchanging Digital Image Information. In: Duerinckx, Andre J., ed. Proceedings of the Society of Photo-Optical Instrumentation Engineers-The International Society for Optical Engineering (SPIE) 1st International Conference and Workshop on Picture Archiving and Communication Systems (PACS) for Medical Applications: Volume 318, Part I; 1982 January 18-21; Newport Beach, CA. Bellingham, WA: SPIE; 1982. 273-277. ISBN: 0-89252-352-2; CODEN: PSISDG.

BETTS, MITCH. 1986. Federal Agencies Driving Market for Optical Disk Storage. Computerworld. 1986 April 14; 13. ISSN: 0010-4841.

BIRENBAUM, ROBERT I.; KING, DAVID R. 1986. VME Imaging Boards Target Machine Vision. Digital Design. 1986 May; 16(4): 28-30, 32. ISSN: 0147-9245.

BOERHOUT, JAN. 1986. Adaptive Filtering of Digital Video Signals for Image Enhancement. Digital Design. 1986 March 25; 16(4): 61-62, 64. ISSN: 0147-9245.

BOWER, JAMES M.; ALVEY, CELINE. 1985. Toward Automation- Implementing Standards in Visual Resource Collections. In: Abstracts, Post Conference Workshop, Art Libraries Society of North America 13th Annual Conference; 1985 February 8-14; Los Angeles, CA. Tucson, AZ: Art Libraries Society of North America; 1985. 39-41.

BOYNE, WALTER J.; OTANO, HERNAN. 1984. Direct Document Capture and Full Text Indexing. An Introduction to the National Air and Space Museum System. Library Hi Tech. 1984; 2(4): 7-14. ISSN: 0737-8831.

BROUAYE, P.; PUDET, T.; VICARD, J. 1984. Managing the Semantic Content of Graphical Data. In: Gardarin, G.; Gelenbe, E., eds. New Applications of Data Bases. London, England: Academic Press; 1984. 63-84. ISBN: 0-12-275550-2.

BROWN, C. M. 1984. Computer Vision and Natural Constraints. Science. 1984 June 22; 224(4655): 1299-1305. ISSN: 0036-8075.

BUSINESS WEEK. 1984. Machines That Can See: Here Comes a New Generation. Business Week. 1984 January. 9; 118-119, 121. ISSN: 0007-7135.

BUSINESS WEEK. 1986. Getting the Right Picture by Videodisk. Business Week. 1986 November 3; 1401. ISSN: 0007-7135.

CAHILL, P. T.; KNEELAND, B.; KNOWLES, R. J. R.; LUNIN, L. F.; TSEN, O. 1983. An Intelligent Data Base for Image Analyses, Storage, and Retrieval in Radiology. In: Williams, Martha E.; Hogan, Thomas H., comps. Proceedings of the 4th National Online Meeting; 1983 April 12-14; New York, NY. Medford, NJ: Learned Information, Inc.; 1983. 83-86. ISBN: 0-938734-05-9.

CANNON, T. M.; HUNT, B. R. 1981. Image Processing by Computer. Scientific American. 1981 October; 245: 214-225. ISSN: 0036-8733.

CASTLEMAN, KENNETH R. 1979. Digital Image Processing. Englewood Cliffs, NJ: Prentice-Hall, Inc.; 1979. 429 p. ISBN: 0-13-212365-7.

CERVA, JOHN R.; STRONG, HARRY M. 1985. Optical Storage Technology in an Integrated Information and Imaging Network. In: Proceedings of the Videodisc, Optical Disk, and CD-ROM Conference & Exposition; 1985 December 9-12; Philadelphia, PA. Westport, CT: Meckler Publishing; 1985. 43-48. ISBN: 0-88736-053-X.

CHANG, N. S.; FU, K. S. 1981. Picture Query Languages for Pictorial Data Base Systems. Computer. 1981 November; 14 (11): 23-33. ISSN: 0018-9162.

CHANG, S. E.; KUNII, T. L. 1981. Pictorial Data Base Systems. IEEE Transactions on Computers. 1981 November; 14 (11): 13-21. ISSN: 0018-9340.

CHANG, S. K.; LIN, B. S.; WALSER, R. 1979. A Generalized Zooming Technique for Pictorial Database Systems. In: Proceedings of the American Federation for Information Processing Societies (AFIPS), National Computer Conference; 1979 June 4-7; New York, NY. Montvale, NJ: AFIPS Press; 1979. 147-156.

CHANG, SHI-KUO; LIU, SHAO-HUNG. 1984. Picture Indexing and Abstraction Techniques for Pictorial Databases. IEEE Transactions on Pattern Analysis and Machine Intelligence. 1984 July; PAMI-6(4): 475-484. ISSN: 0162-8828.

CHANG, SHI-KUO; YANG, CHUNG-CHUN. 1986. A Modular Software for Image Information Systems. In: Wegman, Edward J.; DePriest, Douglas J., eds. Statistical Image Processing and Graphics. New York, NY: Marcel Dekker, Inc.; 1986. 127-144. (From a Workshop, 1983 May; Luray, VA). ISBN: 0-8247-7600-3.

CHASE, MITCHELL. 1986. Pipelined Image Processor IC Speeds Imaging Tasks. Digital Design. 1986 March 25; 16(4): 73-74. ISSN: 0147-9245.

CHENG, GEORGE C.; LEDLEY, ROBERT S. 1968. A Theory of Picture Digitization and Application. In: Chang, George C.; Ledley, Robert S.; Pollock, Donald K.; Rosenfeld, Azriel, eds. Pictorial Pattern Recognition: Proceedings of Symposium on Automatic Photointerpretation; 1967 May 31-June 2; Washington, DC. Washington, DC: Thompson Book Co.; 1968. 329-352. LC: 68-21811.

CHOCK, M.; CARDENAS, A. F.; KLINGER, A. 1984. Database Structure and Manipulation Capabilities of a Picture Database Management System (PIC DMS). IEEE Transactions on Pattern Analysis and Machine Intelligence. 1984 July; PAMI-6(4): 484-492. ISSN: 0162-8828.

COCHRANE, PAULINE A. 1984. Modern Subject Access in the Online Age. American Libraries. 1984 June; 15(6): 438-441, 443. ISSN: 0002-9769.

COHEN, Y.; LANDY, M. S.; PAVEL, M. 1985. Hierarchical Coding of Binary Images. IEEE Transactions on Pattern Analysis and Machine Intelligence. 1985; 7(3): 284-298. ISSN: 0162-8828.

COOK, RICK. 1986. Wanted: A Good Graphics DBMS. Computer Decisions. 1986 October 21; 18(23): 66-68, 69. ISSN: 0010-4558.

COX, GLENDON G.; TEMPLETON, ARCH W.; DWYER, SAMUEL J., III. 1986. Digital Image Management: Networking, Display, and Archiving. Radiologic Clinics of North America. 1986 March; 24(1): 37-54. ISSN: 0033-8389.

CREHANGE, M.; HADDOU, A. AIT; BOUKAKIOU, M.; DAVID, J. M.; FOUCAUT, O.; MAROLDT, J. 1984. Exprim: An Expert System to Aid in Progressive Retrieval from a Pictorial and Descriptive Database. In: Gardarin, G.; Gelenbe, E., eds. New Applications of Data Bases. London, England: Academic Press; 1984. 43-61. ISBN: 0-12-275550-2.

CRISMAN, JILL. 1986. Machine Perception. UNIX Review. 1986 September; 4(9): 36-39. ISSN: 0742-3136.

DANE, WILLIAM J. 1983. Networking and an Art Library. Drexel Library Quarterly. 1983 Summer; 19(3): 52-65. ISSN: 0012-6160.

DAVE, J. V. 1985. Digital Image Manipulation, Analyses and Processing Systems (DIMAPS). A Research Oriented, Experimental Image-Processing System. In: Applications of Artificial Intelligence II: Proceedings of the Society of Photo-Optical Instrumentation Engineers (SPIE): Volume 548. Bellingham, WA: SPIE—The International Society for Optical Engineering; 1985. 100-109. ISSN: 0361-0748.

DAVE, JITENDRA V.; BERNSTEIN, RALPH; KOLSKY, HARWOOD G. 1982. Importance of Higher-Order Components to Multispectral Classification. IBM Journal of Research and Development. 1982 November; 26(6): 715-723. ISSN: 0018-8646.

DAVIS, ANDREW W.; BERMAN, ARI P. 1986. Image Filtering in the Temporal Domain. Digital Design. 1986 November; 65-66, 68-69. ISSN: 0147-9245.

DE LAURIER, NANCY. 1982. Visual Resources: The State of the Art. Art Libraries Journal. 1982 Autumn; 7(3): 7-21. ISSN: 0307-4722.

DEVALK, J. P. S.; VANRIJINS, R. C.; BAKKER, A. R. 1985. Simulation of a Feasible Medical Image Storage Hierarchy Within the Dutch Images Project. Proceedings of the Society of Photo-Optical Instrumentation Engineers. 1985; 529: 240-246. ISSN: 0361-0748.

DICKMAN, JEFFREY L. 1984. An Image Digitising and Storage System for Use in Rock Art Research. Rock Art Research. 1984 May; 1(1): 25-35. ISSN: 0813-0426.

DIGITAL DESIGN. 1986. Pyramid Processing Redefines Machine Vision. Digital Design. 1986 May; 7: 34-36, 39. ISSN: 0147-9245.

DON, HON-SON; FU, KING-SUN. 1985. A Syntactic Method for Image Segmentation and Object Recognition. Pattern Recognition. 1985; 18(1): 73-87. ISSN: 0031-3203.

DON, HON-SON; FU, KING-SUN. 1986. A Parallel Algorithm for Stochastic Image Segmentation. IEEE Transactions on Pattern Analysis and Machine Intelligence. 1986 September; PAMI-8: 594-603. ISSN: 0162-8828.

DOSZKOCS, TAMAS. 1984. [untitled comments]. In: Cochrane, Pauline A. Modern Subject Access in the Online Age. American Libraries. 1984 June; 15(6): 438-441, 443. ISSN: 0002-9769.

DUFF, MEG. 1984. Second International Conference on Automatic Processing of Art History Data: A Report. Art Libraries Journal. 1984 Autumn-Winter; 9: 3-7. ISSN: 0307-4722.

ESKIND, ANDREW. 1985. Videodisc and Computer in Tandem: The Ultimate Finding Aid for Large Collections of Visual Material. In: Abstracts of the Art Libraries Society of North America 13th Annual Conference; 1985 February 8-14; Los Angeles, CA. Tucson, AZ: Art Libraries Society of North America; 1985. 21.

FAWCETT, TREVOR. 1979. Subject Indexing in the Visual Arts. Art Libraries Journal. 1979 Spring; 4: 5-17. ISSN: 0307-4722.

FAWCETT, TREVOR. 1982. Control of Text and Images: Tradition and Innovation. Art Libraries Journal. 1982 Summer; 7(2): 7-16. ISSN: 0307-4722.

FELDMAN, STEVEN. 1985. A Boom is Imminent on the Image-Systems Front. Systems and Software. 1985 September; 35-36, 39.

FIELDS, C. 1981. User Interfaces for Pictorial Databases. In: Proceedings of the 7th International Conference on Very Large Data Bases; 1981 September 9–11; Cannes, France. p. 180. Available from: IEEE, New York, NY.

FISCHLER, M. A. 1982. The SRI Image Understanding Program. In: Image Understanding: Proceedings of a Workshop; 1982 September 15-16; Palo Alto, CA. Springfield, VA: NTIS, 1982. (AD-A120 072; AD-P000 108/1).

FLUTY, STEVE. 1986. A New Image for Office Automation. Journal of Information & Image Management. 1986 November; 19(11): 8, 47. ISSN: 0745-9963.

FREEMAN, ROBERT R. 1985. Global Information Systems and the Application of Environmental Satellite Data for Marine Information Services, A Prototype for New Information Analysis Centers. In: Parkhurst, Carol A., ed. Proceedings of the American Society for Information Science (ASIS) 48th Annual Meeting; 1985 October 20-24; Las Vegas, NV. White Plains, NY: Knowledge Industry Publications, Inc. for ASIS; 1985. 170-181. ISSN: 0044-7870; ISBN: 0-86729-176-1.

FREEMAN, ROBERT R.; SMITH, MONA F. 1986. Environmental Information. In: Williams, Martha E., ed. Annual Review of Information Science and Technology: Volume 21. White Plains, NY: Knowledge Industry Publications, Inc. for ASIS; 1986. 241-305. ISSN: 0066-4200; ISBN: 0-86729-209-1.

FRENTZ, WILL. 1986. Graphics Display Controllers in Image Processing. Digital Design. 1986 March 25; 16(4): 45. ISSN: 0147-9245.

GALE, JOHN C. 1984. Use of Optical Disks for Information Storage and Retrieval. Information Technology and Libraries. 1984 December; 3(4): 379-382. ISSN: 0730-9295.

GARDARIN, G. 1984. Towards the Fifth Generation of Data Management Systems. In: Gardarin, G.; Gelenbe, E., eds. New Applications of Data Bases. London, England: Academic Press; 1984. 3-15. ISBN: 0-12-275550-2.

GELENBE, EROL. 1984. New Applications of Data Bases, Preface. In: Gardarin, G.; Gelenbe, E. eds. New Applications of Data Bases. London, England: Academic Press; 1984. vii-ix. ISBN: 0-12-275550-2.

GEOGRAPHIC TECHNOLOGY, INC. 1986. An Executive Introduction to the Geographic Data Management System. 1986. 7p. Available from: Geographic Technology, Inc., 141-B East Laurel Road, Bellingham, WA 98226.

GITLIN, JOSEPH N. 1986. Teleradiology. Radiologic Clinics of North America. 1986 March; 24(1): 55-68. ISSN: 0033-8389.

GLASS, A. M. 1984. Materials for Optical Information Processing. Science. 1984 November 9; 226 (4675): 657-662. ISSN: 0036-8075.

GONZALEZ, R. C.; WOODS, R. E.; SWAIN, W. T. 1986. Digital Image Processing: An Introduction. Digital Design. 1986 March 25; 16(4): 15-16, 18, 20. ISSN: 0147-9245.

GREENHALGH, MICHAEL. 1982. New Technologies for Data and Image Storage and Their Application to the History of Art. Art Libraries Journal. 1982 Summer; 7: 67-81. ISSN: 0307-4722.

GRIGSBY, MASON. 1985. Micrographic and Optical Disk Document Image Processing in the Automated Office. In: Proceedings of the Videodisc, Optical Disk, and CD-ROM Conference & Exposition; 1985

December 9-12; Philadelphia, PA. Westport, CT: Meckler Publishing; 1985. 77-81. ISBN: 0-88736-053-X.

GRIGSBY, MASON. 1986. Merging Micrographics and MIS. Infosystems. 1986; 5: 94-97. ISSN: 0364-5533.

GROSKY, W. I. 1982. Towards a Data Model for Integrated Pictorial Databases. In: Proceedings of the Workshop on Computer Vision; Representation and Control; 1982 August 23-25; Rindge, N. H. 135-139. Available from: the author, Dept. of Computer Science, Wayne State University, Detroit, MI.

HALL, GENE. 1986. Large-Kernel Convolutions in Image Processing. Digital Design. 1986 August; 16: 46-48. ISSN: 0147-9245.

HAMILTON, WALTER G. 1986. Developing a Document Image Processing System: The Systems Integrator's Role. In: Roth, Judith Paris, comp. Optical Information Systems '86; 1986 December 9-11; Arlington, VA. Westport, CT: Meckler Publishing Corp.; 1986. 142-144. ISBN: 0-88736-113-7.

HANEY, M. J.; JOHNSON, R. L; O'BRIEN, W. D., JR. 1982. On Standards for the Storage of Images and Data. In: Duerinckx, Andre J., ed. Proceedings of the Society of Photo-Optical Instrumentation Engineers-The International Society for Optical Engineering (SPIE) 1st International Conference and Workshop on Picture Archiving and Communication Systems (PACS) for Medical Applications: Volume 318, Part I; 1982 January 18-21; Newport Beach, CA. Bellingham, WA: SPIE; 1982. 294-297.

HARALICK, R. M.; SHAPIRO, L. G. 1985. Image Segmentation Techniques. Computer Vision, Graphics and Image Processing. 1985; 29(1): 100-132. ISSN: 0734-189X.

HELGERSON, LINDA W. 1987. Market Trends for Scanning Systems. Bulletin of the American Society for Information Science. 1987 December-January; 13(2): 16-17. ISSN: 0095-4403.

HELMS, GEORGE C. 1986. Digital Vision: The Camera as an Imaging Peripheral. Journal of Information & Image Management. 1986 November; 19(11): 30-33. ISSN: 0745-9963.

HENDLEY, TONY. 1986. Two Hundred Attend Optical Storage Conference. Optical Information Systems. 1986 July/August; 6(4): 306-309. ISSN: 0886-5809.

HICKS, PATRICIA. 1986. PC Spurs Growth in Image Processing and Media Cybernetics. Washington Technology. 1986 July 10; 1(8): 17. Available from: 1953 Gallows Road, Suite 130, Vienna, VA 22180.

HOFFOS, SIGNE. 1986. Medical and Health Care Applications in the British Market. Optical Information Systems. 1986 July/August; 6(4): 303-305. ISSN: 0886-5809.

HOOTON, WILLIAM L. 1985. The Optical Digital Image Storage System (ODISS) at the National Archives and Records Administration. In: Proceedings of the 1985 Videodisc, Optical Disk, and CD-ROM Conference & Exposition; 1985 December 9-12; Philadelphia, PA. Westport, CT: Meckler Publishing; 1985. 90-94. ISBN: 0-88736-053-X.

HOOTON, WILLIAM L. 1986. An Update on the Optical Digital Image Storage System (ODISS) at the National Archives. In: Roth, Judith Paris, comp. Optical Information Systems '86; 1986 December 9-11; Arlington, VA. Westport, CT: Meckler Publishing Corp.; 1986. 153-157. ISBN: 0-88736-113-7.

HWANG, K.; FU, K. 1983. Integrated Computer Architectures for Image Processing and Database Management. Computer. 1983 January; 16(1): 51–59. ISSN: 0018-9162.

INSURANCE ACCOUNTING SYSTEMS ASSOCIATION (IASA). 1986. Image Processing: Document Imaging for the Future. Journal of Information & Image Management. 1986 November; 19(11): 10–21, 46, 47. ISSN: 0745-9963.

IYENGAR, S. SITHARAMA; MILLER, STEPHAN W. 1986. Efficient Algorithm for Polygon Overlay for Dense Map Image Data Sets. Image and Vision Computing. 1986 August; 4(3): 167–174. ISSN: 0262-8856.

JACK, R. F. 1984. ERDS Main Image File: A Picture Perfect Database for Landsat Imagery and Aerial Photography. Database. 1984 February; 7(1): 35–52. ISSN: 0162-4105.

JAFFE, C. CARL. 1984. Medical Imaging, Vision, and Visual Psychophysics. Medical Radiography and Photography. 1984; 60(1): 1–48. ISSN: 0025-746X.

JAMES, A. EVERETTE, JR.; CARROLL, FRANK; PICKENS, DAVID R. III; CHAPMAN, JOHN C.; ROBINSON, ROSCOE R.; PENDERGRASS, HENRY P.; ZANER, RICHARD. 1986. Medical Image Management. Radiology. 1986 September; 100(3): 847–851. ISSN: 0033-8419.

KAMISHER, LISA M. 1985. The Images System: Videodisc and Database Integration for Architecture. In: Proceedings of the 1985 Videodisc, Optical Disk, and CD-ROM Conference & Exposition; 1985 December 9–12; Philadelphia, PA. Westport, CT: Meckler Publishing; 1985. 104–108. ISBN: 0-88736-053-X.

KIRKPATRICK, NANCY. 1982. Major Issues of the Past 10 Years in Visual Resources Curatorship. Art Libraries Journal. 1982 Winter; 7: 30–35. ISSN: 0307-4722.

KOTELLY, GEORGE V. 1986. U.S. Agency Assesses Disk Storage Industry. Mini-Micro Systems. 1986 February; 19: 5. ISSN: 0364-9342.

KRAYESKI, FELIX. 1984. Transition of an Image System—From Paper to Microfiche to Optical Disk. Journal of Imaging Technology. 1984; 10: 161–162. ISSN: 0747-3583.

KRAYESKI, FELIX. 1985. An Overview of Optical Digital Disk Applications. In: Proceedings of the Videodisc, Optical Disk, and CD-ROM Conference and Exposition; 1985 December 9–12; Philadelphia, PA. Westport, CT: Meckler Publishing; 1985. 109–110. ISBN: 0-88736-053-X.

KRAYESKI, FELIX. 1986. Image Processing and Optical Disk Technology at the Library of Congress Research Service. Optical Information Systems. 1986 March-April; 6(2): 120–122. ISSN: 0886-5809.

KRAYESKI, FELIX; SWORA, TAMARA. 1985. Image Processing and Optical Disk Technology at the Library of Congress and the Congressional Research Service. In: Proceedings of the Videodisc, Optical Disk, and CD-ROM Conference & Exposition; 1985 December 9–12; Philadelphia, PA. Westport, CT: Meckler Publishing; 1985. 111–118. ISBN: 0-88736-053-X.

KUNDEL, H. L. 1986. Visual Perception and Image Display Terminals. Radiologic Clinics of North America. 1986 March; 24(1): 69–78. ISSN: 0033-8389.

KUNT, M.; IKONOMOP, A.; KOCHER, N. 1985. Second Generation Image-Coding Techniques. Proceedings of the IEEE. 1985; 73(4): 549-574. ISSN: 0018-9219.

LAMIELLE, JEAN-CLAUDE; DE HEAULME, MICHEL. 1986. A General Bank of Referential Images Using Laservision Disk in Bio-Medicine. In: Roth, Judith Paris, comp. Optical Information Systems '86; 1986 December 9-11; Arlington, VA. Westport, CT: Meckler Publishing Corp.; 1986. 184-191. ISBN: 0-88736-113-7.

LASERVIEW. 1986. System Overview. LaserData. 1986 July 22; Rev. 1-1. Available from: LaserData, 10 Technology Drive, Lowell, MA 01851.

LIN, B. S.; CHANG, S. K. 1979. Picture Algebra for Interface with Pictorial Database Systems. In: Proceedings of the IEEE Computer Society's 3rd International Computer Software and Applications Conference (COMPSAC); 1979 November 6-8; Chicago, IL. New York, NY: IEEE; 1979. 525-530.

LINDBERG, DONALD A. B. 1982. Computer Networks Within Health Care: The State of the Art in the U.S.A. In: Peterson, H. E.; Isaksson, A. J., eds. Communication Networks in Health Care. Amsterdam, The Netherlands: North Holland Publishing Co.; 1982. 109-120. ISBN: 0-669-02911-4.

LITA NEWSLETTER. 1986. The Potential Impacts of Optical Disk Technology on the Online Catalog: A Discussion. LITA Newsletter. 1986; no. 25: 1-2. ISSN: 0196-1799.

LUNIN, LOIS F.; CAHILL, PATRICK T.; AUH, JONG; LEE, BENJAMIN C. P.; BECKER, DAVID V. 1982. Organizing for Information Inter-action in a Radiology Department: Focus on Image Analysis, Storage and Retrieval. In: Petrarca, Anthony; Taylor, Celianna; Kohn, Robert, eds. Proceedings of the American Society for Information Science (ASIS) 45th Annual Meeting; 1982 October 17-21; Columbus, OH. White Plains, NY: Knowledge Industry Publications, Inc. for ASIS; 1982. 179-181. ISSN: 0044-7870; ISBN: 0-86729-038-2.

LUNIN, LOIS F.; PARIS, JUDITH, eds. 1983. Perspectives on . . . Video-disc and Optical Disk: Technology, Research, and Applications. Intro-duction and Overview. Journal of the American Society for Information Science. 1983 November; 34(6): 406-407. ISSN: 0002-8231.

LYNCH, CLIFFORD A.; BROWNRIGG, EDWIN B. 1985. Document Delivery and Packet Facsimile. In: Parkhurst, Carol A., ed. Proceedings of the American Society for Information Science (ASIS) 48th Annual Meeting; 1985 October 20-24; Las Vegas, NV. White Plains, NY: Knowl-edge Industry Publications, Inc. for ASIS; 1985. 11-14. ISSN: 0044-7870; ISBN: 0-86729-176-1.

LYNCH, CLIFFORD A.; BROWNRIGG, EDWIN B. 1986a. Conservation, Preservation, and Digitization. College and Research Libraries. 1986 July; 47(4): 379-382. ISSN: 0010-0870.

LYNCH, CLIFFORD A.; BROWNRIGG, EDWIN B. 1986b. Electronic Publishing, Electronic Imaging and Document Delivery. Boston, MA: Institute for Graphic Communication, Inc.; 1986 November. 662-667.

LYNCH, CLIFFORD A.; BROWNRIGG, EDWIN B. 1986c. Library Appli-cations of Electronic Imaging Technology. Information Technology and Libraries. 1986 June; 5(2): 100-105. ISSN: 0730-9295.

MACDONALD, BRUCE J. 1986. CDROM Makers Near Consensus on Standards. Mini-Micro Systems. 1986 March; 19: 36, 38, 40. ISSN: 0364-9342.

MANNS, BASIL; SWORA, TAMARA. 1986. Digital Imaging at the Library of Congress. Journal of Information & Image Management. 1986 October; 19(10): 27-32. ISSN: 0745-9963.

MARCEAU, CARLA. 1982. What is a Picture Archiving and Communication System (PACS)? In: Duerinckx, Andre J., ed. Proceedings of the Society of Photo-Optical Instrumentation Engineers–The International Society for Optical Engineering (SPIE) 1st International Conference and Workshop on Picture Archiving and Communication Systems (PACS) for Medical Applications: Volume 318, Part I; 1982 January 18-21; Newport Beach, CA. Bellingham, WA: SPIE; 1982. 24-29. ISBN: 0-89252-352-2; CODEN: PSISDG.

MARKEY, KAREN. 1984. Visual Arts Resources and Computers. In: Williams, Martha E., ed. Annual Review of Information Science and Technology: Volume 19. White Plains, NY: Knowledge Industry Publications, Inc.; 1984. 271-309. ISSN: 0066-4200; ISBN: 0-86729-093-5.

MATULLO, GENE. 1985. Advanced Document Image Processing. In: Proceedings of the 1985 Videodisc, Optical Disk, and CD-ROM Conference & Exposition; 1985 December 9-12; Philadelphia, PA. Westport, CT: Meckler Publishing; 1985. 130-135. ISBN: 0-88736-053-X.

MAURO, JOSEPH. 1986. Image Compression Cuts Imaging Problems Down to Size. Digital Design. 1986 April; 40-43. ISSN: 0147-9245.

MCKEOWN, DAVID M. JR. 1982. Concept Maps. In: Image Understanding: Proceedings of a Workshop; 1982 September 15-16; Palo Alto, CA. AD-A120072.

MCKEOWN, DAVID M. JR. 1983. MAPS: The Organization of a Spatial Database System Using Imagery, Terrain, and Map Data. In: Image Understanding: Proceedings of the 14th Workshop; 1983 June 23; Arlington, VA. 23p. Available from: NTIS, Springfield, VA. AD-A130251.

MCKEOWN, DAVID M. JR. 1984. Digital Cartography and Photo Interpretation From a Database Viewpoint. In: Gardarin, G.; Gelenbe, E., eds. New Applications of Data Bases: Proceedings of the ICOD-2 Workshop; 1983 September 2-3; Cambridge, UK. London, England: Academic Press; 1984. 19-42. ISBN: 0-12-275550-2.

MEIER, A. 1985. A Graph Grammar Approach to Geographical Databases. Information Systems. 1985; 10(1): 9-19. ISSN: 0306-4379.

MEYER-EBRECHT, D.; WENDLER, T. 1983. An Architectural Route Through PACS. Computer. 1983 August; 16(8): 19-28. ISSN: 0018-9162.

MILLER, TIM. 1986. Hybrid Information Systems: There's More to Imaging Than Meets the Eye. Journal of Information & Image Management. 1986 November; 19(11): 27-29. ISSN: 0745-9963.

MODERN OFFICE TECHNOLOGY. 1986. 86: The Year of the Optical Disk. Modern Office Technology. 1986 January; 31(1): 118, 120. ISSN: 0746-3839.

MOORE, CONNIE. 1986a. Image Processing Offers MIS A New View of Information. Computerworld. 1986 June 23; 69-70, 72-75, 79, 82-84. ISSN: 0010-4841.

MOORE, CONNIE. 1986b. Digital Image Processing in Private Industry. Optical Information Systems. 1986 May–June; 6(3): 224–229. ISSN: 0886-5809.

MOORE, FRANK. 1985. The Files Archival Image Storage and Retrieval System. In: Proceedings of the Videodisc, Optical Disk, and CD-ROM Conference & Exposition; 1985 December 9–12; Philadelphia, PA. Westport, CT: Meckler Publishing; 1985. 147–149. ISBN: 0-88736-053-X.

MOORE, FRANK. 1986. The Files Archival Image Storage and Retrieval System (FAISR) at the IRS. Optical Information Systems. 1986 March-April; 6(2): 127–129. ISSN: 0086-5809.

MOSKOWITZ, ROBERT A. 1985. Imaging Systems: Present Imperfect. Computer Decisions. 1985 January 4; 17(11): 36, 40. ISSN: 0010-4558.

MUTRUX, ROBIN; ANDERSON, JAMES D. 1983. Contextual Indexing and Faceted Taxonomic Access System. Drexel Library Quarterly. 1983 Summer; 19: 91–109. ISSN: 0012-6160.

NAGY, G. 1985. Image Database. Image and Vision Computing. 1985; 3(3): 111–117. ISSN: 0262-8856.

NETRAVALI, A. N.; BOWEN, E. G. 1981. A Picture Browsing System. IEEE Transactions on Education. COM-29. 1981 December; 12: 1968–1976. ISSN: 0018-9359.

NOFEL, PETER J. 1986. 40 Million Hits on Optical Disk. Modern Office Technology. 1986 March; 31(3): 84, 86, 88. ISSN: 0746-3839.

NUGENT, WILLIAM R. 1983. Applications of Digital Optical Disks in Library Preservation and Reference. In: Smith, Allen, ed. Proceedings of the American Federation of Information Processing Societies (AFIPS): National Computer Conference: Volume 52; 1983 May 16–19; Anaheim, CA. Montvale, NJ: AFIPS Press; 1983. 771–775. ISBN: 0-88283-039-2.

NUGENT, WILLIAM R. 1984. [untitled comments]. In: Cochrane, Pauline A. Modern Subject Access in the Online Age. American Libraries. 1984 June; 15 (6): 438–441, 443. ISSN: 0002-9769.

NUGENT, WILLIAM R. 1986. Optical Discs—An Emerging Technology for Libraries. IFLA Journal. 1986; 12(3): 175–181. ISSN: 0340-0352.

NUGENT, WILLIAM R.; HARDING, JESSICA R. 1983. Pictures and Productivity: How Image Automation Will Amplify The Output of the Knowledge Worker. In: Vondran, Raymond F.; Caputo, Anne; Wasserman, Carol; Diener, Richard A.V., eds. Proceedings of the American Society for Information Science (ASIS) 46th Annual Meeting; 1983 October 2–6; Washington, DC. White Plains, NY: Knowledge Industry Publications, Inc. for ASIS; 1983. 36–40. ISSN: 0044-7880; ISBN: 0-86729-072-2.

NURCOMBE, VALERIE J. 1984. Databases in Architecture. Art Libraries Journal. 1984 Spring; 9: 44–60. ISSN: 0307-4722.

NYERGES, ALEXANDER LEE. 1982. Museums and the Videodisc Revolution: Cautious Involvement. Videodisc/Videotex. 1982 Fall; 2(4): 267–274. ISSN: 0278-9183.

O'CONNOR, MARY ANN. 1986. Application Development and Optical Media. Optical Information Systems. 1986 January-February; 6(1): 64–67. ISSN: 0886-5809.

OFFICE ADMINISTRATION AND AUTOMATION. 1985. A New Alternative; Image-Based Networks. Office Administration and Automation. 1985 April; 46(4): 46. ISSN: 0886-5809.

OHLGREN, THOMAS H. 1980. Subject Indexing of Visual Resources: A Survey. Visual Resources: An International Journal of Documentation. 1980 Spring; 1(1): 67-73. ISSN: 0197-3762.

OHLGREN, THOMAS H. 1982. Image Analysis and Indexing in North America: A Survey. Art Libraries Journal. 1982 Summer; 7(2): 51-60. ISSN: 0307-4722.

OPTICAL INFORMATION SYSTEMS. 1986a. U.S. Patent and Trademark Office to Automate Using Optical Data Disk Technology. Optical Information Systems. 1986 May-June; 6(3): 179. ISSN: 0086-5809.

OPTICAL INFORMATION SYSTEMS. 1986b. ARTSearch: Helen L. Allen Textile Collection. Videodisc and Retrieval System. Optical Information Systems. 1986 January-February; 6(1): 71-72. ISSN: 0886-5809.

OSBORN, HUGH. 1984. A Look at Videoarchiving. Videodisc and Optical Disk. 1984 November-December; 4(6): 460-467. ISSN: 0742-5740.

OWEN, WAYNE P. 1986. FAX Enters the Digital Era. Telecommunication Products Plus Technology. 1986 May; 4(5): 62-64, 66. ISSN: 0746-6072.

PARKER, ELISABETH BETZ. 1985a. The Library of Congress Non-Print Optical Disk Pilot Program. Information Technology and Libraries. 1985 December; 4: 289-299. ISSN: 0730-9295.

PARKER, ELISABETH BETZ. 1985b. Videodisc Revisited- State of the Art Applications and Current Issues. In: Abstracts of the Art Libraries Society of North America 13th Annual Conference; 1985 February 8-14; Los Angeles, CA. Tucson, AZ: Art Libraries Society of North America; 1985. 22.

PICTURE CONVERSION, INC. 1986. Image Processing on a PC. 1986 November. 1p. Available from: Picture Conversion, Inc., 5109 Leesburg Pike, Six Skyline Place, Suite 212, Falls Church, VA 22041.

POIZNER, STEPHEN L. 1986. Micro Mapping: Data Imaging for Information Managers. Journal of Information & Image Management. 1986 November; 19(11): 35-38. ISSN: 0745-9963.

PRESTON, CRAIG; MOLINARI, FRED. 1986. Applications Proliferate as Digital Image Processing Enters Mainstream Design. Electronic Design. 1986 March 6. ISSN: 0013-4872.

PREWITT, J. M. S.; SELFRIDG, P. G.; ANDERSON, A. C. 1984. Name-Value Pair Specification for Image Data Headers and Logical Standards for Image Data Exchange. Proceedings of the Society of Photo-Optical Instrumentation Engineers. 1984; 515: 452-458. ISSN: 0361-0748.

PRICE, JOSEPH. 1984. The Optical Disk Pilot Program at the Library of Congress. Videodisc and Optical Disk. 1984 November-December; 4(6): 424-432. ISSN: 0742-5740.

PRICE, JOSEPH W. 1985. Optical Disks and Demand Printing Research at the Library of Congress. Information Services & Use. 1985 February; 5(1): 3-20. ISSN: 0167-5265.

PURCELL, PATRICK; OKUN, HENRY. 1983. Information Technology and Visual Images. Two Case Studies. Art Libraries Journal. 1983 Autumn; 8: 43-48. ISSN: 0307-4722.

RAITT, DAVID I. 1985. APOLLO; Document Delivery by Satellite. In: Parkhurst, Carol A., ed. Proceedings of the American Society for Information Science (ASIS) 48th Annual Meeting; 1985 October 20-24; Las

Vegas, NV. White Plains, NY: Knowledge Industry Publications, Inc. for ASIS; 1985. 7-10. ISSN: 0044-7870; ISBN: 0-86729-176-1.

RAMAN, VASUDEVAN; IYENGAR, S. SITHARAMA. 1983. Properties and Applications of Forests of Quadtrees for Pictorial Data Representation. BIT. 1983; 23: 472-486. ISSN: 0006-3835.

REICH, VICTORIA ANN; BETCHER, MELISSA ANN. 1986. Library of Congress Staff Test Optical Disk System. College and Research Libraries. 1986 July; 47(4): 385-389. ISSN: 0010-0870.

RINEHART, MICHAEL. 1982. Art Databases and Art Bibliographies: A Survey. Art Libraries Journal. 1982 Summer; 7: 17-31. ISSN: 0307-4722.

RISHER, CAROL A. 1985. Copyright: An Impediment or an Incentive to Videodisc Creation. In: Abstracts of the Art Libraries Society of North America 13th Annual Conference; 1985 February 8-14; Los Angeles, CA. Tucson, AZ: Art Libraries Society of North America; 1985. 24.

ROBERTS, H. E. 1983. Visual Documentation: Engravings to Videodiscs. Drexel Library Quarterly. 1983 Summer; 19: 18-27. ISSN: 0012-6160.

ROBERTS, H. E. 1985. Visual Resources: Proposals for an Ideal Network. Art Libraries Journal. 1985; 10: 32-41. ISSN: 0307-4722.

ROBINSON, RALPH G. 1983. The Use of Medical Images in Today's Hospitals. In: Dwyer, S.J. ed. Proceedings of the International Society for Optical Engineering (SPIE) 2nd International Conference and Workshop on Picture Archiving and Communication Systems (PACS) for Medical Applications [Tutorial]; 1983 May 8; Kansas City, MO. Bellingham, WA: SPIE; 1983. 1-11. ISBN: 0-89252-453-7; CODEN: PSISDG.

ROSENFELD, AZRIEL. 1969. Picture Processing by Computer. Computing Surveys. 1969 September; 1(3): 147-174. ISSN: 0360-0300.

ROSS, MORRIS. 1986. D.O.E. Funds Art Retrieval. Video Computing. 1985 October/November; 8. ISSN: 0815-628X.

ROUSSOPOULOS, N.; LEIFKER, D. 1985. Direct Spatial Search on Pictorial Databases Using Packed R-Trees. Proceedings of the Association for Computing Machinery, Special Interest Group on Management of Data (ACM-SIGMOD) International Conference on Management of Data; 1985 May 28-31; Austin, TX. SIGMOD Record. 1985 December; 14(4): 17-31. ISSN: 0163-5808.

RYLAND, JANE N. 1985. Storing "Outer Space" Data on Laser Disc. Library Hi Tech. 1985 Winter; 3(4): 77. ISSN: 0737-8831.

S. KLEIN COMPUTER GRAPHICS REVIEW. 1986. No Time for an Identity Crisis. The S. Klein Computer Graphics Review. 1986 Fall; 64, 66-67.

SAFFADY, WILLIAM. 1986. Optical Disks for Data and Document Storage. Westport, CT: Meckler Publishing Corp.; 1986. p. 39. ISBN: 0-88736-065-3.

SARETZKY, GARY. 1985. Bibliographies and Databases for Research on the Preservation of Aural and Graphic Records. Picturescope. 1985 Winter; 31(4): 119-121. ISSN: 0031-9694.

SCHEIBE, PAUL O. 1984. Design Considerations for Multi-Channel Picture Communication Network. In: Dayhoff, Ruth E., ed. Proceedings of the 7th Annual Symposium on Computer Applications in Medical Care; 1983 October 23-26; Washington, DC. Silver Spring, MD: IEEE Computer Society; 1984. Available from: IEEE Computer Society Press,

1109 Spring Street, Silver Spring, MD 20910. 814-815. ISSN: 0195-4210; ISBN: 0-8186-0503-0.

SCHNEIDER, ROGER H. 1982. The Role of Standards in the Development of Systems for Communicating and Archiving Medical Images. In: Duerinckx, Andre, J., ed. Proceedings of the Society of Photo-Optical Instrumentation Engineers–The International Society for Optical Engineering (SPIE) 1st International Conference and Workshop on Picture Archiving and Communication Systems (PACS) for Medical Applications: Volume 318, Part I, 1982 January 18-21; Newport Beach, CA. Bellingham, WA: SPIE; 1982. 270-271. ISBN: 0-89252-352-2; CODEN: PSISDG.

SCHREIBER, NORMAN. 1984. Computerized Photography- New Tools for the Trade. Popular Photography. 1984 March; 91(3): 66-69, 92-93. ISSN: 0032-4582.

SCOTT, DAVID S.; IYENGAR, S. SITHARAMA. 1985. A New Data Structure for Efficient Storing of Images. Pattern Recognition. 1985; 3(3): 211-214. ISSN: 0031-3203.

SCOTT, DAVID S., IYENGAR, S. SITHARAMA. 1986. TID—A Translation Invariant Data Structure for Storing Images. Communications of the Association for Computing Machinery. 1986 May; 29(5): 418-429. ISSN: 0001-0782.

SHERMAN, A.; KOSS, L. G.; ADAMS, S.; SCHREIBER, K.; MOUSSOURIS, H. F., FREED, S. Z.; BARTELS P. H.; WIED, G. L. I. 1981. Bladder-Cancer Diagnosis by Image Analyses of Cells in Voided Urine Using a Small Computer. Analytical and Quantitative Cytology. 1981; 3(3): 239-249. ISSN: 0190-0471.

SIEBERT, JAMES E.; ROSENBAUM, TERRY L.; OOSTERIVIJK, HERMAN. 1985. In: Proceedings of the Society of Photo-Optical Instrumentation Engineers. 1985; 536: 3-10. ISSN: 0361-0748.

SONNEMANN, SABINE S. 1983. The Videodisc as a Library Tool. Special Libraries. 1983 January; 74(1): 7-13. ISSN: 0038-6723.

SORKOW, JANICE. 1983. Videodiscs and Art Documentation. Art Libraries Journal. 1983 Autumn; 8: 27-41. ISSN: 0307-4722.

SOUTHARD, THOMAS E. 1985. Radiographic Image Storage via Laser Optical Disk Technology- A Preliminary Study. Oral Surgery, Oral Medicine, Oral Pathology. 1985; 60(4): 436-439. ISSN: 0030-4220.

STAR, JEFFREY L. 1985. Introduction to Image Processing. BYTE. 1985 February; 12(2): 163-166, 168, 170. ISSN: 0360-5280.

STEFANEK, G.; CHANG, S.-K. 1983. An Automatic Database Generator for a Medical Pictorial Information System. In: COMPSAC 83: Proceedings of the IEEE Computer Society's 7th International Computer Software and Application Conference; 1983 November 7-11; Chicago, IL. IEEE. 229-236. ISBN: 0-8186-0509-X.

STEWART, SCOTT. 1986. The Use of Optical Storage Technology in Health Care and Medical Science. Optical Information Systems. 1986 July–August; 6(4): 298-302. ISSN: 0886-5809.

STRONG, HARRY; LEHMAN, DAVID H. 1983. Videodisc Research at the MITRE Corporation. Journal of the American Society for Information Science. 1983 November; 34(6): 433. ISSN: 0002-8231. ERRATUM. Journal of the American Society for Information Science. 1984 March; 35(2): 138. ISSN: 0002-8231.

STRUM, REBECCA. 1985. High Tech Breakthrough: Interactive Videodisc. Wilson Library Bulletin. 1985 March; 59(7): 450-452. ISSN: 0043-5651.

SUNDERLAND, JOHN. 1982. Image Collections: Libraries, Users, and Their Needs. Art Libraries Journal. 1982 Summer; 7(2): 41-49. ISSN: 0307-4722.

SUSTIK, JOAN M.; BROOKS, TERRENCE A. 1983. Retrieving Information With Interactive Videodiscs. Journal of the American Society for Information Science. 1983 November; 34(6): 424-432. ISSN: 0002-8231.

SUTHASINEKUL, S.; WALKER, F.; COOKSON, J.; RASHIDIAN, M.; THOMA, G. 1985. Prototype System for Electronic Document Image Storage and Retrieval. Journal of Imaging Technology. 1985 October; 11(5): 220-223. ISSN: 0747-3583.

TAMURA, H.; YOKOYA, N. 1984. Image Database Systems. A Survey. Pattern Recognition. 1984; 171(1): 29-43. ISSN: 0031-3203.

TECHNICALITIES. 1986. Videodiscs and the Impending Demise of the Index Table. Technicalities. 1986 June; 6(6). ISSN: 0272-0884.

THIELEN, DAVID N. 1986. The Process of Document and Image Conversion for the Image File System 2000. In: Roth, Judith Paris, comp. Optical Information Systems '86; 1986 December 9-11; Arlington, VA. Westport, CT: Meckler Publishing Corp.; 1986. 231-236. ISBN: 0-88736-113-7.

THOMA, GEORGE R. 1984. Linking Bibliographic Citation Retrieval to an Electronic Document Image Database. In: Flood, Barbara; Witiak, Joanne; Hogan, Thomas H., comps. Proceedings of the American Society for Information Science (ASIS) 47th Annual Meeting; 1984 October 21-25; Philadelphia, PA. White Plains, NY: Knowledge Industry Publications, Inc. for ASIS; 1984. 190-194. ISSN: 0044-7870; ISBN: 0-86729-115-X.

THOMA, GEORGE R. 1985. Electronic Storage and Retrieval of Medical Document Images. In: Proceedings of the 5th International Congress on Medical Librarianship (V. ICML); 1985 September 30-October 4; Tokyo, Japan: Japan Organizing Committee for V. ICML; 1985. 750-756.

THOMA, GEORGE R.; COOKSON, JOHN P.; WALKER, FRANK L. 1986. Integration of an Optical Disk Subsystem Into an Electronic Document Storage and Retrieval System. Optical Information Systems. 1986 March-April; 6(2): 128-129. ISSN: 0886-5809.

THOMA, GEORGE R.; SUTHASINEKUL, S.; WALKER, F. L.; COOKSON, J.; RASHIDIAN, M. 1985. A Prototype System for the Electronic Storage and Retrieval of Document Images. ACM Transactions on Office Automation Systems. 1985; 3(3): 279-291. ISSN: 0734-2047.

TORIWAKE, J.; HASEGAWA, J. 1980. Pictorial Information Retrieval of Chest X-Ray Image Database Using Pattern Recognition Techniques. In: Medinfo 80: Proceedings of the 3rd World Conference on Medical Informatics; 1980 September 29-October 4; Tokyo, Japan. Amsterdam, The Netherlands: North Holland; 1980. Part 2, 1116-1120.

VAN ARSDALE, WILLIAM O.; OSTRYE, ANNE T. 1986. InfoTrac: A Second Opinion. American Libraries. 1986 July/August; 17(7): 514-515. ISSN: 0002-9769.

VANDEN BRINK, JOHN. 1986. What Radiologists Say About PACS. American Journal of Roentgenology. 1986 February; 146(2): 419–420. (Editorial). ISSN: 0190–0471.

VAUGHAN, C. L.; KAMM, T. M.; YASNOFF, W. A.; ANDREWS, L. T.; KLINGLER, J. W. 1985. Image Database Considerations: The Core of PACS. In: IEEE 7th Annual Conference of the Engineering in Medicine and Biology Society. 1985; 1002–1006.

WALTER, G. O. 1984. Optical Digital Storage of Office and Engineering Documents. Journal of Information and Image Management. 1984 April; 17: 27–30+. ISSN: 0745-9963.

WALTER, G. O. 1985. Optical Digital Data Disk Technology for the Management of Engineering Documents. Journal of Information and Image Management. 1985 January; 18: 21–27. ISSN: 0745-9963.

WEBSTER'S SEVENTH NEW COLLEGIATE DICTIONARY. 1967. Springfield, MA: G & C Merriam Company; 1967.

WILSON, ANDREW C. 1986a. Array Processors: The Best Way to Process Images? Digital Design. 1986 January; 16: 47–52. ISSN: 0147-9245.

WILSON, ANDREW C. 1986b. Imaging Rises to Meet User Demands. Digital Design. 1986 February; 16: 23. ISSN: 0147-9245.

WILSON, ANDREW C. 1986c. Standalone Imagers Focus on Workstation Segment. Digital Design. 1986 May; 16: 17–18. ISSN: 0147-9245.

WILSON, ANDREW C. 1986d. Graphics Boards Tackle Imaging. Digital Design. 1986 June; 16: 84–88. ISSN: 0147-9245.

WILSON, ANDREW C. 1986e. AT&T Makes the Right Choice for Imaging Support. Digital Design. 1986 September; 16: 40–42. ISSN: 0147-9245.

WILSON, ANDREW C. 1986f. Color Cell Compression Reduces Images to 2 Bits/Pixel. Digital Design. 1986 October; 16(11): 23. ISSN: 0147-9245.

YAMAGUCHI, K.; OHBO, N.; KUNII, T. L.; KITAGAWA, H.; HARADA, M. 1980. Elf; Extended Relational Model for Large, Flexible Picture Databases. In: Proceedings of the Workshop on Picture Data Description and Management; 1980 August 27–28; Asilomar, CA. 95–100.

YANG, CHUNG-CHUN; CHANG, SHI-KUO. 1984. Picture Encoding Techniques for a Pictorial Database. In: 1st International Conference on Computers and Applications; 1984; New York, NY. Silver Spring, MD: IEEE Computer Society Press; 1984. 777–786.

YANG, C. C.; HSIEH, C. Y., CHANG, S. K. 1986. Picture Information Measure and Its Applications in Pictorial Database Management. Washington, DC: Naval Research Laboratory; 1986 October 8. (NRL Report 9004). Available from: Naval Research Laboratory, Code 5380, Washington, DC 20375.

III

Applications

Section III includes chapters entitled "End-User Searching of Bibliographic Databases" by William H. Mischo and Jounghyoun Lee of the University of Illinois at Urbana-Champaign, "Systems that Inform: Emerging Trends in Library Automation and Network Development" by Ward Shaw and Patricia B. Culkin of the Colorado Alliance of Research Libraries, and "Agricultural Information Systems and Services" by Robyn C. Frank of the National Agricultural Library.

In their chapter, "End-User Searching of Bibliographic Databases" Mischo and Lee define end users, in opposition to intermediary searchers, as searchers who also use the information that is the result of the search. Among the most heavily used end-user computer interfaces for information retrieval is the online catalog. Mischo and Lee discuss a variety of studies dealing with such interfaces. They briefly review the history of online database searching going back to the early 1970s and bring the reader up to date with an overview of recent work on end-user searching and software search aids such as interfaces, gateways, and transparent systems. Mischo and Lee investigate the problems of training end users, and within that framework they cover command language training, the resultant new intermediaries, and the use of front-end software. End-user search services in libraries are discussed, and the end-user searches are characterized.

In their library automation and network chapter Ward Shaw and Patricia B. Culkin examine the last several years of the literature of library automation, and they discuss how automated library systems and computer-based information services are being transformed into much more powerful "systems that inform." Five major topics are used to structure the developments: 1) the revitalization of local systems development and the changing nature of vendor/customer relationships; 2) the emergence of local application networks; 3) the use of library systems as channels to local resources, to external systems, to nonbibliographic data, and to electronic publishing; 4) the professional leadership; and 5) the direct influence on learning.

In her chapter on "Agricultural Information Systems and Services" Robyn C. Frank notes that because agriculture is a broad, multidisciplinary subject, which includes many of the life, physical, and social sciences, users of agricultural information are a very diverse group. Users of agricultural information range from policy makers and researchers to journalists and consumers. This chapter traces the history of agricultural information in the United States and highlights some of the key reference tools. The "big three" major agricultural databases– AGRICOLA, AGRIS, and CAB–are discussed and compared. Agricultural research project databanks are highlighted. Searching by various agricultural disciplines is explored to illustrate the breadth of subject coverage. Food and feed composition databanks are featured as examples of numeric databanks.

Computers, and in particular microcomputers, are used by all segments of the agricultural community including extension personnel and farmers. Videodisks, CD-ROMs, and expert systems are emerging technologies that are being tapped by information professionals in agriculture. Frank notes that greater emphasis needs to be placed on determining the information needs of specific clientele groups because not all groups have the same needs. A marketing approach should then be taken to develop products and services for current as well as potential users of agricultural information.

7 End-User Searching of Bibliographic Databases

WILLIAM H. MISCHO
University of Illinois, Urbana-Champaign

JOUNGHYOUN LEE
University of Illinois, Urbana-Champaign

INTRODUCTION

One of the most widely reported developments in the library and information science literature over the past several years has been the growing interest in end-user or direct patron access to online bibliographic databases. This increased interest in end-user searching has been driven by technology. The past several years have witnessed the development and evaluation of online catalogs, the continued proliferation of online databases, the introduction of new online search services, the increased availability of microcomputer workstations, and the development of optical information technologies.

End users have been defined as "processors of information who use information sources directly" (BUNTROCK & VALICENTI, p. 203) and "persons who end up using information they have sought, by whatever means" (JANKE, 1984a, p. 152). End users are the "information consumers," the patrons of libraries and information centers. LEVY & HAWKINS note the traditional division of information seekers into intermediaries, who retrieve information as a service to others, and end users, whose information-seeking activity is done in support of other functions. ARNOLD (1987), for reasons detailed in this review, proposes a definition of end users as the individuals who ask the questions, while special librarian refers to individuals who conduct online searches for others and *new intermediary* describes individuals who perform some of the same functions as special librarians but have non-library responsibilities and work outside the library setting.

The practice whereby end users perform their own online searching to retrieve bibliographic information has been referred to in a number of ways,

Annual Review of Information Science and Technology (*ARIST*), Volume 22, 1987
Martha E. Williams, Editor
Published for the American Society for Information Science (ASIS)
by Elsevier Science Publishers B.V.

including "nonmediated searching" and "client searching" (JANKE, 1985a). This review uses the terminology end-user searching. Grammatically, "end user" is the noun form of the term and "end-user" the adjective form. However, in the literature there has been no consistency, and even the term "enduser" has been used. For this review, we have used the *ARIST* convention of hyphenating the adjective but not the noun.

Recent literature expresses a conviction that there is a need for enhanced and coordinated access to rapidly growing collections, particularly in areas traditionally lacking bibliographic control, such as periodicals. Libraries are exploring four methods of providing increased access to the periodical literature: 1) loading periodical citation databases into local online catalogs; 2) providing end-user searching services or training programs; 3) utilizing optical disk databases; and 4) bringing databases in-house for in-house processing.

Clearly, important parallels exist among end-user searching of online catalogs, remote bibliographic databases, and CD-ROM (compact disk, read-only memory) indexes. One would expect that insights obtained in the study of any of these areas would have implications in the design and use of the others. Because of the breadth of the literature, this review focuses on end-user searching of remote bibliographic databases and emphasizes selected studies that seem to represent the field best.

ONLINE CATALOG USE

Online catalogs are, by definition, bibliographic access tools for end users. For many library patrons, online catalogs represent the most visible and heavily used end-user computer interface for information retrieval. For example, some 25 million library patrons use the bar-coded library cards of the turnkey vendor CLSI (MILLER, 1986). The design, implementation, and evaluation of online catalogs have been the subject of several recent comprehensive reviews (HILDRETH, 1982, 1985; MATTHEWS; MATTHEWS ET AL.). This chapter briefly mentions the major points relating to end-user access identified from the online catalog research.

The Council on Library Resources (CLR) sponsored various user studies of online catalogs through questionnaires, transaction log analysis, and focused group interviews (HILDRETH, 1984; MARKEY, 1983). These studies found that:

- Subject searching is the predominant mode of searching; in some libraries it accounts for more than one-half of all searches;
- Most catalog users want materials on a topic;
- Catalog users report the most problems with subject searching;
- Users approach online catalogs expecting to find enhanced access to a broader field of materials, particularly periodicals, than that in the card catalog;
- Catalog users place the highest priority for improvements on various subject-search enhancements;

- One-third to one-half of searches result in nothing retrieved;
- Subject searches using keywords with a non-Boolean strategy or with search arguments providing a partial match with controlled vocabulary terms often produce a large number of citations;
- Systems with keyword searching receive more subject searching.

The results of the CLR studies have been interpreted as a mandate for enhanced subject access.

The CLR survey results and other studies indicate an overwhelming acceptance of online catalogs and high user satisfaction. However, NIELSEN (1986) points to studies showing that judgments made from questionnaire responses can be suspect and notes that transaction log data and other evidence indicate that online catalog user satisfaction does not imply that successful searches are being conducted.

The importance of the user–system interface in the online catalog, particularly in subject searching, has been emphasized (HILDRETH, 1982; MARKEY, 1984). VIGIL (1986) in last year's *ARIST* summarized the current software interface research and noted the problems of subject searching in online catalogs and other bibliographic databases.

Efforts to apply expert system techniques to online catalogs have been reported. DOSZKOCS described the CITE NLM (National Library of Medicine) public access catalog, which offers natural language query input, closest-match search strategy, ranked document output, and the use of dynamic user feedback. Improvements to the interface in the MELVYL online catalog have been described by BRANDRIFF & LYNCH.

Several comprehensive plans for improved subject access in online catalogs have been advanced (BATES, 1986a; MARKEY, 1984). These techniques emphasize increasing the catalog entry vocabulary, access points, and search strategy manipulation. HILDRETH (1987) has presented a general prescription for improved user access in online catalogs.

One technique to increase local access to the periodical literature is to load periodical citation databases into online catalogs (HILDRETH, 1987; QUINT). There are presently demonstration projects at various academic libraries to load certain databases—Current Contents, ERIC, MEDLINE, or those of Information Access Corp. (IAC)—into local online catalogs. BROERING has described the miniMEDLINE system, a subset of MEDLINE that runs in an online catalog environment. The library of the Washington University medical school has tested the use of Current Contents in their local online catalog called BACS (CRAWFORD ET AL.). Using BRS (Bibliographic Retrieval Services) software, Georgia Institute of Technology has loaded several IAC databases into a distributed online catalog (*INFORMATION TODAY*, 1987b). Another project is investigating providing access to remote bibliographic databases as a search option in an online catalog through an intelligent microcomputer interface (MISCHO).

HISTORICAL DEVELOPMENT OF DATABASE SEARCHING

Since online searching of commercial and governmental bibliographic data-bases began in the early 1970s, it is primarily trained intermediaries who have had direct interaction with these systems (M. EISENBERG; MARRON & FIFE). The role of the end user has been to supply information for search-strategy formulation and modification and to interact with the intermediary searcher. As MCCARN has pointed out, this has been the catch-22 of infor-mation retrieval—i.e., that online systems designed to provide enhanced access for users can only be accessed through intermediaries. HAWKINS (1981) summarized the situation in 1981, "Most bibliographic online searches today are done by an intermediary rather than by the end user for good reasons" (p. 179).

The problems with searching these retrieval systems in their native mode on the commercial and governmental database vendor systems (these are called "Databanks" by PEMBERTON in contrast to the general understand-ing of a databank as being a database containing numeric data) include: 1) dissimilarities among the different vendors; 2) the bewildering number of available databases (within a general area and even within a single subject area such as business); 3) the complexities of search-strategy formulation and logic; 4) the variations in command languages from vendor to vendor; 5) the lack of standardization in database search elements; and 6) the substantial costs of online searching and printing.

The prevailing view in the 1970s, supported by several studies, indicated that the cost of searching could be significantly reduced by the use of inter-mediaries (MARTIN; P. W. WILLIAMS, 1977). BARRACLOUGH noted that because early experiences with online systems indicated that intermediaries were the primary users, the vendors developed the system features along the lines desired by intermediaries, thus perpetuating the status quo.

Earlier *ARIST* reviews have suggested methods of facilitating direct end-user searching by coordinated training methods and promotion efforts (MCCARN; WANGER). MEADOW (1979), in an oft-cited article, argues through analogy that end-user searching would increase and become common just as computer programming had proliferated with the development of high-level programming languages. LOWRY concluded that increased emphasis on user training and education, promotion efforts, and the development of software interfaces to online systems would result in a shifting from inter-mediary to end-user searching.

Several recent bibliographies of end-user searching have been published (DES CHENE; JANKE, 1984b; JANKE, 1985a; LYON; WOOD). SEWELL & TEITELBAUM review their findings from a ten-year study of end-user search-ing in NLM by health science professionals and provide an extensive bibliog-raphy of end-user searching. BORGMAN, in an excellent review of the literature on user behavior in online catalogs and bibliographic retrieval systems, suggests an agenda for further research in the area.

There has clearly been a renewed interest in end-user searching of biblio-graphic databases and a concomitant growth in the literature over the past six years. In summary, the various reasons cited for this increased interest include:

- The continued exponential growth of information and the demonstrated value of online information retrieval;
- The wide availability of online full-text databases (such as LEXIS) directed at professional users;
- The proliferation of microcomputer workstations with communications capabilities in both the workplace and home settings;
- The emphasis on computer literacy in education, office automation, professional occupations, and recreation;
- The inauguration of nonpeak-time, less expensive, more user-friendly search systems by BRS, DIALOG, and STN;
- The growing awareness among the end-user population of the existence of online database services because of articles in the popular and specialized literature, conference exhibits, and vigorous promotion by database producers and vendors;
- The growing familiarity by library users with online catalogs and, by extension, online bibliographic databases;
- The increase of workloads for intermediaries;
- The development of research and commercial front-end and gateway software packages to facilitate online searching by untrained users.

ARNOLD (1986) recalled the prevailing optimistic view of end-user searching that existed in 1981 (p. 5): "End users are buying PCs in record numbers and will become online searchers; end users are less price sensitive than information specialists; end users won't require how-to training; end users know what they want; and when end users do their own searches, they will be able to save time and money." There now exists a clearer picture of the role of end-user searching in information services.

OVERVIEW OF RECENT WORK ON END-USER SEARCHING

The potential end-user searcher has been exposed to a marketing blitz of overly simplistic, partly accurate articles in the popular and professional literature (DIODATO; OJALA, 1985). The database vendors stay very visible; Mead and DIALOG run advertisements in the *Wall Street Journal.* The vendors are marketing and demonstrating their products at all the appropriate professional society meetings. The online vendors, database producers, and search software developers have targeted the vast personal computer market, focusing primarily on the information needs of the professional worker in both the academic and corporate settings. To a much lesser extent, the literature reports efforts, primarily via front-end microcomputer software and factual databases, to penetrate the home computer market.

Fueling this effort is the information industry's desire to continue a growth rate that has found revenues increasing 90% from 1982 to 1984 (M. E. WILLIAMS, 1986a).

Libraries and information centers, responding to growing user interest and aware of the changing information technologies, have become actively involved in promoting end-user searching with their designated clientele. Examining the results reported in the literature, recent studies show that end-user searching falls into the two categories discussed below.

Formal Training to Targeted Clientele

This category involves formal training programs for online searching by library and information center staff for targeted clientele. The end users are typically taught to search in command language. One goal is often to extend online searching into the work environment of the professional. Trained end users are expected to conduct searches in their work environment. These studies usually involve vendor systems such as DIALOG, BRS, SDC (System Development Corp.), MEDLINE and STN because of their wide availability and the search power of the native-mode systems. Also included in this category are programs designed to train specific professional user groups (e.g., health science professionals) to access, in the workplace, the "user-friendly" versions of the vendor systems, such as BRS/AFTER DARK (JANKE, 1983), BRS/BREAKTHRU (HOOK), BRS/Saunders Colleague (C. A. BAKER; CLANCY), KNOWLEDGE INDEX (KAPLAN; OJALA, 1983), and DIALOG Business Connection (O'LEARY, 1986). While the programs involving the traditional systems seemingly duplicate earlier attempts to train end users, they represent the application of modern training methods directed to, ostensibly, users who are more information conscious.

End-User Search Services in Libraries

End-user search services in libraries can perhaps best be described as walk-in search services. These services involve the establishment or demonstration of end-user search centers, almost exclusively in university libraries. These search facilities are offered as an extension of library reference services to the general population served and typically offer access to the after-hours search services. Several programs utilize front-end search software for their clientele. The distinction between these programs and the programs discussed above is that the library-based programs cannot be described as outreach directed to a clientele in a work environment and may not even use formal training programs.

SOFTWARE SEARCH AIDS

One of the most active areas of development in information technology has been in the growth of software search aids for accessing information retrieval systems. The development of software search aids has received much impetus from the wide use of microcomputers in database searching. These aids are designed to facilitate online searching by simplifying and automating the search process, with an eye to saving costs and improving search results. This general class of software has been referred to in many ways, including database search assistance software, interface software, expert gateways, and transparent systems. M. E. WILLIAMS (1985) introduced a classification that groups search software packages into three types: 1) gateways, 2) intermediary systems, and 3) front ends. HAWKINS & LEVY (1985) further defined and consolidated the terminology by dividing software for online searching into gateway software and front-end software. This terminology has been adapted by other writers (COONS, 1986a; FENICHEL, 1986; TOLIVER, 1986).

In this classification scheme, a gateway is defined as a software interface that automatically logs on to a database vendor via a communications network; it may also provide for uploading of search statements and downloading of citations. Front-end software, as defined, extends the features of a gateway by simplifying and performing some of the steps of the search process. Front ends attempt to make the online retrieval service transparent to the user by, typically, assisting in or providing database selection and translating the user search request into the language of the vendor. These definitions provide coherence to the terminology, but the distinctions between the two software classes can become blurred. For example, Pro-Search is a front end in its high-level mode but a gateway in native mode. Likewise, SearchMaster is defined as a gateway but becomes a front end when using software scripts.

Other definitions have been offered. M. E. WILLIAMS (1986a) defines an intermediary system as a front-end package with comprehensive search negotiation features that is used to access a specific database or family of databases. HUSHON and HUSHON & CONRY also note this distinction by dividing search aid software into two categories: gateway packages (which include front ends) that interface to a single host system, and those that interface to multiple host systems. The search software is then further refined by an indication of an audience level of end user or information specialist.

There is also a tendency to refer to front-end systems residing on vendor mainframes (e.g., BRS/AFTER DARK, KNOWLEDGE INDEX) as user-friendly systems (FENICHEL, 1986; KING & BRUEGGEMAN).

Listings of available front-end and gateway software packages are included in HAWKINS & LEVY (1986a), COONS (1986a), KING & BRUEGGEMAN, M. E. WILLIAMS (1986b), and HUSHON. The reader should consult these references for a comparative analysis of the features of this proliferating software market.

The available gateway software is predominately microcomputer-based and offers various features designed to facilitate searching for the trained searcher. For example, in addition to automatic log-on, SearchWorks provides uploading, accounting modules, and post-processing capabilities, such as templates, sorting, and ranking of citations (COONS, 1986b).

Work on front-end packages has progressed from predecessor research programs to today's commercial systems. The INTREX Project at MIT (1965–

1973) was one of the earliest attempts to provide end users with a user-friendly interface to an online periodical database (OVERHAGE & REINTJES). This demonstration project led to the CONIT interface, which emphasized the searching of heterogeneous commercial retrieval systems with a common command language. Early on, the CONIT research pinpointed search strategy formulation as the most critical element in end-user retrieval success (MARCUS & REINTJES). The linear progression from CONIT to IIDA to OL'Sam (TOLIVER, 1982) to Sci-Mate is discussed by KEHOE.

Front-end software can reside in four locations: 1) on a microcomputer; 2) on a vendor mainframe (BRS/AFTER DARK, KNOWLEDGE INDEX); 3) on remote dial-up computers (EasyNet); and 4) on a local mainframe with direct access. TOLIVER (1986) and P. W. WILLIAMS (1985) discuss the advantages and disadvantages of these respective locations, and both note the flexibility of microcomputer front ends.

Two early microcomputer front-end packages, In-Search and Sci-Mate, were designed specifically for the end-user and scientific/professional markets. In 1984 Menlo Corp. released In-Search, a front end to DIALOG featuring menus, windows, and online tutorials and bluesheets. NEWLIN candidly describes the marketing problems and deteriorating relationship with DIALOG that led to Menlo's discontinuation of the product. In-Search's successor, Pro-Search, which is aimed at the professional searcher market, was later taken over by Personal Bibliographic Software.

ISI's Sci-Mate front end has been marketed primarily to researchers but has been used in several libraries and end-user centers. The early versions lacked the capability of storing search strategy offline and several studies have found that the online menus can be time consuming and unwieldy (BRODY ET AL.; ROUSSEAU & BRONARS; RUDIN ET AL.; P. W. WILLIAMS, 1985).

Information Access Corp. (IAC) markets Search Helper as a library end-user package. This software, which accesses only IAC databases, has been criticized for its lack of an "or" Boolean operator and limited print capabilities (ENSOR & CURTIS; KLEINER). Users of Search Helper at the Carnegie-Mellon University libraries found the system valuable and easy to use, with 50% of the users indicating that one-half or less of the retrieved citations were relevant (PISCIOTTA ET AL.). Several public libraries have offered patrons the Search Helper package (ELLEFSEN; ELMORE; QUEENS BOROUGH CENTRAL LIBRARY STAFF). Elmore found that patrons were not performing successful searches and recommended that librarians act as intermediaries.

Dial access to remote front ends has also generated a great deal of attention. HOROWITZ ET AL. described the PaperChase front-end software that simplifies access to the medical literature. ROBERTS compared Telebase System's EasyNet front end with the BCN (Business Computer Network) (O'LEARY, 1985a) gateway, concluding that EasyNet offered many advantages. HAWKINS & LEVY (1986b) described the introduction of EasyNet to AT&T Bell Laboratories end users. While BCN has gone out of business, EasyNet in its various forms—Einstein, InfoMaster, IQuest, Alanet Plus, SearchLink, and AccuSearch—has gained wide acceptance (*INFORMATION TODAY*, 1987a).

The database vendors are also developing menu-driven, user-friendly front ends residing on the vendor's mainframe. BRS/AFTER DARK and DIALOG's KNOWLEDGE INDEX are general-purpose front ends that provide access to a broad subject range of the vendor-mounted databases. DIALOG has recently introduced two front-end services aimed at vertical market segments, DIALOG Business Connection and DIALOG Medical Connection, which allow access to selected databases (*INFORMATION TODAY*, 1987c). O'LEARY (1986) maintains that remote front ends, such as EasyNet and Dialog Business Connection, "foreshadow the end of disk-based [microcomputer] frontend search assistance interfaces" (p. 20). However, there are numerous customized microcomputer front ends being developed to meet local needs, such as OAK (Online Access to Knowledge) (MEADOW, 1986), ILLINOIS SEARCH AID (MISCHO), Grateful Med (SNOW ET AL.), and OASIS (P. W. WILLIAMS, 1984). SearchMaster, with its provision for programmed scripts, can be modified to meet a specific application. LEVY & HAWKINS point out the need for specialized front ends that are customized to subject-oriented databases, such as CA Search.

There is a growing consensus that customized front-end software can provide end users with facilitated and sophisticated access to online bibliographic databases. HAWKINS & LEVY (1985) state that front-end systems "are not merely a passing fad, but are a significant step in the evolution of online searching systems and one that opens a vast new level of information provision" (p. 32). The success of EasyNet, with its value-added interface, over the gateway package BCN illustrates the value and perceived need for front ends to the various system command languages.

As KEHOE and FENICHEL (1986) have pointed out, the major problem with the present front-end packages is that they do not adequately aid the user in search-strategy formulation or assist in search navigation. These are obvious areas for the application of artificial intelligence (AI) and expert system techniques (SMITH). Current research on AI techniques in end-user systems was summarized at a recent conference (JACOBSON & WITGES).

There continues to be great interest in developing front-end software for general-purpose searching and for database or subject-specific fields. Speaking of front-end software development, SUMMIT comments, "we have yet to see the VISICALC of the online retrieval world, but it may be just around the corner" (p. 63).

TRAINING TARGETED END-USER GROUPS

Command Language Training

In examining the results reported in studies concerning the training of end users in command language searching, several trends can be observed. One almost universal observation is that a high level of initial enthusiasm for direct online searching among the end-user target population does not translate into a continuing commitment. The studies indicate that a substantial percentage of the trained end users do not continue searching after training.

RICHARDSON, reporting on an end-user study involving 20 volunteer engineers/scientists, indicated that only six remained relatively active searchers. User searching was characterized by high initial enthusiasm, followed by a period of settling out, leaving a final cadre of proficient users. However, overall levels of usage were "astonishingly low" particularly since the first seven months of searching were absolutely free of charge.

Two studies at Kodak Research Laboratories used end-user volunteers to search chemical databases via DIALOG, SDC, and CAS ONLINE (HAINES; HAINES ET AL.). A total of 38% and 32%, respectively, of the trained end-users remained as regular or frequent searchers. These studies involved approximately 75 scientists from a staff of 1,000 professional employees.

KIRK describes a multipart training program for end users at Amoco and emphasizes that "only the most motivated and/or most computer-oriented of the end-users completing Course I or II—which represent about one-third of our classroom participants—decide to request their own SDC or DIALOG password and become bonafide online searchers" (p. 22).

WALTON & DEDERT reported that 50% of the chemists trained to access relevant databases directly engaged in their own searching several months after the training sessions. However, a follow-up article (WALTON) revealed that the long-term success rate for the earlier study was actually 0%—i.e., no one in the trained group continued searching.

FLYNN, describing an end-user program at 3M Co., found that about 25% of attendees at open presentations on searching later indicated an interest in searching.

SEWELL & TEITELBAUM found in their multiyear study that "at least a third of scientists and practitioners will continue to use online databases" after training (p. 243).

BODTKE-ROBERTS, after training ten members of the life sciences faculty already involved with database searching through intermediaries, reported that 40% continued searching four months after training.

In a study of patent attorneys with DIALOG passwords, VOLLARO & HAWKINS observed the settling out effect reported by Richardson, noting that only 55% of the attorneys with passwords relied exclusively on their own searching.

LEIPZIG ET AL. reported on end-user searching activities in a pharmaceutical company. They noted that while the original number of trained MEDLINE and SDC end-user searchers in their organization had all but disappeared, the use of an in-house document storage and retrieval system, with more user-friendly access protocols, had grown.

SNOW reported the results of an NLM follow-up on training health-care professionals in performing their own online searches. Of the 1,000 individuals trained, less than 50% had gone on to apply for a password, and most of those who trained searched less than one hour per month.

The significant dropout rate documented in the training of searchers using the native-mode systems can also be observed in several experiments involving specific library clientele who were trained in the user-friendly systems.

POISSON reported two studies of end-user training of medical staff on BRS/AFTER DARK and MEDLINE at New York Hospital-Cornell Medical Center in which 50% of the staff attended at least one class, but only 8% and 7% became frequent searchers.

After training more than 150 health sciences faculty, staff, and students at the University of Minnesota in BRS Colleague and AFTER DARK, GLASGOW & FOREMAN found that 48% were not yet signed up to search and 44% of those not searching had decided against doing their own searching.

Other authors have supplied anecdotal evidence concerning the difficulties of training and retaining end-user searchers. THOMPSON sees a disparity between the predictions in the literature of increased end-user searching and the observations of practitioners. She reports that seminars on online searching for agricultural researchers have not induced people to search on their own.

CASE ET AL. studied the information-seeking behavior of energy researchers and concluded, "it may be that only the most innovative and motivated of end-users will elect to access databases directly" (p. 306).

Several reasons have been cited for the poor success rates in training scientists, researchers, and business professionals in using command-mode online systems: 1) the inconvenience of searching; 2) infrequency of searching; 3) convenience of using intermediaries; 4) previously documented problems of difficulties with command language; 5) difficulties with microcomputer searching protocols; 6) other pressing demands on users' time; and 7) the high cost of searching.

In summary, LEIPZIG ET AL. found that end users who are trained to use native-mode systems like to think online (thus increasing search costs), resist using a thesaurus, dislike typing, have trouble remembering commands, and prefer to delegate complicated searches to information professionals.

New Intermediaries

These efforts by special libraries, database producers, and database vendors to train specific end users have been directed at heterogeneous groups using various training methods. VIGIL (1984), noting that numerous end-user studies have shown that relatively high percentages of trained searchers have not continued their searching, suggested that it would be more effective to limit training to individuals with the initiative to search. Some of the recent literature on training end users has indicated a shift in focus from the originally targeted clientele of end users to a largely different group.

ARNOLD (1985; 1986; 1987) after much examination of the end-user market, concludes that the great expectations of increased end-user searching have not become reality; that attracting end users is a "difficult, costly, and time-consuming job" (1986, p. 5). He notes that the individuals attending training seminars on end-user searching are often intermediaries with a non-librarian job title rather than actual end users. He concludes that these "new intermediaries," who perform the role of designated searchers for other end users, will form the largest potential market for increased end-user searching. In fact, many of the articles describing programs for training end-user searchers address training of "new intermediaries" (FOGEL & ZIGMUND; JANKE, 1985b; KUPFERBERG). One reported danger in the reliance on programs that train these new intermediaries or designated searchers for a group is that when these individuals leave, searching activities may cease for the entire group (BODTKE-ROBERTS; WALTON).

It should also be noted that the nonbibliographic online systems and full-text databases designed for and tailored to specific end-user markets have experienced some of the same user training problems. In 1986 the Dow Jones News/Retrieval Service changed from counting the number of total passwords issued to reporting monthly password use to determine the number of customers. While the Mead Data Central LEXIS law databases have enjoyed enormous success—Mead doubled the number of its LEXIS customers in the single year 1983 (PEMBERTON & EMARD)—the system has been criticized for its search complexity (HUNTER, 1983). HARMAN notes that many lawyers are unwilling to learn the searching protocols and will delegate searches to traditional intermediaries or designated searchers. She also reports almost universal dissatisfaction among Reuters's journalism staff with the NEWSBANK full-text database. Finally, Mead's NEXIS system has recently been criticized by librarian intermediaries for unannounced changes, complexity of use, and lack of needed documentation (MILLER, 1987). Thus, there is evidence that within the factual and full-text markets, there is increased searching by new intermediaries or nonlibrarians acting as designated searchers for other end users.

Intermediate Level of Training

Another approach being taken to increase the efficacy of end-user searching is to train end users to utilize a subset of the complete command set. WARD & OSEGUEDA advocate training student users to perform simple searches in specific databases. MARTIN & DUTTON trained users in a large corporate setting to perform simple searches on one host system (DATA–STAR). SEWELL & TEITELBAUM routinely teach only the *AND* logical operator, limiting training in the other operators to selected end users.

Similarly, the philosophy of some of the front-end packages is to provide ease of use in return for less sophisticated searches using a subset of system command features.

Although higher success rates appear to be reported in situations in which intermediate-level searches of one primary database are involved, this would appear to work to the disadvantage of multidisciplinary workers and users in fields with multiple discrete databases, such as business or engineering.

Use of Front-End Software

Another clearly discernible reaction to the problems of training end users has been a closer examination of the role of front-end interface software in end-user searching. TATALIAS surveyed researchers at MITRE Corp. who were currently using online databases via intermediaries. She found that 65% of the respondents indicated a desire to do at least some end-user searching. Their requirements for a search system included menu choices (73%) and database(s) suggestions (78%).

HAWKINS & LEVY (1986b) surveyed users of the AT&T Bell Laboratories mediated search service. They found that while only 15% of the users had done their own online searches, 81% were interested in becoming searchers and 89% of those interested wished to conduct searches from their offices. A trial of the EasyNet front-end service showed that users were willing and able to utilize a front-end system for database searching.

An article by SEEFELDT & THOMAS presents a microcosm of end-user searching in the research environment. Efforts to train end users at Borg-Warner Corp. have resulted in: 1) a significant user dropout rate; 2) a core group of searchers performing straightforward searches on a few databases; 3) the training of researchers who function as new intermediaries by searching for other members of a group; and 4) an increasing number of end users utilizing front-end search software.

In another classic example, WALTON reported on the experiences of Exxon over a seven-year period. After testing an early software search aid (IIDA), the information center staff began training programs for end-user searching in native mode. After experiencing a high dropout rate and abandoning in-house training, the information center turned to a microcomputer front-end package, SearchMaster from SDC. This software allows the creation of customized script files for searching specific databases in a highly structured manner. The SearchMaster programs, which are used for author searches, reference verification, and routine subject searches, have been widely accepted and regularly used.

The development of the OAK system for energy researchers chronicles a series of steps in which users concerned with complicated search protocols and unwilling to use command-driven systems were able to utilize a customized front-end package (CASE ET AL.; MEADOW, 1986).

END-USER SEARCH SERVICES IN LIBRARIES

Many university libraries report the establishment of drop-in end-user searching centers, often as an extension of a fee-based online searching program. JANKE (1984b), describing the Online After Six service at the University of Ottawa, lists 37 other institutions offering end-user search services. For an ARL (Association of Research Libraries) Spec Kit prepared in 1986 (THOMAS), 23 university libraries were contacted, and all but one either reported an active end-user searching program or one in the planning stages.

These end-user search services typically use the user-friendly after-hours services and/or front-end microcomputer software. Several large-scale operational services and demonstration projects have been reported in the literature. Almost in contrast to the studies involving targeted user groups, these university open-ended end-user programs report an enthusiastic user response and high user satisfaction.

The Texas A&M University Library offered an experimental end-user search service in 1983/1984 (DODD ET AL.). The study compared 705 searches performed on BRS/AFTER DARK and IAC's front-end software Search Helper. User satisfaction with both services was very high (for AFTER DARK, 1.9 on a five-point Likert scale with 1 representing the highest ranking). The library chose to retain BRS/AFTER DARK and began offering this service on a no-fee sign-up basis in September 1984 (JAROS ET AL.). Between September 1984 and December 1985 a total of 6,358 end-user searches were performed—an average of 398 searches per month—with graduate students comprising over 60% of the users.

The Online After Six service utilizing BRS/AFTER DARK was inaugurated at the University of Ottawa in July 1983 after a pilot project involving 25 end users (JANKE, 1983). By April 1984, some 227 end-user searches had been run under a charging scheme whereby end users pay $2 per five minutes of connect time. The largest group of searchers was graduate students, comprising 44% of the total.

The University of Michigan offered BRS/AFTER DARK at no cost in the fall of 1984 (CROOKS). Data collected from 450 users showed that 69% found significant relevant references and that only 4% considered the system difficult to use. Graduate students were the largest search group, comprising 46.6% of searchers surveyed.

In 1985 the University of Pittsburgh established an experimental end-user search facility to evaluate several available search systems and software packages (BRODY ET AL.). Between February and December 1985, a total of 5237 users performed 25,445 searches at no charge for most of the period. User questionnaires administered to a sample of the search population showed 56% of the users to be graduate students. The user surveys also indicated that 66% of the users felt "successful" in their ability to use the search systems, averaging 2.17 on a Likert-type scale. An even higher percentage indicated a "positive" attitude toward the total computer searching experience.

SLINGLUFF ET AL. described an end-user search service offered at a nominal cost at the University of Maryland Health Science Library using BRS/AFTER DARK. In a follow-up article, SIMON reported that 88% of 96 users surveyed between September 1984 and September 1985 considered their searches at least partially successful.

HALPERIN & PAGELL described the free self-service searching program at the Lippincott Library of the Wharton School of Business utilizing BRS/AFTER DARK and the Wiley Executive Information Service on BRS. They found that 75% of the users were master's or doctoral students. A questionnaire, administered after completed searches, showed that users were very satisfied with the results of the search (median of 1 on a 1–5 Likert scale) and felt that the material retrieved was useful (median of 1 on Likert).

PENHALE & TAYLOR report on end-user searching introduced into a bibliographic instruction program at Earlham College. In a comparison with manual searching, students were able to retrieve as many relevant citations in 20 minutes of online time as they did in two hours of manual searching. Of the students tested 82% thought that the BRS/AFTER DARK system was easy to use, and 88% felt that online searching should be widely taught as part of a library instruction curriculum.

Other small-scale demonstration projects have yielded similar results. In a study involving 20 subjects (including 12 graduate students and three faculty) at the University of Wisconsin-Stout, TRZEBIATOWSKI found that 85% of the participants rated the results of their searches of BRS/AFTER DARK as either excellent or good. No subjects rated the results as poor.

In an experimental study of 22 AFTER DARK searchers at Pennsylvania State University, FRIEND found users "unanimously enthusiastic about performing their own searches" (p. 140).

As discussed previously, front-end software is also being used in end-user search centers. In the Pittsburgh laboratory, searchers made heavy use of Search Helper, In-Search (until its demise), and to a lesser extent, Sci-Mate (BRODY ET AL.). Texas A&M tested but decided not to use Search Helper (DODD ET AL.) and Sci-Mate (ROUSSEAU & BRONARS). Several other university libraries have tested Search Helper (ENSOR & CURTIS; KLEINER). Public libraries report very little activity in remote end-user database searching, save for the three libraries reporting the use of Search Helper.

An obvious factor in the popularity of these search services is that they provided searches free of charge or at small fees. However, the large number of users and enthusiastic survey responses clearly indicate a high level of interest in end-user searching within academic institutions. These are users already familiar with the printed indexing tools, and many are also using online catalogs. Thus, this user group might already expect to be provided enhanced online access. End-user searching seems to hold particular attraction for graduate students, who are the researchers and professionals of tomorrow. Academic library reference staff have noted that graduate students often act as new intermediaries for their research groups.

Several studies suggest that a high percentage of the users of end-user search services are not regular users of associated mediated search services. It would appear that these two types of search services attract different customers. BRODY ET AL. at Pittsburgh found through observation and survey results that "many Lab users had not requested searches through librarians" (p. 44); many indicated no knowledge of the mediated service. Of the users of the Michigan end-user service 62% had not had a previous search performed with an intermediary (CROOKS). A higher percentage of those not willing to pay for an end-user search had not had a mediated search done, suggesting that a fee schedule discourages potential users. SIMON found that 66% of the users of the end-user service had never used the library's mediated service. KRAVITZ & WESTLING note that 61% of the PaperChase users had never requested a mediated search. POISSON found that 80% of end users were nonusers of search services. GLASGOW & FOREMAN observed that, while 80% of the end users had requested a search from the library, most of the individuals using the end-user service had not been frequent users of the library's mediated service.

At the same time, various studies report a decrease in simple, verification searches and a slowdown in the rate of growth of their mediated search services (HAINES; JANKE, 1984b; KRAVITZ & WESTLING; WALTON).

CHARACTERIZATION OF END-USER SEARCHES

Effectiveness of Searchers

The literature reporting end-user searching indicates significant activity but mixed results in attracting users. These studies indicate limited success in direct training by library staff of specific clientele in their working environment. However, the library-based end-user search services, particularly those offering searches at no cost, report widespread user enthusiasm.

One obvious area of interest is the question of the effectiveness of end-user searches. A great deal has been learned about the individual searching behavior of end users.

In studies describing the training of targeted professional groups, perhaps because of the low user retention rates, little has been reported about user satisfaction or searcher performance. HAINES reported that 76.5% of sampled searchers were at least moderately satisfied with their results.

The majority of the library-based end-user studies were conducted under controlled conditions with subjects given post-search questionnaires. These studies offer an interesting parallel with the evaluation studies of online catalogs. Similar to those studies, while end-users indicate overwhelming satisfaction with searches, analysis of search results using standard retrieval measures shows that end users are not performing particularly effective searches.

Several studies have attempted to evaluate end user searching by comparing the results of searches performed by end users with the results of searches on the same topic conducted by a trained intermediary. Using subjects from an academic library environment, HURT found that end users were significantly more satisfied when searches were performed by intermediaries rather than by themselves. The preference for intermediaries appeared to be associated with perceived difficulties in system search protocols.

KIRBY & MILLER (1985; 1986) asked end-user searchers to compare their searches of BRS Colleague with the results of searches on the same topic conducted by the authors. In 60% of the searches the end user stated that the search by the intermediary was better. Of the Colleague searches initially judged successful by the end user, 54% were missing important articles discovered during the follow-up search. Further analysis of the searches showed that most end-user retrieval failures were due to problems with search strategy.

PENHALE & TAYLOR compared student and librarian online search results. They found that in the same amount of search time the librarians retrieved five times as many highly relevant citations and about two times as many moderately relevant citations as did the students. The investigators observed that the major problem faced by student searchers is the development of good search strategy. They also found that student searchers used fewer search terms and print commands than librarians did and did not use many alternative ways to narrow or broaden searches.

JANKE (1984b) noted a "fundamental qualitative difference" (p. 20) in the online searches conducted with intermediaries and those done by end users themselves. At least 50% of the users surveyed had difficulty getting satisfactory results.

In the study of NLM searchers conducted by SEWELL & TEITELBAUM, 56% of the end users in the group who both searched by themselves and used intermediaries felt that the mediated searches were of higher quality, and only 21% of this group preferred their own searches.

POISSON examined several searches by medical end users and found that in 60% of the cases users had retrieved only 12% to 22% of the relevant articles identified by expert searchers. There is also much anecdotal evidence and observation showing that end users have particular difficulties with search-strategy formulation and the use of Boolean logic.

Numerous authors report end users as having significant difficulty with the proper application of Boolean logic (BORGMAN; FRIEND; HALPERIN & PAGELL; LEIPZIG ET AL.; MADER & PARK; SLINGLUFF ET AL.; TRZEBIATOWSKI). JANKE (1984b) remarked "the most frequently heard complaint was this: end users all too often are not fully aware of the application of Boolean logic in formulating an effective online search strategy" (p. 21). Several evaluation surveys indicate that the use of Boolean operators is viewed as the most difficult aspect of retrieval. The crux of the problem appears to center around the contradictory use of the grammatical *AND* and the logical *AND*. The problem of confusion between the Boolean operators *OR* and *AND* was identified in the MIT INTREX Project at MIT in the late 1960s (KUGEL).

Numerous articles detail user searching difficulties and report that users request help and require staff assistance. SEWELL & TEITELBAUM found in transaction logs that about one-half to two-thirds of their regular searchers had some technical problem in their searches, and almost 30% of these searches had significant uncorrected logic or command failures.

Experience in the large end-user searching program at Texas A&M (JAROS ET AL.) has shown that "despite these services' [AFTER DARK, KNOWL-EDGE INDEX] claim to be user friendly, patrons still required substantial instruction and assistance" (p. 225). In their preliminary study (DODD ET AL.), library staff attendants provided assistance an average of 1.6 times during each AFTER DARK session. ROUSSEAU & BRONARS found that patrons required assistance more than five times in 80.8% of the search sessions using Sci-Mate and in 29.4% of the sessions using Tech Data (the AFTER DARK software).

In the Wisconsin study described by Trzebiatowski, users were given an orientation and training session before their search and were asked to complete a post-search questionnaire. Questionnaire results indicated that only 5% of the subjects considered the training session "not necessary"; 50% of the users responded that they had encountered problems with search strategy. An analysis of the search commands and strategy revealed numerous problems, including duplicate search statements, incorrect Boolean operator processing order, difficulties with concept combinations, jumping from specific to broad search strategy, and confusion about print options. Of the 20 searchers, three used truncation correctly, and only three performed their entire searches without seeking help. The average time spent by librarians with end users in the study was one hour and ten minutes.

FRIEND required the experimental end-user group to attend several training sessions and have their search strategy approved prior to searching. Investigators found that end users had problems with choice of terminology, Boolean operators, search-strategy modification, and the printing of documents. A post-session questionnaire showed that 80% of the respondents felt that a librarian with both AFTER DARK and subject expertise needed to be available for assistance while users were learning. Once they were experienced, 48% of the subjects indicated that someone with at least BRS training was required for assistance. Every subject who had returned to run a new search needed to meet with a trained searcher beforehand to review system commands and logic.

At the University of Michigan, CROOKS found that 85% of the users needed library staff assistance during their search; 76% of survey respondents preferred to perform searches by themselves but with library staff present to assist.

Questionnaire responses and staff observations at the Pittsburgh end-user center showed that choosing, combining, and refining search terms was the most difficult aspect of the search process (BRODY ET AL.). Approximately 50% of the survey respondents indicated that they requested aid with the selection of search terms and with system commands. Interviews with laboratory staff, who were constantly present, revealed, for example, that most users required help to print their search results.

In the public library setting, ELMORE found that users of Search Helper were unable to utilize the package adequately, and searches had to be performed for patrons by librarians. Other studies also strongly suggest that end users experience difficulties with the searching process. A University of Ottawa questionnaire showed that 80% of end users prefer a librarian be on hand for assistance (JANKE, 1984b). SLINGLUFF ET AL. found that users had difficulty in constructing search strategies and applying Boolean logic. PENHALE & TAYLOR found that 82% of the subjects surveyed said they wanted help from a librarian on the mechanics of searching, search strategy, or both.

All of the end-user search centers report that operating the service in a library setting requires a significant investment in staff to monitor the service and assist users (BRODY ET AL.; HALPERIN & PAGELL; JANKE, 1985a; JAROS ET AL.; PEISCHL & MONTGOMERY; SLINGLUFF ET AL.).

It is interesting to examine the results from these end-user behavior studies with the literature on the searching behavior of intermediaries. Analysis of the results of numerous studies on searcher behavior indicates that experienced searchers perform more complicated and better searches but that poor searches occur at every level of experience (BORGMAN; FENICHEL, 1980). It appears that search-strategy formulation is a problem at all levels of searching. There is evidence that new users can learn to perform simple searches after fairly brief training (FENICHEL, 1981). A complicating factor in interpreting this research is the wide variation found in searching behavior at every level. TRIVISON ET AL. noted that "serious questions can be raised about how much we know about what is really going on in search processes" (p. 341). In light of these results, it should come as no surprise that some end users have search difficulties. It should also be remembered that even the low-frequency intermediate searchers classified as novice searchers in these studies have more training and experience than the typical end-user searcher.

Costs of Searching

Because of their infrequent searching, many end users will remain novice searchers. Survey results indicate that end users will avail themselves of search services fairly infrequently: 80% less often than once a month (JANKE, 1984b); 72% once a month or less (FRIEND); 64% once or twice a month (HAWKINS & LEVY, 1987); and 74% once a month or less when charged a fee (CROOKS).

The cost of searching can be an important factor in user searching frequency, both in the library setting and in the workplace. BODTKE-ROBERTS found that search activity dropped more than 90% after the free trial period had ended. In the Michigan study (CROOKS), the survey indicated that 99% of the participants would use the service if it were free, but only 56% would use it if a $10 fee were imposed. The KLEINER trial of Search Helper showed a "startling increase in use" after the service was made available free of charge, even though the search fee had been only $2. In the Pittsburgh study (BRODY ET AL.), a user fee of $5 per half hour was instituted for the last month of the study. Use was still regarded as high (496 users), but database use dropped 79% from the previous month.

Numerous other authors have noted the cost factor and the financial ramifications that increased end-user searching can have on libraries or corporations sponsoring end-user searching (ARNOLD, 1986; KOSMIN; PEISCHL & MONTGOMERY). Clearly, end users and novice searchers spend more time online during a search session than intermediaries and their searches therefore will be more expensive (FENICHEL, 1981; JANKE, 1984b; PENHALE & TAYLOR; TRZEBIATOWSKI; VOLLARO & HAWKINS).

Training Methods

The area of training users of online systems has been a topic of much discussion. There is relevant information in the literature on training intermediaries, online catalog users, and end-user searchers. There is evidence that people who receive some training in online catalog use report higher rates of success and satisfaction (MATTHEWS & LAWRENCE). It is also clear, however, that online catalog users are disinclined to seek training, attend training seminars, or even consult the Library of Congress Subject Headings list (B. BAKER; MARKEY, 1983).

BORGMAN summarized the research on training across all types of information retrieval systems: "frequent use of a system and a database leads to better use, and some training is better than no training. These results are not yet very strong" (p. 394). Because of the wide variation in searching behavior among searchers with the same level of experience, the effects of training methods can be difficult to measure.

In the end-user studies, the usual training method was the group lecture with a demonstration search, followed by hands-on experience for users. The amount of time spent on instruction and the detail presented varied. Several studies explored more than one approach.

SEWELL & TEITELBAUM prefer a one-on-one training method and are confident that this method has contributed to the success enjoyed by their users. Evaluative surveys have confirmed that one-on-one individual instruction is also preferred by users. These authors have prepared a search manual emphasizing the problem searching areas they have identified.

SIMON found that the formal hour-long user education programs held for health science end users were attended by only two or three people. A survey question asking users to rank training methods showed "learn by using" as the most popular, with individual instruction second, CAI third, and workshops and reading manuals far behind. This again illustrates the role of the interface in system–user interaction in situations where users are anxious to dispense with the training and begin searching.

POISSON found that the end users who were trained directly by librarians rather than by other end users or in group training sessions performed more effective searches.

The investigators in the Texas A&M search service compared a slide/tape program, a printed manual, and a CAI program in an effort to arrive at the most satisfactory instructional method (JAROS ET AL.). While CAI was favored by users, analysis of the results indicated the printed manual was the most effective. All new users are required to read the manual and have their search strategy approved by library staff before going online.

GORDON reviewed various training methods for end users and suggested that the best way to teach online searching is the one-on-one method in a comfortable setting. Because this method is usually unrealistic, innovative training methods need to be explored.

The INTREX Project staff found that an online CAI approach that took the user "by the hand" through the procedures required to extract information from the system was the most effective (OVERHAGE & REINTJES).

Numerous authors have suggested the use of CAI in training end users (VIGIL, 1986). A software package designed by GROTOPHORST has been adapted for training end users on the user-friendly vendor systems.

A number of software packages have been developed to train and assist novices and library school students in online searching (BOYCE ET AL.; CARUSO; EISENBERG ET AL.; HENDRY ET AL.; STIRLING). BOYCE has recently noted that much effort has gone into designing these CAI programs for online searching but that virtually no effort has been put into determining whether they actually improve instruction. Nevertheless, the potential for flexible and sophisticated CAI programs designed for end-user searching certainly exists within the front-end interfaces. This is an area that will receive increased attention.

While end-user survey respondents indicate a desire for training, users tend not to use the training materials and search tools available to them. CROOKS reports that 62% of Michigan users surveyed expressed a desire for training sessions. However, only 48% read any of the available manuals or search guides. HAWKINS & LEVY (1986b), when surveying the participants in an Easynet trial, found that one-half of the users wanted no training at all. Librarians at the Pittsburgh end-user laboratory held formal bibliographic instruction sessions that were attended by a total of 227 people, a small percentage of the 5,237 searchers who used the laboratory. Carnegie-Mellon Search Helper users did not consult written instructions and ranked the system manual as the least desirable source for help, behind asking a friend, online help, library staff, and asking anyone available (PISCIOTTA ET AL.). Again, insight here can be gained from the INTREX Project, which found that all manual tools, from the full manual through abbreviated manuals to wall charts, were seldom used (OVERHAGE & REINTJES).

Because of end-user problems with the mechanics of search-strategy formulation and Boolean logic and the questions surrounding methods of training, several of the library-based programs require end users to complete a presearch strategy form and have it approved by a trained searcher (JAROS ET AL.; SLINGLUFF ET AL.; THOMAS). JANKE (1985a) has established a formal presearch (and post-search) counseling service and believes that the service is "nothing less than key to the client searcher's performance online" (p. 16). In this way, one-on-one training and individual review of the user's search strategy can be accomplished.

These studies reveal that when asked to compare the results of searches performed by themselves with those done by intermediaries, end users typically indicate that the intermediary search contains more relevant citations. However, this may not be a concern for those end users seeking only a few relevant citations (JANKE, 1984b; KIRBY & MILLER, 1985). These demonstration projects and surveys show that users have a strong interest in conducting their own searches; however, cost is an important factor and will heavily influence actual user searching activity.

ROLE OF LIBRARIANS

Survey articles have appeared over the past six years examining the general issues and questions surrounding end-user searching (FAIBISOFF & HURYCH; OJALA, 1985; 1986). Nearly every author reporting an end-user study conjectures on the philosophical and policy questions connected with direct user searching. Of particular concern has been the role of intermediary librarians vis-a-vis end users and the implications of increased direct end-user searching (DALRYMPLE; JACK; JANKE, 1985b; PEART). NIELSEN (1980) suggested that online searching, which up to that point had been regarded as a professionalizing force for librarianship, could lead to disadvantages for librarians if trends in end-user access continued. At the other extreme, SUMMIT & MEADOW predicted that for librarians involved in searching, their roles, salaries, and stature in society would rise.

NICHOLAS & HARMAN have noted that "there is scant evidence to indicate that the prospect of becoming an end-user is attracting, in droves, users" (p. 182). As end-user searching has grown and developed, most writers have taken the cautious position put forth by DES CHENE and OJALA

(1985) that, while the role of librarians has and will continue to change, the function of the intermediary will remain. Authors point to additional advisory and consulting roles, particularly with the new intermediaries (JANKE, 1985b).

OPTICAL DISK TECHNOLOGIES

CD-ROM technology offers on-disk access to bibliographic databases without connect time costs, telecommunication fees, and printing charges. All would agree with M. E. WILLIAMS (1986a) that "the excitement and interest generated by this new technology is probably greater than that associated with any technology in the online database field since the introduction of the personal computer" (p. 2). However, the impact of this technology on the online field is unknown, and the first commercial product releases do not make future predictions easier.

While IAC's optical disk database, InfoTrac, was greeted with enthusiasm by users and library staff, several authors have criticized its lack of keyword searching and reliance on LCSH (Library of Congress Subject Headings) for subject access (HALL ET AL; HAWKINS, 1986; VAN ARSDALE & OSTRYE). The early optical products have contained small subsets of databases, have not provided sophisticated retrieval techniques, and have long response times. There are questions concerning the aggregate costs of subscriptions, particularly in academic libraries where one can visualize a "CD arcade" of different databases.

It is not clear that the large, complex scientific databases will lend themselves to the CD-ROM environment. The DEC (Digital Equipment Corp.) CD-ROM system, featuring various engineering and chemical databases, was voted first-place winner of the Optical Memory News 1985 Distinguished Achievement Award For Excellence in Optical Technology. Several months later, DEC discontinued its optical disk line.

Very little in the way of user studies has been done for the CD products. In a letter, LYLE indicates that Clemson University, after experimenting with optical databases, believes that "end-user searching has proven to be the most useful for students and faculty" (p. 9). HALL ET AL. expressed concern with the quality of the searches and found that students used InfoTrac when they would have been better served with another index. DEWEY reports that a study of the CD-ROM ISI Science Citation Index found that slowness of the system was a problem, that users seem to need training, and that the system proved not as easy to use as the users thought it should be. Dewey also reported on a study of SilverPlatter's ERIC database in which 84% of respondents preferred using the CD-ROM to the online database. In this study, staff indicated that most users needed some guidance in using the system. DENNIS reported that 67% of the users of Compact Cambridge/Medline at Hahnemann University were satisfied with the results of their search, but 33% recommended that the library not purchase the system. TOOEY found responses to Silver-Platter's PsycLIT at the University of Maryland overwhelmingly positive. It should be clear from online catalog and end-user database searching studies that many users will enthusiastically search CD databases, but the retrieval effectiveness of these searches needs to be examined.

CONCLUSION

The literature reveals an enormous amount of interest in end-user database searching. The primary areas of study are online catalog use, end-user searching of remote databases, and the use of optical disk databases. User studies of online catalogs show a need for improved and enhanced subject access, including access to the periodical literature. Searching of remote bibliographic databases has historically been the province of trained intermediaries. Recently, there has been a renewed interest in direct end-user searching with the introduction of user-friendly vendor search systems, the wide availability of microcomputers, and the development of front-end and gateway software.

End-user searching activities can be grouped into two categories: 1) training professional users to search in the work environment; and 2) establishing end-user search centers in primarily academic libraries. Efforts to train professional users in command language searching reveal a high dropout rate of trained searchers. This has been attributed to the inconvenience and infrequency of searching, the problems of command language, the cost of searching, and other demands on users' time. Studies have shown that a new class of non-librarian searchers, or new intermediaries, is being trained. Several front-end search packages to facilitate access to online systems and databases have gained support, and developmental efforts on front ends continue.

End-user search services in academic libraries have reported high user satisfaction with the after-hours user-friendly systems. The primary clientele for these search centers has been graduate students. It would appear that a high percentage of the users of end-user services are not regular users of the library-mediated search services.

Analysis of end-user search results reveals that end users are not performing particularly effective searches. Comparisons of end-user searches with repeat searches by trained intermediaries show that the intermediaries retrieve more relevant citations and spend less time online. End users have particular difficulty with search-strategy formulation and Boolean logic and often seek help from library staff during their search.

Studies suggest that searching costs influence search frequency. Various training methods for end-user searching have been explored, with one-on-one training and CAI appearing to be the most effective.

Optical disk technology offers enhanced access without connect time and telecommunication costs. The future impact of this technology is not clear.

End users are enthusiastic about inexpensive information retrieval systems, such as online catalogs, user-friendly online search systems, and optical disk databases, that can be searched with little training. However, they will not retrieve as many relevant citations or perform as efficiently as trained intermediaries.

BIBLIOGRAPHY

ARNOLD, STEPHEN E. 1985. Hard Lessons About End Users. In: Williams, Martha E.; Hogan, Thomas H., comps. Proceedings of the 6th National Online Meeting; 1985 April 30–May 2; New York, NY. Medford, NJ: Learned Information, Inc.; 1985. 11–18. ISBN: 0-938734-09-1.

ARNOLD, STEPHEN E. 1986. End Users: Old Myths and New Realities. In: Williams, Martha E.; Hogan, Thomas H., comps. Proceedings of the 7th National Online Meeting; 1986 May 6–8; New York, NY. Medford, NJ: Learned Information, Inc.; 1986. 5–10. ISBN: 0-938734-12-1.

ARNOLD, STEPHEN E. 1987. End-Users: Dreams or Dollars. Online. 1987 January; 11(1): 71–81. ISSN: 0146-5422.

BAKER, BETSY. 1986. A New Direction for Online Catalog Instruction. Information Technology and Libraries. 1986 March; 5(1): 35–41. ISSN: 0730-9295.

BAKER, CAROLE A. 1984. COLLEAGUE: A Comprehensive Online Medical Library for the End User. Medical Reference Services Quarterly. 1984 Winter; 3(4): 13–26. ISSN: 0276-3869.

BARBER, A. STEPHANIE; BARRACLOUGH, ELIZABETH D.; GRAY, W. A. 1973. On-Line Information Retrieval as a Scientists Tool. Information Storage and Retrieval. 1973 August; 9(8): 429–440. ISSN: 0020-0271.

BARRACLOUGH, ELIZABETH D. 1977. Progress in Documentation: On-Line Searching in Information Retrieval. Journal of Documentation. 1977 September; 33(3): 220–238. ISSN: 0022-0418.

BATES, MARCIA. 1986a. Subject Access in Online Catalogs: A Design Model. Journal of the American Society for Information Science. 1986 November; 37(6): 357–376. ISSN: 0002-8231.

BATES, MARCIA. 1986b. Terminological Assistance for the Online Subject Searcher. In: Jacobson, Carol E.; Witges, Shirley A., comps. Proceedings of the 2nd Conference on Computer Interfaces and Intermediaries for Information Retrieval; 1986 May 28–31; Boston, MA. Alexandria, VA: Defense Technical Information Center; 1986. 285–293. (DTIC/TR-86/5).

BODTKE-ROBERTS, ALICE. 1983. Faculty End-User Searching of BIOSIS. In: Williams, Martha E.; Hogan, Thomas H., comps. Proceedings of the 4th National Online Meeting; 1983 April 12–14; New York, NY. Medford, NJ: Learned Information, Inc.; 1983. 45–56. ISBN: 0-938734-05-9.

BORGMAN, CHRISTINE L. 1986. Why are Online Catalogs Hard to Use? Lessons Learned from Information-Retrieval Studies. Journal of the American Society for Information Science. 1986 November; 37(6): 387–400. ISSN: 0002-8231.

BOYCE, BERT R. 1987. Computer-Assisted Instruction for Online Searchers. Bulletin of the American Society for Information Science. 1987 April/May; 13(4): 34. ISSN: 0095-4403.

BOYCE, BERT R.; MARTIN, DAVID; FRANCIS, BARBARA; SIEVERT, MARY ELLEN. 1984. The DAPPOR Answer Evaluation Program. Information Technology and Libraries. 1984 September; 3(3): 306–309. ISSN: 0730-9295.

BRANDRIFF, ROBERT K.; LYNCH, CLIFFORD, A. 1985. The Evolution of the User Interface in the MELVYL Online Catalog, 1980–1985. In: Parkhurst, Carol A., ed. ASIS '85: Proceedings of the American Society for Information Science (ASIS) 48th Annual Meeting; 1985 October 20–24; Las Vegas, NV. White Plains, NY: Knowledge Industry Publications, Inc.; 1985. 102–105. ISSN: 0044-7870; ISBN: 0-867291-76-1.

BRODY, FERN; WHITMORE, MARILYN; MCCORMICK, GREG. 1986. End User Searching: An Experiment at the University of Pittsburgh. Pittsburgh, PA: University Library System, University of Pittsburgh; 1986 May.

BROERING, NAOMI C. 1985. The MiniMEDLINE SYSTEM: A Library-Based End-User Search System. Bulletin of the Medical Library Association. 1985 April; 73(2): 138–145. ISSN: 0025-7338.

BRUNNING, DENNIS R.; STEWART, DOUG. 1986. Review of Searcher's Tool Kit. Information Technology and Libraries. 1986 December; 5(4): 363–366. ISSN: 0730-9295.

BUNTROCK, ROBERT E.; VALICENTI, ALDONA K. 1985. End-Users and Chemical Information. Journal of Chemical Information and Computer Sciences. 1985 August; 25(3): 203–207. ISSN: 0095-2338.

BYERLY, GREG. 1983. Online Searching: A Dictionary and Bibliographic Guide. Littleton, CO: Libraries Unlimited; 1983. 288p. ISBN: 0-872873-81-1.

CARUSO, ELAINE. 1981. TRAINER. Online. 1981 January; 5(1): 36–38. ISSN: 0146-5422.

CASE, DONALD; BORGMAN, CHRISTINE L.; MEADOW, CHARLES T. 1986. End-User Information-Seeking in the Energy Field: Implications for End-User Access to DOE/RECON Databases. Information Processing & Management. 1986; 22(4): 299–308. ISSN: 0306-4573.

CLANCY, STEPHEN. 1985. BRS/Saunders Colleague: An Information Service for Medical Professionals. Database. 1985 June; 8(2): 108–121. ISSN: 0162-4105.

COONS, BILL. 1986a. Frontiers in Front-Ends and Gateways. In: Proceedings of the ONLINE '86 Conference; 1986 November 4–6; Chicago, IL. Weston, CT: Online Inc.; 1986. 30–35.

COONS, BILL. 1986b. SearchWorks: Does it Really Work? Database. 1986 December; 9(6): 62–68. ISSN: 0162-4105.

CRAWFORD, SUSAN; HALBROOK, BARBARA; IGIELNIK, SIMON. 1986. Testing the Use of a Medical Center-Wide Online Current Awareness Service: BACS/Current Contents. In: Hurd, Julie M., ed. ASIS '86: Proceedings of the American Society for Information Science (ASIS) 49th Annual Meeting; 1986 September 28–October 2; Chicago, IL. Medford, NJ: Learned Information, Inc.; 1986. 335–340. ISSN: 0044-7870; ISBN: 0-938734-14-8.

CROOKS, JAMES E. 1985. End-User Searching at the University of Michigan Library. In: Williams, Martha E.; Hogan, Thomas H., comps. Proceedings of the 6th National Online Meeting; 1985 April 30–May 2; New York, NY. Medford, NJ: Learned Information, Inc.; 1985. 99–110. (With handout at the meeting). ISBN: 0-938734-09-1.

DALRYMPLE, PRUDENCE W. 1984. Closing the Gap: The Role of the Librarian in Online Searching. RQ. 1984 Winter; 24(20): 177–185. ISSN: 0033-7072.

DENNIS, SHARON. 1987. Medical University Library Evaluates Medline CD-ROM. Information Today. 1987 May; 4(5): 1. ISSN: 8755-6286.

DES CHENE, DORICE. 1985. Online Searching by End Users. RQ. 1985 Fall; 25(1): 89–95. ISSN: 0033-7072.

DEWEY, BARBARA I. 1987. Report of the SIG/ED Session on Academic
End-User Access to Databases for the SIG/ED Newsletter. In: Summary
of Papers Read at the American Society for Information Science (ASIS)
Mid-Year Meeting; 1987 May 17-20; Kings Island, Ohio. 2p.
DICKSON, JEAN. 1984. An Analysis of User Errors in Searching an Online
Catalog. Cataloging & Classification Quarterly. 1984 Spring; 4(3): 19-
38. ISSN: 0163-9374.
DIODATO, VIRGIL. 1984. Popular Magazines Discuss Online Information
Retrieval. Online. 1984 May; 8(3): 24-29. ISSN: 0146-5422.
DODD, JANE; GILREATH, CHARLES; HUTCHINS, GERALDINE. 1985.
A Comparison of Two End-User Operated Search Systems: Final Report.
Washington, DC: Office of Management Studies, Association of Research
Libraries; 1985. ERIC: ED 255224.
DOSZKOCS, TAMAS E. 1983. CITE NLM: Natural-Language Searching in
an Online Catalog. Information Technology and Libraries. 1983
December; 2(4): 364-380. ISSN: 0730-9295.
EISENBERG, LAURA J.; STANDING, ROY A.; TIDBALL, CHARLES S.;
LEITER, JOSEPH. 1978. MEDLEARN: A Computer-Assisted Instruc-
tion (CAI) Program for MEDLARS. Bulletin of the Medical Library
Association. 1978 January; 66(1): 6-13. ISSN: 0025-7338.
EISENBERG, MICHAEL. 1983. The Direct Use of Online Bibliographic
Information Systems by Untrained End Users: A Review of Research.
Syracuse, NY: National Institute of Education; 1983. 40p. ERIC:
ED 238440.
ELLEFSEN, DAVID. 1985. Automated Periodical Reference Service. In-
formation Technology and Libraries. 1985 December; 4(4): 353-355.
ISSN: 0730-9295.
ELMORE, BARBARA. 1985. End-User Searching in a Public Library. In:
Online '85: Conference Proceedings; 1985 November 4-6; New York,
NY. Weston, CT: Online Inc.; 1985. 98-101.
ENSOR, PAT; CURTIS, RICHARD A. 1984. Search Helper: Low-Cost On-
line Searching in an Academic Library. RQ. 1984 Spring; 23(3): 327-
331. ISSN: 0033-7072.
EVANS, NANCY; PISCIOTTA, HENRY. 1985. Search Helper: Testing
Acceptance of a Gateway Software System. In: Williams, Martha E.;
Hogan, Thomas H., comps. Proceedings of the 6th National Online
Meeting; 1985 April 30-May 2; New York, NY. Medford, NJ: Learned
Information, Inc.; 1985. 131-136. ISBN: 0-938734-09-1.
FAIBISOFF, SYLVIA G.; HURYCH, JITKA. 1981. Is There a Future for
the End User in Online Bibliographic Searching? Special Libraries. 1981
October; 72(4): 347-355. ISSN: 0038-6723.
FENICHEL, CAROL HANSEN. 1980. The Process of Searching Online
Bibliographic Databases: A Review of Research. Library Research.
1980 Summer; 2(2): 107-127. ISSN: 0164-0763.
FENICHEL, CAROL HANSEN. 1981. Online Searching: Measures that
Discriminate among Users with Different Types of Experiences. Journal
of the American Society for Information Science. 1981 January; 32(1):
23-32. ISSN: 0002-8231.
FENICHEL, CAROL HANSEN. 1986. Intermediary Information Systems:
The Range of Options from the User's Point of View. In: Jacobson, Carol
E.; Witges, Shirley A., comps. Proceedings of the 2nd Conference on

Computer Interfaces and Intermediaries for Information Retrieval; 1986 May 28-31; Boston, MA. Alexandria, VA: Defense Technical Information Center; 1986. 9-16. (DTIC/TR-86/5).

FLYNN, KAREN L. 1985. The 3M Experience: Use of External Databases in a Large Diversified Company. Special Libraries. 1985 Spring; 76(2): 81-87. ISSN: 0038-6723.

FOGEL, LAURENCE D.; ZIGMUND, CLAIRE F. 1985. End User vs. Intermediary: A Personal Perspective. In: Williams, Martha E.; Hogan, Thomas H., comps. Proceedings of the 6th National Online Meeting; 1985 April 30-May 2; New York, NY. Medford, NJ: Learned Information, Inc.; 1985. 153-159. ISBN: 0-938734-09-1.

FOX, EDWARD. 1986. A Design for Intelligent Retrieval: The CODER System. In: Jacobson, Carol E.; Witges, Shirley A., comps. Proceedings of the 2nd Conference on Computer Interfaces and Intermediaries for Information Retrieval; 1986 May 28-31; Boston, MA. Alexandria, VA: Defense Technical Information Center; 1986. 135-154. (DTIC/TR-86/5).

FRIEND, LINDA. 1985. Independence at the Terminal: Training Student End Users to Do Online Literature Searching. Journal of Academic Librarianship. 1985 July; 11(3): 136-141. ISSN: 0099-1333.

GLASGOW, VICKI L.; FOREMAN, GERTRUDE. 1986. U-Search: A Program to Teach End User Searching at an Academic Health Sciences Library. In: Wood, M. Sandra; Horak, Ellen Brassil; Snow, Bonnie, eds. End User Searching in the Health Sciences. New York, NY: Haworth Press, Inc.; 1986. 137-148. ISBN: 0-86656-465-9.

GORDON, DENA W. 1983. Online Training for the End User or Information Consumer. Paper presented at the Mid-Year Meeting of the American Society for Information Science (ASIS); 1983 May 22-25; Lexington, KY. 13p. ERIC: ED 245697.

GROTOPHORST, CLYDE W. 1984. Training University Faculty as End-use Searchers: A CAI Approach. In: Williams, Martha E.; Hogan, Thomas H., comps. Proceedings of the 5th National Online Meeting; 1984 April 10-12; New York, NY. Medford, NJ: Learned Information Inc.; 1984. 77-82. ISBN: 0-938734-07-5.

HAINES, JUDITH S. 1982. Experiences in Training End-User Searchers. Online. 1982 November; 6(6): 14-23. ISSN: 0146-5422.

HAINES, JUDITH S.; NAJJAR, ROBERT C.; WEHNER, KAREN. 1986. In-House Training of Chemists in Searching CAS Online Databases. In: Williams, Martha E.; Hogan, Thomas H., comps. Proceedings of the 7th National Online Meeting; 1986 May 6-8; New York, NY. Medford, NJ: Learned Information, Inc.; 1986. 157-161. ISBN: 0-938734-12-1.

HALL, CYNTHIA; TALAN, HARRIET; PEASE, BARBARA. 1987. Info-Trac in Academic Libraries: What's Missing in the New Technology? Database. 1987 February; 10(1): 52-56. ISSN: 0162-4105.

HALPERIN, MICHAEL; PAGELL, RUTH A. 1985. Free 'Do-It-Yourself' Online Searching... What to Expect. Online. 1985 March; 9(2): 82-84. ISSN: 0146-5422.

HARMAN, JENNIFER. 1986. Reuters: A Survey of End-User Searching. ASLIB Proceedings. 1986 January; 38(1): 35-42. ISSN: 0001-253X.

HAWKINS, DONALD T. 1981. Online Information Retrieval Systems. In: Williams, Martha E., ed. Annual Review of Information Science and Technology: Volume 16. White Plains, NY: Knowledge Industry Publications, Inc. for the American Society for Information Science; 1981. 171-208. ISSN: 0066-4200; ISBN: 0-914236-90-3.

HAWKINS, DONALD T. 1986. A Trial of the InfoTrac System in a Technical Library Environment. In: Proceedings of the ONLINE '86 Conference; 1986 November 4-6; Chicago, IL. Weston, CT: Online Inc.; 1986. 103-105.

HAWKINS, DONALD T.; LEVY, LOUISE R. 1985. Front End Software for Online Database Searching. Part 1: Definitions, System Features, and Evaluation. Online. 1985 November; 9(6): 30-37. ISSN: 0146-5422.

HAWKINS, DONALD T.; LEVY, LOUISE R. 1986a. Front End Software for Online Database Searching. Part 3: Product Selection Chart and Bibliography. Online. 1986 May; 10(3): 49-58. ISSN: 0146-5422.

HAWKINS, DONALD T.; LEVY, LOUISE R. 1986b. Introduction of On-line Searching to End Users at AT&T Bell Laboratories. In: Williams, Martha E.; Hogan, Thomas H., comps. Proceedings of the 7th National Online Meeting; 1986 May 6-8; New York, NY. Medford, NJ: Learned Information, Inc.; 1986. 1-4. ISBN: 0-938734-12-1.

HAWKINS, DONALD T.; LEVY, LOUISE R. 1987. A Year's Experience with End User Searching. In: Williams, Martha E.; Hogan, Thomas H., comps. Proceedings of the 8th National Online Meeting; 1987 May 5-7; New York, NY. Medford, NJ: Learned Information, Inc.; 1987. 155-159. ISBN: 0-938734-17-2.

HENDRY, IAN G.; WILLETT, PETER; WOOD, FRANCES E. 1986. INSTRUCT: A Teaching Package for Experimental Methods in Information Retrieval. Part I. The Users' View. Program. 1986 July; 20(3): 245-263. ISSN: 0033-0337.

HILDRETH, CHARLES R. 1982. Online Public Access Catalogs: The User Interface. Dublin, OH: OCLC; 1982. 263p. ISBN: 0-933418-34-5.

HILDRETH, CHARLES R. 1984. Pursuing the Ideal: Generations of Online Catalogs. In: Aveney, Brian; Butler, Brett, eds. Online Catalogs, Online Reference: Converging Trends: Proceedings of a Library and Information Technology Association Preconference Institute; 1983 June 23-24; Los Angeles, CA. Chicago, IL: American Library Association; 1984. 31-56. ISBN: 0-838933-08-4.

HILDRETH, CHARLES R. 1985. Online Public Access Catalogs. In: Williams, Martha E., ed. Annual Review of Information Science and Technology: Volume 20. White Plains, NY: Knowledge Industry Publications, Inc. for the American Society for Information Science; 1985. 233-285. ISSN: 0066-4200; ISBN: 0-86729-175-3.

HILDRETH, CHARLES R. 1987. Beyond Boolean: Designing the Next Generation of Online Catalogs. Library Trends. 1987 Spring; 35(4): 647-658. ISSN: 0024-2594.

HOOK, SARA ANNE. 1986. BRS/BRKTHRU: A Happy Medium. Online. 1986 January; 10(1): 97-101. ISSN: 0146-5422.

HOROWITZ, GARY L.; JACKSON, JEROME D.; BLEICH, HOWARD L. 1983. PaperChase; Self-Service Bibliographic Retrieval. Journal of the American Medical Association. 1983 November 11; 250(18): 2494-2499. ISSN: 0098-7484.

HUNTER, JANNE A. 1983. What Did You Say the End-User Was Going to Do at the Terminal, and How Much Is It Going to Cost? In: Williams, Martha E.; Hogan, Thomas H., comps. Proceedings of the 4th National Online Meeting; 1983 April 12-14; New York, NY. Medford, NJ: Learned Information, Inc.; 1983. 223-229. ISSN: 0-938734-05-9.

HUNTER, JANNE A. 1984. When Your Patrons Want to Search--The Library as Advisor to End-Users. . . A Compendium of Advice and Tips. Online. 1984 May; 8(3): 36-41. ISSN: 0146-5422.

HURT, C. D. 1983. Intermediaries, Self-Searching and Satisfaction. In: Williams, Martha E.; Hogan, Thomas H., comps. Proceedings of the 4th National Online Meeting; 1983 April 12-14; New York, NY. Medford, NJ: Learned Information, Inc.; 1983. 231-238. ISBN: 0-938734-05-9.

HUSHON, JUDITH. 1986. How Micro-CSIN, A New Gateway to Online Systems, Stacks Up. In: Williams, Martha E.; Hogan, Thomas H., comps. Proceedings of the 7th National Online Meeting; 1986 May 6-8; New York, NY. Medford, NJ: Learned Information, Inc.; 1986. 203-210. ISBN: 0-938734-12-1.

HUSHON, JUDITH M.; CONRY, THOMAS J. 1986. The Evolution of Gateway Technology. In: Jacobson, Carol E.; Witges, Shirley A., comps. Proceedings of the 2nd Conference on Computer Interfaces and Intermediaries for Information Retrieval; 1986 May 28-31; Boston, MA. Alexandria, VA: Defense Technical Information Center; 1986. 181-201. (DTIC/TR-86/5).

INFORMATION TODAY. 1987a. AccuSearch to Market EasyNet to Libraries. Information Today. 1987 February; 4(2): 10. ISSN: 8755-6286.

INFORMATION TODAY. 1987b. Georgia Tech Wires Campus for PC Access to Library. Information Today. 1987 May; 4(5): 23. ISSN: 8755-6286.

INFORMATION TODAY. 1987c. Online Dose for Doctors: Dialog Medical Connection. Information Today. 1987 May; 4(5): 1. ISSN: 8755-6286.

JACK, ROBERT F. 1986. Why Searchers Hate End-Users. Database End-User. 1986 March; 2(3): 23-25. ISSN: 0882-326X.

JACOBSON, CAROL E.; WITGES, SHIRLEY A., comps. 1986. Proceedings of the 2nd Conference on Computer Interfaces and Intermediaries for Information Retrieval; 1986 May 28-31; Boston, MA. Alexandria, VA: Defense Technical Information Center; 1986. 371p. (DTIC/TR-86/5).

JANKE, RICHARD V. 1983. BRS/After Dark: The Birth of Online Self-Service. Online. 1983 September; 7(5): 12-29. ISSN: 0146-5422.

JANKE, RICHARD V. 1984a. Just What Is an End User? In: Online '84 Conference Proceedings; 1984 October 29-30; San Francisco, CA. Weston, CT: Online Inc.; 1984. 148-154. ISSN: 0146-5422.

JANKE, RICHARD V. 1984b. Online After Six: End-User Searching Comes of Age. Online. 1984 November; 8(6): 15-29. ISSN: 0146-5422.

JANKE, RICHARD V. 1985a. Presearch Counseling for Client Searchers (End-Users). Online. 1985 September; 9(5): 13-26. ISSN: 0146-5422.

JANKE, RICHARD V. 1985b. Client Searchers and Intermediaries: The New Online Partnership. In: Online '85 Conference Proceedings; 1985 November 4-6; New York, NY. Weston, CT: Online Inc.; 1985. 165-171.

JAROS, JOE; ANDERS, VICKI; HUTCHINS, GERI. 1986. Subsidized End-User Searching in an Academic Library. In: Williams, Martha E.; Hogan, Thomas H., comps. Proceedings of the 7th National Online Meeting; 1986 May 6-8; New York, NY. Medford, NJ: Learned Information, Inc.; 1986. 223-229. ISBN: 0-938734-12-1.

KAPLAN, ROBIN. 1985. Knowledge Index: A Review. Database. 1985 June; 8(2): 122-128. ISSN: 0162-4105.

KEHOE, CYNTHIA A. 1985. Interfaces and Expert Systems for Online Retrieval. Online Review. 1985 December; 9(6): 489-505. ISSN: 0309-314X.

KING, JOSEPH; BRUEGGEMAN, PETER. 1986. Frontenders, Gateways, User-Friendly Systems, or Whatever You Want to Call Them. Database End-User. 1986 June; 2(6): 17-21. ISSN: 0882-326X.

KIRBY, MARTHA; MILLER, NAOMI. 1985. MEDLINE Searching on BRS Colleague: Search Success of Untrained End Users in a Medical School and Hospital. In: Williams, Martha E.; Hogan, Thomas H., comps. Proceedings of the 6th National Online Meeting; 1985 April 30-May 2; New York, NY. Medford, NJ: Learned Information, Inc.; 1985. 255-263. ISBN: 0-938734-09-1.

KIRBY, MARTHA; MILLER, NAOMI. 1986. MEDLINE Searching on Colleague: Reasons for Failure or Success of Untrained End Users. Medical Reference Services Quarterly. 1986 Fall; 5(3): 17-34. ISSN: 0276-3869.

KIRK, CHERYL L. 1986. End-User Training at the Amoco Research Center. Special Libraries. 1986 Winter; 77(1): 20-27. ISSN: 0038-6723.

KLEINER, JANE P. 1985. User Searching: A Public Access Approach to Search Helper. RQ. 1985 Summer; 24(4): 442-451. ISSN: 0033-7072.

KOSMIN, LINDA J. 1986. Economic Pitfalls of Hi-Tech End User Searching. In: Williams, Martha E.; Hogan, Thomas H., comps. Proceedings of the 7th National Online Meeting; 1986 May 6-8; New York, NY. Medford, NJ: Learned Information, Inc.; 1986. 257-262. ISBN: 0-938734-12-1.

KRAVITZ, RHONDA A. RIOS; WESTLING, ELLEN R. 1986. Implementing End User Systems at the Massachusetts General Hospital Health Sciences Libraries. In: Wood, M. Sandra; Horak, Ellen Brassil; Snow, Bonnie, eds. End User Searching in the Health Sciences. New York, NY: Haworth Press, Inc.; 1986. 149-162. ISBN: 0-86656-465-9.

KRISMANN, CAROL. 1986. End-User Searching of Online Databases: An Overview. Colorado Libraries. 1986 March; 12(1): 16-22. ISSN: 0147-9733.

KUGEL, PETER. 1971. Dirty Boole? Journal of the American Society for Information Science. 1971 July-August; 22(4): 293-294. ISSN: 0002-8231.

KUPFERBERG, NATALIE. 1986. End-Users: How Are They Doing? A Librarian Interviews Six "Do-It-Yourself" Searchers. Online. 1986 March; 10(2): 24-28. ISSN: 0146-5422.

LANCASTER, F. W.; RAPPORT, R. L.; PENRY, J. K. 1972. Evaluating the Effectiveness of an On-Line Natural Language Retrieval System. Information Storage and Retrieval. 1972; 8(5): 223-245. ISSN: 0020-0271.

LEIPZIG, NANCY; KOZAK, MARLENE GALANTE; SCHWARTZ, RONALD A. 1983. Experiences with End-User Searching at a Pharmaceutical Company. In: Williams, Martha E.; Hogan, Thomas H., comps. Proceedings of the 4th National Online Meeting; 1983 April 12-14; New York, NY. Medford, NJ: Learned Information, Inc.; 1983. 325-332. ISBN: 0-938734-05-9.

LEVY, LOUISE R. 1984. Gateway Software: Is It for You? Online. 1984 November; 8(6): 67–69. ISSN: 0146-5422.

LEVY, LOUISE R.; HAWKINS, DONALD T. 1986. Front End Software for Online Database Searching. Part 2: The Marketplace. Online. 1986 January; 10(2): 33–40. ISSN: 0146-5422.

LINDER, GLORIA A.; LENON, RICHARD A.; SU, VALERIE; WIBLE, JOSEPH G.; STANGL, PETER. 1986. Training the End User: The Stanford University Medical Center Experience. In: Wood, M. Sandra; Horak, Ellen Brassil; Snow, Bonnie, eds. End User Searching in the Health Sciences. New York, NY: Haworth Press, Inc.; 1986. 113–126. ISBN: 0-86656-465-9.

LOWRY, GLENN R. 1981. Training of Users of Online Services: A Survey of the Literature. Science & Technology Libraries. 1981 Spring; 1(3): 27–40. ISSN: 0194-262X.

LYLE, MARTHA. 1987. Letter to the Editor. The Ei Insider. 1987 Spring; 5: 9.

LYON, SALLY. 1984. End-User Searching on Online Databases: A Selective Annotated Bibliography. Library Hi Tech. 1984 Summer; 2(2): 47–50. ISSN: 0737-8831.

MADER, SHARON; PARK, ELIZABETH H. 1985. BRS/After Dark: A Review. Reference Services Review. 1985 Spring; 13(1): 25–28. ISSN: 0090-7324.

MANCALL, JACQUELINE C. 1983. Training Students to Search Online: Rationale, Process, and Implications. Drexel Library Quarterly. 1983 Winter; 20(1): 64–84. ISSN: 0012-6160.

MARCUS, R. S.; REINTJES, J. F. 1981. A Translating Computer Interface for End-User Operation of Heterogeneous Retrieval Systems. I. Design. Journal of the American Society for Information Science. 1981 July; 32(4): 287–303. ISSN: 0002-8231.

MARKEY, KAREN. 1983. Thus Spake the OPAC User. Information Technology and Libraries. 1983 December; 2(4): 381–387. ISSN: 0730-9295.

MARKEY, KAREN. 1984. Subject Searching in Library Catalogs: Before and After the Introduction of Online Catalogs. Dublin, OH: OCLC; 1984. 174p. ISBN: 0-933418-54-X.

MARRON, BEATRICE; FIFE, DENNIS. 1976. Online Systems—Techniques and Services. In: Williams, Martha E., ed. Annual Review of Information Science and Technology: Volume 11. Washington, DC: American Society for Information Science; 1976. 163–210. ISSN: 0066-4200; ISBN: 0-87715-212-8.

MARTIN, JANE F.; DUTTON, BRIAN G. 1985. Online End-User Training: Experiences in a Large Industrial Organization. Program. 1985 October; 19(4): 351–358. ISSN: 0033-0337.

MARTIN, THOMAS H. 1975. Reflections upon the State-of-the-Art in Interactive Retrieval. In: Information Systems and Networks: 11th Annual Symposium; 1974 March 27–29; Los Angeles, CA. Westport, CT: Greenwood Press; 1975. 77–83. ISBN: 0-8371-7717-0.

MATTHEWS, JOSEPH R. 1985. Public Access to Online Catalogs. New York, NY: Neal-Schuman; 1985. 497p. ISBN: 0-918212-89-8.

MATTHEWS, JOSEPH R.; LAWRENCE, GARY S. 1984. Further Analysis of the CLR Online Catalog Project. Information Technology and Libraries. 1984 December; 3(4): 354–376. ISSN: 0730-9295.

MATTHEWS, JOSEPH R.; LAWRENCE, GARY S.; FERGUSON, DOUGLAS K. 1983. Using Online Catalogs: A Nationwide Survey. New York, NY: Neal-Schuman; 1983. 255p. ISBN: 0-918212-96-6.

MCCANDLESS, PATRICIA; CHAPLAN, MARGARET; CLARK, BARTON M.; KOHL, DAVID F.; WALLACE, DEE; WERT, LUCILLE. 1985. The Invisible User: User Needs Assessment for Library Public Services. Washington, DC: Office of Management Studies, Association of Research Libraries; 1985. 47p. ERIC: ED 255227.

MCCARN, DAVIS B. 1978. Online Systems—Techniques and Services. In: Williams, Martha E., ed. Annual Review of Information Science and Technology: Volume 13. White Plains, NY: Knowledge Industry Publications, Inc. for the American Society for Information Science; 1978. 85-124. ISSN: 0066-4200; ISBN: 0-914236-21-0.

MEADOW, CHARLES T. 1979. Online Searching and Computer Programming: Some Behavioral Similarities (or. . . Why End Users Will Eventually Take Over the Terminal). Online. 1979 January; 3(1): 49-52. ISSN: 0146-5422.

MEADOW, CHARLES T. 1986. OAK—A New Approach to User Search Assistance. In: Jacobson, Carol E.; Witges, Shirley A., comps. Proceedings of the 2nd Conference on Computer Interfaces and Intermediaries for Information Retrieval; 1986 May 28-31; Boston, MA. Alexandria, VA: Defense Technical Information Center; 1986. 215-224. (DTIC/TR-86/5).

MILLER, TIM. 1986. SilverPlatter: Dishing up Data for Libraries. Information Today. 1986 June; 3(6): 24. ISSN: 8755-6286.

MILLER, TIM. 1987. Congressional Research Service: Vanguard of Online Database Users. Information Today. 1987 May; 4(5): 7, 46-47. ISSN: 8755-6286.

MISCHO, WILLIAM H. 1986. End-User Searching in an Online Catalog Environment. In: Jacobson, Carol E.; Witges, Shirley A., comps. Proceedings of the 2nd Conference on Computer Interfaces and Intermediaries for Information Retrieval; 1986 May 28-31; Boston, MA. Alexandria, VA: Defense Technical Information Center; 1986. 241-262. (DTIC/TR-86/5).

NEWLIN, BARBARA B. 1985. In-Search: The Design and Evolution of an End User Interface to DIALOG. In: Williams, Martha E.; Hogan, Thomas H., comps. Proceedings of the 6th National Online Meeting; 1985 April 30-May 2; New York, NY. Medford, NJ: Learned Information, Inc.; 1985. 313-319. ISBN: 0-938734-09-1.

NICHOLAS, DAVID; HARMAN, JENNIFER. 1985. The End-User: An Assessment and Review of the Literature. Social Science Information Studies. 1985 October; 5(4): 173-184. ISSN: 0143-6236.

NIELSEN, BRIAN. 1980. Online Bibliographic Searching and the Deprofessionalization of Librarianship. Online Review. 1980 September; 4(3): 215-224. ISSN: 0309-314X.

NIELSEN, BRIAN. 1986. What They Say They Do and What They Do: Assessing Online Catalog Use Instruction through Transaction Monitoring. Information Technology and Libraries. 1986 March; 5(1): 28-34. ISSN: 0730-9295.

O'LEARY, MICK. 1985a. Business Computer Network: A 'Gateway' to Multiple Databanks. Online. 1985 May; 9(3): 118-122. ISSN: 0146-5422.

O'LEARY, MICK. 1985b. Easynet: Doing It All for the End User. Online. 1985 July; 9(4): 106-113. ISSN: 0146-5422.
O'LEARY, MICK. 1986. Dialog Business Connection: Dialog for the End-User. Online. 1986 September; 10(5): 15-24. ISSN: 0146-5422.
OJALA, MARYDEE. 1983. Knowledge Index: A Review. Online. 1983 September; 7(5): 31-33. ISSN: 0146-5422.
OJALA, MARYDEE. 1985. End-User Searching and Its Implications for Librarians. Special Libraries. 1985; 76(2): 93-99. ISSN: 0038-6723.
OJALA, MARYDEE. 1986. Views on End-User Searching. Journal of the American Society for Information Science. 1986 July; 37(4): 197-203. ISSN: 0002-8231.
OVERHAGE, CARL F. J.; REINTJES, J. FRANCIS. 1974. Project Intrex: A General Review. Information Storage and Retrieval. 1974 May/June; 10(5/6): 157-188. ISSN: 0020-0271.
PEART, PETER A. 1985. Online Retrieval: Intermediaries vs. End Users. In: Williams, Martha E.; Hogan, Thomas H., comps. Proceedings of the 6th National Online Meeting; 1985 April 30-May 2; New York, NY. Medford, NJ: Learned Information, Inc.; 1985. 357-363. ISBN: 0-938734-09-1.
PEISCHL, THOMAS M.; MONTGOMERY, MARILYN. 1986. Back to the Warehouse or Some Implications on End User Searching in Libraries. In: Williams, Martha E.; Hogan, Thomas H., comps. Proceedings of the 7th National Online Meeting; 1986 May 6-8; New York, NY. Medford, NJ: Learned Information, Inc.; 1986. 347-352. ISBN: 0-938734-12-1.
PEMBERTON, JEFFREY K. 1985. Databank. Online. 1985 May; 9(3): 95. ISSN: 0146-5422.
PEMBERTON, JEFFREY K.; EMARD, JEAN-PAUL. 1984. What's Happening at Mead Data Central... Online. 1984 July; 8(4): 13-19. ISSN: 0146-5422.
PENHALE, SARA, J.; TAYLOR, NANCY. 1986. Integrating End-User Searching into a Bibliographic Instruction Program. RQ. 1986 Winter; 26(2): 212-220. ISSN: 0033-7072.
PISCIOTTA, HENRY; EVANS, NANCY; ALBRIGHT, MARILYN. 1984. Search Helper: Sancho Panza or Mephistopheles? Library Hi Tech. 1984; 2(3): 25-32. ISSN: 0737-8831.
POISSON, ELLEN H. 1986. End-User Searching in Medicine. Bulletin of the Medical Library Association. 1986 October; 74(4): 293-299. ISSN: 0025-7338.
PREECE, SCOTT E.; WILLIAMS, MARTHA E. 1980. Software for the Searcher's Workbench. In: Benenfeld, Alan R.; Kazlauskas, Edward John, eds. Communicating Information: Proceedings of the American Society for Information Science (ASIS) 43rd Annual Meeting: Volume 17; 1980 October 5-10; Anaheim, CA. White Plains, NY: Knowledge Industry Publications for ASIS; 1980. 403-405. ISSN: 0044-7870.
QUEENS BOROUGH CENTRAL LIBRARY STAFF. 1985. The Queens Borough Experience. Online. 1985 November; 9(6): 53-56. ISSN: 0146-5422.
QUINT, BARBARA. 1987. Journal Article Coverage in Online Library Catalogs: The Next Stage for Online Databases? Online. 1987 January; 11(1): 87-90. ISSN: 0146-5422.
RICHARDSON, ROBERT J. 1981. End-User Online Searching in a High-Technology Engineering Environment. Online. 1981 October; 5(5): 44-57. ISSN: 0146-5422.

ROBERTS, L. KAY. 1986. Evaluation of the EasyNet Gateway. In: Williams, Martha E.; Hogan, Thomas H., comps. Proceedings of the 7th National Online Meeting; 1986 May 6-8; New York, NY. Medford, NJ: Learned Information, Inc.; 1986. 375-381. ISBN: 0-938734-12-1.

ROUSSEAU, ROSEMARY; BRONARS, LORI. 1985. Engineers' Use of Microcomputers for Literature Searching: A Comparison of Sci-Mate & Tech Data Software. Paper presented at the American Society for Engineering Education 93rd Annual Meeting; 1985 June 16-19; Atlanta, GA. 7p.

RUDIN, JOAN; HAUSELE, NANCY; STOLLAK, JAY. 1985. Comparison of In-Search, Scimate and an Intelligent Terminal Emulator in Biomedical Literature Searching. In: Williams, Martha E.; Hogan, Thomas H. comps. Proceedings of the 6th National Online Meeting; 1985 April 30-May 2; New York, NY. Medford, NJ: Learned Information, Inc.; 1985. 403-414. ISBN: 0-938734-09-1.

SEEFELDT, ROBERTA B.; THOMAS, SUSAN. 1986. Second Generation End User Searching in a Corporate Environment, Part 1. In: Proceedings of the ONLINE '86 Conference; 1986 November 4-6; Chicago, IL. Weston, CT: Online Inc.; 1986. 227-231.

SEWELL, WINIFRED. 1986. Overview of End User Searching in the Health Sciences: An Opinion Paper. In: Wood, M. Sandra; Horak, Ellen Brassil; Snow, Bonnie, eds. End User Searching in the Health Sciences. New York, NY: Haworth Press, Inc.; 1986. 3-14. ISBN: 0-86656-465-9.

SEWELL, WINIFRED; TEITELBAUM, SANDRA. 1986. Observations of End-User Online Searching Behavior over Eleven Years. Journal of the American Society for Information Science. 1986 July; 37(4): 234-245. ISSN: 0002-8231.

SHAPIRO, BETH J. 1986. Laserdisks in Academic Libraries. . .Is InfoTrac the Answer? In: Proceedings of the ONLINE '86 Conference; 1986 November 4-6; Chicago, IL. Weston, CT: Online Inc.; 1986. 232-237.

SHEDLOCK, JAMES. 1986. End User Search Systems: An Overview. In: Wood, M. Sandra; Horak, Ellen Brassil; Snow, Bonnie, eds. End User Searching in the Health Sciences. New York, NY: Haworth Press, Inc.; 1986. 65-84. ISBN: 0-86656-465-9.

SIMON, MARJORIE. 1986. The BRS/AFTER DARK Search Service in a Health Sciences Library. In: Wood, M. Sandra; Horak, Ellen Brassil; Snow, Bonnie, eds. End User Searching in the Health Sciences Library. New York, NY: Haworth Press, Inc.; 1986. 163-178. ISBN: 0-86656-465-9.

SLINGLUFF, DEBORAH; LEV, YVONNE; EISAN, ANDREW. 1985. An End-User Search Service in an Academic Health Sciences Library. Medical Reference Services Quarterly. 1985 Spring; 4(1): 11-21. ISSN: 0276-3869.

SMITH, LINDA C. 1980. Implications of Artificial Intelligence for End-User Use of Online Systems. Online Review. 1980 December; 4(4): 383-391. ISSN: 0309-314X.

SNOW, BONNIE. 1986. Summary of Address: Highlights from New York End User Searching: Summary of the July 2nd STS Program. STS Signal: The Newsletter of the ACRL Science and Technology Section. 1986 Fall; 1(2): 2-3. ISSN: 0888-6563.

SNOW, BONNIE; CORBETT, ANN L.; BRAHMI, FRANCES A. 1986. Grateful Med: NLM's Front End Software. Database. 1986 December; 9(6): 94-99. ISSN: 0162-4105.

STEFFEN, SUSAN SWORDS. 1986. College Faculty Goes Online: Training Faculty End Users. Journal of Academic Librarianship. 1986 July; 12(3): 147-151. ISSN: 0099-1333.

STIRLING, KEITH H. 1984. A Micro-Based Emulator for On-Line Search Services. Drexel Library Quarterly. 1984 Fall; 20(4): 87-97. ISSN: 0012-6160.

STOUT, CATHERYNE; MARCINKO, THOMAS. 1983. Sci-Mate: A Menu-Driven Universal Online Searcher and Personal Data Manager. Online. 1983 September; 7(5): 112-116. ISSN: 0146-5422.

SUMMIT, ROGER K. 1987. Online Information: A Ten-Year Perspective and Outlook. Online. 1987 January; 11(1): 61-64. ISSN: 0146-5422.

SUMMIT, ROGER K.; MEADOW, CHARLES T. 1985. Emerging Trends in the Online Industry. Special Libraries. 1985 Spring; 76(2): 88-92. ISSN: 0038-6723.

TATALIAS, JEAN. 1985. Attitudes and Expectations of Potential End-User Online Searchers. In: Williams, Martha E.; Hogan, Thomas H., comps. Proceedings of the 6th National Online Meeting; 1985 April 30-May 2; New York, NY. Medford, NJ: Learned Information, Inc.; 1985. 457-462. ISBN: 0-938734-09-1.

TENOPIR, CAROL. 1984. Database Access Software. Library Journal. 1984 October; 109(16): 1828-1829. ISSN: 0363-0277.

TENOPIR, CAROL. 1985. Systems for End Users: Are There End Users for the Systems? Library Journal. 1985 June; 110(11): 40-41. ISSN: 0363-0277.

THOMAS, SARAH E., comp. 1986. End-User Searching Services, Spec Kit 122. Washington, DC: Office of Management Studies, Association of Research Libraries; 1986. 112p. ISSN: 0160-3582.

THOMPSON, BENNA BRODSKY. 1983. The Linear File: Future Direct Users of Sci-Tech Electronic Databases. Database. 1983 June; 6(2): 6-9. ISSN: 0162-4105.

TOLIVER, DAVID E. 1982. OL'Sam: An Intelligent Front End for Bibliographic Information Retrieval. Information Technology and Libraries. 1982 December; 1(4): 317-326. ISSN: 0730-9295.

TOLIVER, DAVID E. 1986. Whether and Whither Micro-Based Front-Ends. In: Jacobson, Carol E.; Witges, Shirley A., comps. Proceedings of the 2nd Conference on Computer Interfaces and Intermediaries for Information Retrieval; 1986 May 28-31; Boston, MA. Alexandria, VA: Defense Technical Information Center; 1986. 225-234. (DTIC/TR-86/5).

TOOEY, MARY JOAN. 1986. PsycLIT on CD-ROM: The UMAB Experience OR Serving Psychological Abstracts to Library Users on a SilverPlatter. In: Proceedings of the ONLINE '86 Conference; 1986 November 4-6; Chicago, IL. Weston, CT: Online Inc.; 1986. 246-260.

TOROK, ANDREW; HURYCH, JITKA. 1986. End User Online Searching among University Faculty. In: Hurd, Julie M., ed. ASIS '86: Proceedings of the American Society for Information Science (ASIS) 49th Annual Meeting; 1986 September 28-October 2; Chicago, IL. Medford, NJ: Learned Information, Inc.; 1986. 335-340. ISSN: 0044-7870; ISBN: 0-938734-14-8.

TRIVISON, DONNA; CHAMIS, ALICE Y.; SARACEVIC, TEFKO; KANTOR, PAUL. 1986. Effectiveness and Efficiency of Searchers in Online Searching: Preliminary Results from a Study of Information Seeking and Retrieving. In: Hurd, Julie M., ed. ASIS '86: Proceedings of the

American Society for Information Science (ASIS) 49th Annual Meeting; 1986 September 28–October 2; Chicago, IL. Medford, NJ: Learned Information, Inc.; 1986. 341–349. ISSN: 0044-7870; ISBN: 0-938734-14-8.

TRZEBIATOWSKI, ELAINE. 1984. End-User Study on BRS/After Dark. RQ. 1984 Summer; 23(4): 446–450. ISSN: 0033-7072.

VAN ARSDALE, WILLIAM O.; OSTRYE, ANNE T. 1986. InfoTrac: A Second Opinion. American Libraries. 1986 July/August; 17(7): 514–515. ISSN: 0002-9769.

VIGIL, PETER J. 1984. End-User Training: The Systems Approach. In: Williams, Martha E.; Hogan, Thomas H., comps. Proceedings of the 5th National Online Meeting; 1984 April 10–12; New York, NY. Medford, NJ: Learned Information Inc.; 1984. 419–424. ISBN: 0-938734-07-5.

VIGIL, PETER J. 1986. The Software Interface. In: Williams, Martha E., ed. Annual Review of Information Science and Technology: Volume 21. White Plains, NY: Knowledge Industry Publications, Inc. for the American Society for Information Science; 1986. 63–86. ISSN: 0066-4200; ISBN: 0-86729-209-1.

VOLLARO, ALICE J.; HAWKINS, DONALD T. 1986. End-User Searching in a Large Library Network: A Case Study of Patent Attorneys. Online. 1986 July; 10(4): 67–72. ISSN: 0146-5422.

WALTON, KENNETH R. 1986. SearchMaster: Programmed for the End-User. Online. 1986 September; 10(5): 70–79. ISSN: 0146-5422.

WALTON, KENNETH R.; DEDERT, P. L. 1983. Experiences at Exxon in Training End-Users to Search Technical Databases Online. Online. 1983 September; 7(5): 42–50. ISSN: 0146-5422.

WANGER, JUDITH. 1979. Education and Training for Online Systems. In: Williams, Martha E., ed. Annual Review of Information Science and Technology: Volume 14. White Plains, NY: Knowledge Industry Publications, Inc. for the American Society for Information Science; 1979. 219–245. ISSN: 0066-4200; ISBN: 0-914236-44-X.

WARD, SANDRA N.; OSEGUEDA, LAURA M. 1984. Teaching University Student End-Users about Online Searching. Science & Technology Libraries. 1984 Fall; 5(1): 17–31. ISSN: 0194-262X.

WIEDERHOLD, GIO. 1986. Structural versus Application Knowledge for Improved Database Interfaces. In: Jacobson, Carol E.; Witges, Shirley A., comps. Proceedings of the 2nd Conference on Computer Interfaces and Intermediaries for Information Retrieval; 1986 May 28–31; Boston, MA. Alexandria, VA: Defense Technical Information Center; 1986. 17–96. (DTIC/TR-86/5).

WILLIAMS, MARTHA E. 1985. Highlights of the Online Database Field—Gateways, Front Ends and Intermediary Systems. In: Williams, Martha E.; Hogan, Thomas H., comps. Proceedings of the 6th National Online Meeting; 1985 April 30–May 2; New York, NY. Medford, NJ: Learned Information, Inc.; 1985. 11–18. ISBN: 0-938734-09-1.

WILLIAMS, MARTHA E. 1986a. Highlights of the Online Database Field: CD-ROM and New Technologies vs. Online. In: Williams, Martha E.; Hogan, Thomas H., comps. Proceedings of the 7th National Online Meeting; 1986 May 6–8; New York, NY. Medford, NJ: Learned Information, Inc.; 1986. 1–4. ISBN: 0-938734-12-1.

WILLIAMS, MARTHA E.   1986b.   Transparent Information Systems through Gateways, Front Ends, Intermediaries, and Interfaces.   Journal of the American Society for Information Science.   1986 July; 37(4): 204–214.   ISSN: 0002-8231.

WILLIAMS, PHIL W.   1977.   The Role and Cost Effectiveness of the Intermediary.   In: Proceedings of the 1st International On-Line Meeting; 1977 December 13–15; London, England.   Oxford, England: Learned Information, Ltd.; 1977.   53–63.   ISBN: 0-904933-10-5.

WILLIAMS, PHIL W.   1984.   User Trials of the OASIS Search System.   In: Williams, Martha E.; Hogan, Thomas H., comps.   Proceedings of the 5th National Online Meeting; 1984 April 10–12; New York, NY.   Medford, NJ: Learned Information Inc.; 1984.   437–452.   ISBN: 0-938734-07-5.

WILLIAMS, PHIL W.   1985.   How Do We Help the End User?   In: Williams, Martha E.; Hogan, Thomas H., comps.   Proceedings of the 6th National Online Meeting; 1985 April 30–May 2; New York, NY.   Medford, NJ: Learned Information, Inc.; 1985.   495–505.   ISBN: 0-938734-09-1.

WOOD, M. SANDRA.   1986.   End User Searching: A Selected Annotated Bibliography.   In: Wood, M. Sandra; Horak, Ellen Brassil; Snow, Bonnie, eds.   End User Searching in the Health Sciences.   New York, NY: Haworth Press, Inc.; 1986.   213–274.   ISBN: 0-86656-465-9.

Systems That Inform: Emerging Trends in Library Automation and Network Development

8

WARD SHAW
Colorado Alliance of Research Libraries

PATRICIA B. CULKIN
Colorado Alliance of Research Libraries

INTRODUCTION

The literature of library automation has focused until recently on those computer-based systems that handle library management activities. An "automated" library was one that could identify the status of any item in a collection through a computer query and could generate appropriate notices, reports, and statistics regarding the use of that collection. Circulation, acquisitions, cataloging, serials control, and interlibrary loan were all functions deemed appropriate for automating, and a complete system was one in which these functions were integrated into a communicating whole. Automated library systems were considered different from "information" or search systems. KOENIG, FAYEN (1983), and KESNER & JONES describe specific stages or phases of library automation in the introductions to their works. DE GENNARO distinguishes between library automation and information automation.

Public access catalogs (PACs), online search services, and the newer, micro-based videodisk services have all spawned a literature unique to themselves. A glance through the contents of early 1980s *ARIST* volumes confirms this. However, there is clearly some indication in the current literature that there is a convergence of thinking about how these various computer-based capabilities should be packaged for the end user. The components of the package include: 1) traditional automated library systems, 2) PACs, 3) online search services, 4) electronic publishing, 5) electronic delivery, 6) micro-based videodisk services, 7) nonbibliographic databases, and 8) personal computers and scholars' workstations. These phenomena, among others, are now being viewed by information professionals as potential parts of the whole. The view

Annual Review of Information Science and Technology (ARIST), Volume 22, 1987
Martha E. Williams, Editor
Published for the American Society for Information Science (ASIS)
by Elsevier Science Publishers B.V.

is conceptual and not physical, and because it is grounded on an understanding of how gateways, mainframe interconnections, and established telecommunications networks are being combined in current applications, it is more realistic than that reflected in some earlier writing about the computer revolution. DE GENNARO (p. 40) claims that "there are signs now that advances in information automation are being transferred to library automation and are speeding its development. The two streams are rapidly converging."

The notion that an end user (meaning library user rather than librarian) should be able to sit down at a system and interact with it, moving through (up, down, and around) both local and distant resources, and emerging with a question answered, a question refined, a new question, a document retrieved, a document ordered, and best of all, a sense of accomplishment and of having learned something, is much less radical than it would have been earlier in the decade. User-transparent applications, whereby an end user has access to a whole variety of products and services through a generic program, are already a reality. KEHOE (p. 489) reviews the history of online interfaces and discusses how these have led to the development of artificial intelligence (AI) programs and expert systems to "act as consultants in specific domains that usually require human expertise."

Perhaps more surprising is the realization, from a review of the literature, that these applications are being designed, engineered, and supported locally and that rich networks are being built from the local bases to the remote sources in response to specific local needs. Far from being moribund, the local library is positioned to be a primary provider of these "systems that inform." The largest barrier appears to be—again from the literature—the quality of leadership the profession brings to its consequent and inevitable redefinition.

This paper discusses how automated library systems and computer-based information services are being transformed into much more powerful systems that inform by examining five topics:

- The revitalization of local systems development and the changing nature of vendor/customer relationships;
- The emergence of local application networks;
- The library system as a channel to local resources, external systems, nonbibliographic data, and electronic publishing;
- The professional leadership; and
- The direct influence on learning.

SCOPE AND COMMENTARY

The literature included in this review covers the period 1982–1986—i.e., from the last *ARIST* chapter on automation (LUNDEEN & DAVIS) to February 1987. Because this chapter attempts to expand the concept of library automation, the bibliography includes material traditionally associated with other areas of information research. It is derived from four types of writers: philosophers, chroniclers, practitioners, and researchers.

Philosophers or commentators attempt to relate automation efforts to broader institutional goals. Their publications usually refer to efforts of the past and express opinions ranging from disdain to cautious optimism for the results of current effort. Their tone ranges from gentle humor to outright sarcasm. Basically they usually try to point out the folly of relying on computers for miracles while allowing that "they're o.k. in their place." DE GENNARO and CART are examples of this approach.

Chroniclers do surveys and make lists of things. Knowing that there are X more databases to search in the social sciences this year than last, or that there are Y different circulation systems of varying levels of functionality in use in the northern United States, or that there are Z new CD-ROM microcomputer packages on the market is clearly useful in some contexts but does not contribute much to the understanding of information systems. These two types of writing account for a major portion of the total bibliography. The ASSOCIATION OF RESEARCH LIBRARIES, OFFICE OF MANAGEMENT STUDIES (1984; 1986), BELLARDO & STEPHENSON, MATTHEWS (1985), and NEUFELD & CORNOG are examples of list-and-survey literature.

Practitioners' writing describes experience with particular applications or with particular hardware. Taken individually, the pieces usually describe the merits of a particular system or application with only passing reference to the larger issues of design and utility. Viewed collectively, however, practitioners' writing creates an interesting portrait of current activity and provides the building blocks required to create conceptual models. CULKIN & SHAW, DISKIN & MICHALAK, ERNEST & MONATH, and SYBROWSKY & WILSON as well as many papers from conference proceedings represent this category.

Researchers do not have as large a presence in the literature as is probably desirable, yet what there is lends strong credence to the notion that "systems that inform" are developing at a fast and impressive rate and that librarians and information professionals may be positioned to affect and advance the learning process. BATES, BAWDEN, BELLARDO, BORGMAN, BROOKS ET AL., MARKEY, OCLC, INCORPORATED, and SARACEVIC ET AL., among others, report research that advances the concept of systems that inform.

THE REVITALIZATION OF LOCAL SYSTEMS DEVELOPMENT AND THE CHANGING NATURE OF VENDOR/CUSTOMER RELATIONSHIPS

Many of the early library automation efforts were disasters. KOENIG (p. 50) refers to the early era as a time when "intention egregiously outran accomplishment, and indeed, where not only intention, but prediction and hope, were reported as accomplishment." He cites the experiences of the Chicago Circle, the Florida Atlantic, and MIT (Project Intrex) as being good examples of "great hope and only modest achievement." FAYEN (1983, p. 118) describes a particularly dramatic example of failure in her case study section, although she notes that "to protect the identities of the participants, no actual names are used." DE GENNARO (p. 37) says "there is a vast gap between promise and reality when it comes to integrated online library systems."

These bad experiences led to distrust of local capability and to the view that acquisition of turnkey systems was a better guarantee of functionality. In fact, most of the recent literature on system procurement either does not consider local development as an option or argues strongly against it. Describing choices for online PAC design, MATTHEWS (1985, p. 21) says:

> Despite . . . limitations, most libraries with operational online catalogs have implemented turn-key systems. This trend will probably continue because of the high costs and risks of developing one's own system and the rapid advances currently being made by vendors in developing comprehensive or full-feature online catalogs.

The rise of commercially vended turnkey systems has fostered library dependence on the consultant/RFP (request for proposal)/contract method of procurement. In this scenario, libraries obtain the services of a consultant to assess their needs and operations, generate a formal request-for-proposal (RFP) document, distribute the RFP to the vendor market, analyze and evaluate the responses, select a vendor, and (usually) negotiate a contract. As REYNOLDS (p. 230) says, "When a library is purchasing a large scale system, or creating a multi-institutional network, experienced consultants can often provide a degree of insight and skill that is worth their fees." Of the RFP document, he says (p. 243), "It can establish an analytical framework within which different systems are evaluated and a selection is made, and can further serve as the groundwork for negotiating a final contract with the vendor whose system is chosen." The fact that consultants and even vendors make consistent and numerous contributions to the library automation literature attests to the degree to which their presence in the selection dynamic has become established.

Formalizing the selection process and targeting it to the commercial sector was intended to minimize risk and contain cost. It appeared as if the method should ensure for the library generic functionality for predictable dollars and remove responsibility for development and maintenance from library staff. Although this sometimes worked, it has often not prevented commercial systems from being underconfigured or lacking in functionality.

The complexities of library system design and development are closely related to the institutional complexity and functional ambitions of the customer. Failure of commercial systems is nearly directly proportional to the size of the library. Turnkey systems work in smaller libraries, especially if the systems are confined to simple management functions, but for larger libraries, with thousands of users and hundreds of thousands of volumes, which have complex needs for control, delivery, and feedback, they are not usually up to the task.

The problem is compounded by the fact that systems staffs of vendors often understand their system well but don't equally grasp the intricacies of library processing. McGee (in DRABENSTOTT, 1985c, p. 110) notes:

> Even today, when libraries with unusual requirements attempt to obtain suitable systems from the turnkey market, there is some-

times a reluctance or inability on the part of vendors to appreciate the differences between their products and the libraries' stated needs. . . . Vendors do not fully anticipate how some capabilities and features will be used. . . . Besides failing to understand libraries' requirements, vendors often develop capabilities that are convenient to provide, rather than truly responsive to genuine library requirements.

Library systems are complex systems operating in complex institutional environments. Failure to understand this complexity has often made the partnership between libraries and vendors uneasy and unsatisfactory. The considerable instability in the vendor community, particularly among those companies attempting to address the large system market, provides substantial credence to the contention that complex library systems are not easy for vendors to deliver.

Vendor difficulties in the large-end market have been well publicized. The Dataphase experience at Chicago Public Library is probably the most dramatic example. Few would argue that the especially public nature of the disagreements between the two parties contributed strongly to the Dataphase decision to divest itself of its large-end ALIS III sytem to UTLAS. Other vendors, however, are also subject to reorganization, divestiture, or dissolution. Review of the "News" and "Bulletin Board" sections in library periodicals such as *LIBRARY JOURNAL* and *AMERICAN LIBRARIES* over recent months reveals many instances of corporate reorganization.

According to Boss (in DRABENSTOTT, 1985b, p. 95), criteria for judging vendor viability should include: "number of installed systems, number of sales in the past year, income in the past year, profitability, number of programmer/analysts (in full-time equivalents or FTE), and the software's degree of completeness." According to these criteria, he finds that only four vendors—of a total of 37 in April 1986 (MATTHEWS, 1986)—can pass these tests.

REYNOLDS writes about the problems of libraries distrusting vendors and vendors going out of business as being real barriers to libraries in trying to choose and maintain a system. He says (p. 165): "Vendors have complained that librarians often do not say what they really mean and do not have a realistic understanding of the expenses involved in developing automation systems, in introducing enhancements, and so on. Librarians have tended to view vendors as having obsessive interest in profits and as being prone to promising anything and everything to realize a sale."

The well-publicized difficulties of library system vendors have contributed to libraries' hesitation and fear of the substantial commitments automation requires. CART (p. 38) says: "I am terrified because of library automation's incredible complexity and the resulting confusion that swirls angrily about it like the waters around Scylla; I am terrified because of the indecipherable cant in which its disciples converse; I am terrified because of its colossal cost; and I am terrified because of the commitment it demands." It is becoming clear that commercially vended turnkey systems are not panaceas. Their emergence has spawned a complicated selection process that guarantees not much more than basic functionality and often inhibits customized enhancement and easy interface with other local information sources.

THE EMERGENCE OF LOCAL APPLICATION NETWORKS

As a direct result of these phenomena, libraries are again looking to themselves or to partnership arrangements with library-oriented organizations for the development of local systems. Some of the most functionally stable yet dynamic systems are those that have been developed by and for libraries at the local library level. These local systems are successful because they are built to local specification and in response to a particular information environment. Some of these systems include:

BACS	Washington University School of Medicine
BOBCAT	New York University
CARL	Colorado Alliance of Research Libraries
CARNEGIE-MELLON	Carnegie-Mellon University
ILS	National Library of Medicine's Lister Hill National Center for Biomedical Communications
LCS	University of Illinois
LIAS	Pennsylvania State University
Maggie III	Pike's Peak Library District
MELVYL	University of California System
MSUS/PALS	Minnesota State University (Mankato State)
NOTIS	Northwestern University
OSU	Ohio State University
PHOENIX	University of New Brunswick
SULIRS	Syracuse University Libraries Information Retrieval System
TLS	Claremont Colleges
VTLS	Virginia Technical Institute

FAYEN (1983), MATTHEWS (1985), POTTER, and REYNOLDS contain descriptions of most of these applications.

Some of these systems are now being marketed to other libraries directly or through third-party licenses, often with customized options and features that accommodate local policy. Others still serve only their primary clientele. They tend to be very different from one another in functionality and aesthetics. They also tend to be very different from their own original forms. Most have needed to be changed, and in some cases to be reinvented as their influence and effect have become better understood. HILDRETH (1984) introduces the notion of successive generations of development and has identified three generations to date. MATTHEWS (1985) gives detailed component descriptions and screen examples that illustrate the variety of design approaches. FAYEN (1983) lists many of the factors involved in choosing and evaluating an online catalog.

The significant common factor uniting most local systems is that they were developed originally in response to particular perceived local need, whether that need was to find a substitute for or continuation of a card or microform catalog (almost all), to consolidate the holdings of a group of libraries (CARL, MELVYL, LCS), or to facilitate sharing of management information such as data on circulation and serials with end users. CULKIN & SHAW speak of creating a "single research resource for their member institutions." GORMAN (1984, p. 153) notes that an online catalog is more than "just a catalog" because it "is often concerned with holdings of more than one library, always allows access to the records it contains by more than the standard access points, and, most crucially, always gives the status of the item sought." The endurance and transportability of locally developed systems suggest that local need is a galvanizing force in its own right- powerful enough to fuel the invention of functional systems and powerful enough to support continuing incorporation of ideas and technology.

ATKINSON (p. 23) was concerned early in the 1970s that a library system "should be one which would speak to the problems of its users rather than simply the problems of the library." In this, he was fairly prophetic for that time, and his concern has become the principle that has guided the development of the better information systems. All of the systems mentioned above were intended ultimately to serve the end user better. FAYEN (1983, p. 21) declares that a library's "underlying design philosophy must regard the online catalog primarily as a tool for the library's users." Even if early-generation end-user service took the form only of interactive searching of bibliographic databases and delivery of holdings and status information, it was still a major step beyond the creation of technical support systems. It is now becoming clear that local systems can be much more than online catalogs of library holdings.

Dowlin's Maggie III in Colorado Springs was one of the first applications to incorporate the notion that the library system should support access to community information services from outside the traditional library domain. DOWLIN (p. 149) incorporated this vision into the library's 1979 long-range plan. The plan specified that the mission of the library is:

- To serve as a resource center for published materials;
- To serve as a community information center, linking organizations and individuals in the community; and
- To serve as a community communication center, facilitating communication between individuals and organizations.

Dowlin was quick to recognize that vendors could not assist with these goals— i.e., they could not expand their systems into a general purpose computer system for the library. He adapted instead the CARL system to Maggie III, using it to manage data from a variety of sources, and it is still evolving today. The online catalog is one of ten databases on the Maggie III menu, the others ranging from an events calendar to a day care referral service to a local authors database to an urban planning documents database. Future plans include the creation of a public domain electronic mail service that businesses, city agencies, and individual citizens can use.

The report by CULKIN & SHAW on The Colorado Alliance of Research Libraries (CARL) experience is also relevant. CARL is a consortium of six research libraries in metropolitan Denver, Colorado, which decided in the mid-1970s to create a single research resource for the publics they serve. Thinking at that time dictated that CARL should contract with an established vendor for delivery of a PAC or, at the very least, it should enter into a joint venture with a vendor to develop a suitable product.

This they did, and four years later the only tangible result of the partnership was the existence of some mainframe hardware and a few terminals; there was no database and no software. At that point, CARL decided to develop its own system. That system now features a PAC, a database containing 2.2 million bibliographic records, a government documents database, several nonbibliographic databases, and fully integrated circulation, acquisitions, and bibliographic maintenance subsystems.

The primary network contains over 450 terminals, and the university members offer additional access through their broadband networks to campus locations and dial-up users. Between 25,000 and 30,000 users access the CARL system daily. It has been cloned twice in Colorado, at the Boulder Public Library and at the Pike's Peak Library in Colorado Springs (the third generation of Maggie). It is in the process of being cloned twice more, for MARMOT, a consortium of 200 libraries on Colorado's western slope, and for Arizona State University in Tempe. CARL supports direct computer-to-computer links to the clones, meaning that a large number of libraries in Colorado are directly interconnected.

FAYEN (1983, p. 114) describes the evolution of the Syracuse University Libraries Information Retrieval System (SULIRS) from its beginnings as an acquisitions/cataloging/batch circulation system in the 1960s and early 1970s to a partially integrated bibliographic circulation status system in 1980 to a fully interactive catalog by 1981. She notes that:

> The Syracuse University Libraries continue to make enhancements to SULIRS to assist online catalog users. . . . Campus-wide access through dial-up phone lines has been added. . . These enhancements are consistent with the ultimate objective of the Syracuse University Libraries, which is, according to Gregory N. Bullard, associate director for technical and automated services, "to put a terminal in every office and wherever students ongregate in the university, thus making the catalog widely available."

DISKIN & MICHALAK (p. 9) explain how Carnegie-Mellon Libraries have developed software to allow users to access information about library services through the online PAC. The software, called the Information Function (IF), was developed by C-MU staff and runs in conjunction with OCLC's LS/2000 system. Both the public catalog and the IF are available to the university community through the campus broadband network. IF is intended to benefit the university community in the following ways:

- IF will provide patrons and staff fast access, local or remote, to basic information about the libraries. Over time, IF can become the first choice for a large proportion of the community seeking to learn current hours, policies, special services, and appropriate staff contacts for reference and other assistance.
- Use of IF will free staff from some part of the normal and time-consuming routine of providing standard information about library operations and services.
- IF will keep patrons current with library news, recent acquisitions, technological developments, and policies and services.
- IF will provide online access to standard bibliographies and reference guides, prepared in-house as part of the library publications program, that are currently available in printed form on site during library hours. Patrons can download or print such materials for later reference.

Locally developed systems are clearly leading to the creation of truly informing systems. These can be extremely complex systems technologically, intellectually, and culturally. They need to be invented, managed, and enhanced by those who understand the local environment. Political, economic, technical, and intellectual factors all play a part in successful information system development. NIELSEN (p. 83) notes that "the new movement toward localization of information processing is analogous to community development—the conscious effort to organize local citizenry in pursuit of a common purpose, and in the process perhaps regain some aspects of community."

Principles to guide the formation of appropriate coalitions are still being developed. In short, it is too early to think of these systems as being formulaic or packageable, and until they are, design and development must emerge from local initiative. NIELSEN (p. 86) states that "Visionary leadership focusing on local needs, local resources, and local talent can make the next decade considerably brighter."

The vitality of local systems is well recognized, and thus much attention is being given to linking and system interconnection. The Library of Congress (LC) is participating with the major utilities, Western Library Network (WLN), OCLC, and Research Libraries Information Network (RLIN) in the Linked Systems Project (LSP). LSP's goal is to develop a standard and economical means of tying each of these major computer bases together. The impetus for doing this is to create a linked national resource from existing components and to develop generic communications protocols that will allow disparate computers to communicate quickly and easily.

MCCOY (p. 33) summarizes the progress, promise, and realities of LSP and points to its significance in the evolution of systems:

LSP protocols join other library standards (MARC and *AACR2*, for example) which are the essential enabling ingredients of

library cooperation, and without which shared cataloging, interlibrary loan networks, and the other cooperative activities we
take for granted would not be possible. . . . Standard protocols
which permit dissimilar computer systems to communicate with
each other are a new requirement, the result of new technologies,
but they are just as essential to library cooperation as the earlier
standards.

LYNCH (p. 78) reviews the history and scope of LSP and warns of potential
performance and network conversion issues but concludes that library automation cannot occur without it:

The linked systems protocol is essential to the evolution of library
automation both locally and nationally. Adoption and wide
implementation of the linked systems protocol standard will be a
major step forward in broadening access to all kinds of information resources, in sharing existing resources between organizations
and in integrating together currently insular databases.

The next step will be to use the LSP protocols to link the networks to
local systems. This will allow processing that has only local relevance to
remain resident (and shareable) locally while information that needs to be
available regionally or nationally can be passed to the utility level and shared
across the broader networks.

Underlying the goal of providing links between systems are many issues
relating to various standards as described by AVRAM, DENENBERG,
HILDRETH (1986), and others. Avram discusses the role of local systems,
how and with whom to link, and linking standards and database ownership
issues. Denenberg discusses the need for standards in computer-to-computer
communication, gives examples of technical issues, and illustrates the framework of the Open Systems Interconnection (OSI) reference model. Hildreth
describes proposed standards for a "Common Command Language for Online,
Interactive Information Retrieval" that have been developed by Committee G
of the National Information Standards Organization (NISO) and are now
being reviewed for adoption.

Many commentators are writing of the tremendous potential of public
access library systems to provide coordinated access to information resources
beyond bibliographic records that describe library holdings. Dowlin's Maggie
III is one of the earliest pioneers of this effort. GORMAN (1984, p. 156)
notes that:

For the first time, our systems will be conditioned not by our
categorizations and limitations but by the availability of relevant
information—no matter what the form of publication described,
no matter what the source of information might be. . . . Freed
from the chimera of the single omnipotent system, we can range
among the many different kinds of bibliographic, processing,

and data systems without altering the nature of those systems and guided by the friendly interface programs.

For POTTER (p. 130) the catalog is a "lens for exploring a larger...universe than any one library could ever hope to contain."

THE LIBRARY SYSTEM AS CHANNEL TO LOCAL RESOURCES

The fact that local data reside locally also means that they are available for sharing across local area networks (LANs). Some institutions (e.g., Carnegie-Mellon, CARL, Syracuse University, New York University) are beginning to offer access to their libraries' PACs on these networks as well as access to various other nonbibliographic types of information. Locally developed systems lend themselves more easily to this kind of link than proprietary, vended systems because the software is under the institution's control.

MORAN describes Brown University as one of the so-called "Star Wars" universities. Brown has launched a ten-year program to design and install 10,000 scholar's workstations. She notes (p. 25): "When the workstations are fully operational, many services that are now available only in libraries will be available to users of the workstations, including access to the library's catalog and other bibliographic and nonbibliographic databases." She also mentions Clarkson College as a good example of the automation possible in the libraries of smaller institutions.

The bidirectional nature of these links is a portentous feature. Not only are local library data being accessed by users of the network but nonlibrary network data can be accessed from the library terminals. This situation will allow libraries to nurture a strong "habit of use" among all types of users who have not been exposed or predisposed to use online services but who may benefit greatly from the resource.

NEFF, writing from a computing background, and MOLHOLT (1985) and DOUGHERTY & LOUGEE, writing from library backgrounds, note how the roles of computer centers and libraries are converging. Each makes the case that the services and expertise of each agency can directly benefit the other in building future integrated information support systems. Libraries will be wise to forge working relationships with agencies that have technological and networking expertise if their distribution channels for information services are to be broad, effective, and democratic.

THE LIBRARY SYSTEM AS CHANNEL TO EXTERNAL SYSTEMS

If full channeling from local to external systems is to occur, more research, experimentation, and real-life applications need to be executed. Still, much progress has been made since the early 1970s. Transparency and interconnection are concepts that have begun to enter the layman's understanding of computer systems and to affect his requirements for easy access. "Gateway" is a term approaching buzzword status. These terms are becoming part of the vernacular because commercial applications are being well promoted in the public domain and the prevalence of those applications is fueling an understanding of the meaning and potential of true, functional transparency. KEHOE provides a useful review of the current status and future directions of these techniques.

For WILLIAMS (1986, p. 205) "transparent information retrieval" occurs when a user obtains an answer to a query "without seeing the complexity of the intervening transactions that take place between the posing of the query and the provision of the final results." As she notes:

> The variety and variability in databases, search systems, and the techniques needed to exploit them. . .is not new. It was observed in the early 1970's. . . . For many years researchers have been working to simplify the access to, and use of, online systems. Many of the research efforts were known by other names or had different labels, and, then as now, there was a terminology problem. Terms that have been used are: front end, interface, intermediary system, post processor, gateway, and transparent system. All of these have the same general objective of making the complexities of online searching transparent to the user, and in that sense all can be said to fall under the rubric of transparent systems.

While "transparency" describes the effect of the concept, "interconnection" describes the mechanics. As various online products were developed, from PACs to search services to personal computer software, there was an initial and loud cry for standards. In addition to telecommunications and networking standards described above, this has also included requirements for standardized messaging or command languages, standardized search functions, and standardized file structures. However, standardization is difficult to impose on a young and experimental discipline, where the solutions are even now not well understood or agreed on. As WILLIAMS (1986, p. 205) says:

> The variety and variability in databases, search systems, and the techniques needed to exploit them. . .has prompted many to suggest that the solution should come from standardization. But, as we all know, standards take many years to develop and their use in matters such as information resources and systems cannot be imposed. Also, the variety is so great that it might not even be possible to develop a single standard.

While efforts to standardize do exist, they are concentrated now more often and more usefully on the interconnection of disparate systems. Most hardware manufacturers support interconnection software that allows machines by the same manufacturer to talk to each other. Most are also supporting products that allow communication with other brands of hardware. TURTLE provides background on the Open Systems Interconnection (OSI) reference model and a description of the upper-level OSI protocols based on the model. The work on OSI, TCP/IP, Ethernet, and many other protocols is beginning to standardize the communication between machines and applications, allowing users to get to remote data quickly and easily. Often it is impossible for a user to know, unless the software makes a point of telling him, exactly which computer he is operating in during a given activity.

The results of early efforts to achieve transparency are now incorporated in many different commercial products (further discussed in the chapter by MISCHO in this volume). EasyNet, InfoMaster, Pro-Search, BRS After Dark, Dialoglin', SearchMaster, Intelligent Gateway, Sci-Mate, and Search Helper are but a few of the front end systems that have been invented to simplify searching for individuals in business, government, academia, and the public domain. Campus local area networks, research networks such as Arpanet, and bulletin board networks are others that provide front end access to various kinds of data. OJALA, describing front end systems for end users, points out that Comp-U-Store can be accessed from CompuServe and Dow Jones News/Retrieval and that the OAG (Official Airline Guide) is available through a gateway (a system that passes a search request on to another system for processing) on several different systems.

Now that some of these capabilities have been developed and their potential is better understood, the profession needs to consider how to take the best advantage of transparency. One problem is that current distribution mechanisms offer these resources to only a small percentage of potential users. Only highly motivated or well-funded searchers can discover and take advantage of the full range of available services.

People with personal computers and modems, certain businesses, and most general-purpose libraries promote the services. However, in most libraries, logistics and economic factors tend to discourage broad use. A user usually has to find the service, make an appointment to do a search, and have a search "interview" before proceeding. In addition, he often has to let an intermediary perform the search, as professional wisdom says that direct searching is too difficult for laymen. MORAN (p. 18) confirms this: "Because of the expense of online searches and because the systems are still difficult for the new and infrequent user, the normal method of providing access to online searching in academic libraries is to use a trained librarian searcher to act as an intermediary in performing the search." OJALA (p. 202) describes how direct user searching is gaining on intermediary searching but holds that the "demographics of end user searchers restrict them to the professional/managerial/technical types with access to technology."

SARACEVIC ET AL. challenge this approach with studies of the elements involved in information search and retrieval, particularly in relation to the cognitive decisions and human interactions involved. BROOKS ET AL. have done considerable research on the components of the information provision mechanism (IPM) by studying "real life human user-human intermediary information interactions."

Economic factors also discourage in-depth searching. Most online search vendors charge by elapsed time plus number of hits. As users become anxious about the cost, they will abort a search before taking full advantage of the resource. ROOSE points out that libraries typically have a different attitude toward online searching, singling it out as a "big fee" service while charging nothing or only minimal fees for other services. She suggests that this attitude should be subject to redefinition. OJALA (p. 202) notes that, "In reality, people tend to perceive the price as being too much to pay for these services."

Libraries, by virtue of their traditional mission of service and education combined with their knowledge of local and remote system capability, are well positioned to change and even take control of the distribution model for computer-based information. As ROCHELL (p. 46) says,

> when academic libraries got around to embracing the new tech-
> nology, they designed systems uniquely capable of interconnec-
> tion and interface—systems that are specifically equipped for out-
> reach and resource sharing. . . . While this is significant, it is
> hardly surprising. After all, it is what we do. Libraries are, as
> much as anything, the vehicles of information exchange. And
> our position at the center of the university's new information
> environment is totally in line with our traditional mission and
> our role as collectors, organizers, and neutral providers.

Not only is the academic library at the center of the university's information environment, it is also often the center of its social and cultural environment. The same can be said of public libraries in their own contexts, especially those that have a dynamic and aggressive program of outreach and community service. Because library traffic is heavy and diverse, it presents an opportunity to showcase and distribute unified information services. It does not preclude or supplant home and office use but rather opens the market by advertising capability to large segments of the population.

Librarians also have considerable professional value to add. Not all computerized databases are useful. Some are not even intellectually responsible. By centering access to network services in libraries, the traditional contributions of evaluating, selecting, and organizing information are much more likely to be possible. OJALA (p. 202) notes that librarians "have an important role to play. . .both in educating end users about database accuracy and in pressuring database producers to provide good data." NIELSEN (p. 84) says that librarians "owe it to [the public] to share our considerable knowledge of how to evaluate competing systems." SPIGAI (p. 102) predicts that some librarians will become "knowledge engineers" but doesn't think that "database evaluation is in the wind any more than it has been for some of the print tools. I don't think it's something that we function very well at."

Vendors who want to create popular, useful products will need to pay attention to those who create network access. ROSZAK (1986a, p. 175) supports the notion of the library as channel, arguing from the other direction—i.e., as a frequent and committed user of libraries:

> If the equipment for computerized reference facilities were con-
> centrated in local libraries or, better still for reasons of economy,

if every local library were linked to a generously funded regional reference center, this would be the fastest and cheapest way for the general public to gain open access to whatever benefits the Information Age may have to offer. Private, profit-making computer-based information services (of which there are a growing number) are not viable substitutes for what the library can provide as a public reference service, if it is given the chance to show what it can do. Such businesses simply take the service out of the public domain.

THE LIBRARY SYSTEM AS CHANNEL TO NONBIBLIOGRAPHIC DATA

Several authors (e.g., DOWLIN and DISKIN & MICHALAK) write about the inclusion of nonbibliographic data in library systems. Most of the databases that are being accessed through library-supported online systems are bibliographic—i e., they contain records that describe, index, and summarize the contents of documents but do not contain the text of the documents. Some systems do allow the user to order the full text (for a price), but for the most part the user must rely on some other combination of mechanics and legwork to retrieve the document.

Library systems, however, have the capability of supporting access to full text as well as to surrogate records. Further, they can juxtapose traditional library documents (e.g., journal articles, book chapters, encyclopedia articles, statistical information, technical reports) with more general information (e.g., bus schedules, campus directories, consultant's directories, airline reservations, airline departure/arrival information, bulletin boards, course information), thus simplifying the logistics of information gathering for the user and cross-promoting the different kinds of information to vastly different audiences. SANTOSUOSSO (p. 107) notes that:

> A growing number of encyclopedias, directories, and full text reference works are now available. Libraries are the primary market for these sources. Libraries seeking ways to maximize investment in hardware and avoid duplication of print sources may use dedicated reference terminals to access these databases, or may download these files into the online catalog.

A senior citizen attempting to find a bus route to a shopping center might use the same system to find out if a current best seller is available at the local branch library. A doctoral candidate might take a break from research to find out what's showing at the local movie house. A teacher facing certification can consult course offerings and schedules at the local university. A middle-school student can find encyclopedia articles to help him get started on his first term paper.

Because full-text data can be fully and precisely indexed, users can find answers to specific questions or can browse through information of either particular or peripheral interest to them--they can maneuver their way through the various data as need or curiosity dictate. The fact that the user drives the process creates a major advantage over one-way-only information broadcasting in which a user is subject to a serial, prepackaged presentation of data. BINDER describes current videotex and teletext technologies and their potential for libraries.

Online full-text databases have been offered for the most part in microcomputer CD-ROM configurations. These configurations are limited both by their high cost and their finite size and therefore have not served to promulgate full text well. Many of these technologies can, however, be presented to users in coordinated ways through general library systems.

THE LIBRARY SYSTEM AS CHANNEL TO ELECTRONIC PUBLISHING

If libraries are going to become channels to information, they may be able to affect its actual form and delivery. DOUGHERTY & LOUGEE (p. 44) recognize that electronic publishing creates new responsibilities for librarians: "the critical path must involve librarians staying in the mainstream of activities regardless of the rate of change, the competition, or the technologies that emerge- for it is only in the mainstream that librarians can attempt to predict and control the future course of scholarly communication." Electronic publishing, like transparent searching, is a concept that has been recognized for decades, but it has only lately become a reality. Like transparency, it is a young and experimental area of research. Electronic publishing effort should be designed to complement the educational process, but poorly implemented, it can oversimplify and almost trivialize that process, as illustrated by the debate about the merits of the microcomputer based CD-ROM systems. ERNEST & MONATH, VAN ARSDALE, and CARNEY consider many of these issues, using IAC's (Information Access Corp.) InfoTrac system as an example. Summarizing, we can say that on the one hand, the systems are user friendly, appropriately high-tech, and convenient to use. They can even supply the user with a printed list of citations from which to launch his research. On the other hand, their scope and content are limited by a profit-driven need for inexpensive and simple packaging. Nor are such systems dynamic. They are limited to read-only memory, and updates are slow and costly. In addition to the subscription cost, someone must pay the costs for microcomputer, monitor, disk player, and printer equipment, and the result is an expensive learning tool whose static nature cannot take an inspired and curious user to any level beyond its own disk. While enormously useful for certain applications, CD-ROM systems are clearly not a panacea, either economically or intellectually. GURNSEY (p. ix-x) sums up some of the professional ambivalence about electronic publishing.

> Being close to the topic, I find my own attitude to electronic publishing is positive, but fairly complex. On the one hand, I sincerely believe it offers massive potential to enhance the responsibilities of the information professional. On the other hand, I believe a great deal of what is going on at present is blatant

technology push—and has little market relevance. These attitudes are not in conflict. It is not surprising that with such a fast emerging area some speculative and even ridiculous products and services are attempted.

There is no question that access to full text is next on the continuum for interconnected, transparent systems and that electronic media will be the basis for that text. It remains to the information professionals to design appropriate structures for offering that type of data, structures that lead, encourage, and inspire yet do not confuse.

THE PROFESSIONAL LEADERSHIP

These developing informing systems offer the possibility for libraries as institutions to lead in the democratization and effective application of information. For that to happen, however, the profession and its leaders must assume an active role in redefining their institutions and systems in the broadest sense. That may be happening, but it is not yet widely reflected in the literature.

If the role of libraries is, as ROCHELL says, to be "collectors, organizers, and neutral providers," then the profession has the inside track on leadership. Libraries have been collecting and organizing for years and doing it well, but the profession has concentrated so hard on achieving neutrality and preserving a status quo that its members are often not present in institutional decision-making contexts. GURNSEY (p. ix) writes:

Unfortunately, the attitudes of the information professionals often do not facilitate developments. Instead of being willing to assess and evaluate the newer technologies, and so accept a leading role, we are often too critical and obstructive. In this way, instead of being a major influence on change, we stand largely outside an area which is central to our very existence.

Thus it is not surprising that librarianship ranks among the lowest of the professions in surveys of salary, status, and stress. FAYEN (1986, p. 242) points out that "librarians are among the best-educated but lowest-paid professionals in the work force today."

The expectation in recent decades, however, has been that the profession has more to offer society than warehousing expertise, that it has a responsibility to manage access and retrieval and to focus its professional effort on the productive support and guidance of the end user. Librarians, in general, have accepted this definition of their mission but unfortunately do not see themselves as positioned to fulfill it. They are one of the later links in the database use chain outlined by WILLIAMS (1986, p. 210) as follows:

1. Generator/author
2. Primary publisher

3. Secondary publisher
4. Tertiary publisher   (online vendor)   added value
5. Quaternary publisher   (gateway system;   added cost;
    intermediary   decreased
    system)   proprietorship
6. Information broker   (intermediary   decreased
    organization)   control
7. Intermediary searcher
8. User (end user)
9. Ultimate user

Libraries are number six in the chain. The only product decision left to them is to buy or not to buy. They have little or no influence at levels three, four, and five, where products are designed and generated, although in some cases, the more talented of the profession have left the traditional structure and joined organizations higher in the chain that have a dynamic role in production.

Meanwhile, and partly as a result, leadership in many libraries has become defensive. It has come to mean negotiating the bureaucratic structure, hopefully without disaster to programs or personal credibility. Preserving materials budgets, protecting faculty status, and retaining a voice in institutional management have become measures of success whereas they should be routine administrative tasks.

In many cases, this translates into resistance to change—because new products and services do not integrate easily into established routines—and to backing away from cooperative efforts. As GORMAN (1986, p. 327) says, "technological difficulties have led to a lack of faith in cooperative endeavors" and have fostered a fortress mentality. BERRY & WILLIAMS (p. 43) write:

> there will always be some library staff members who generally feel threatened by any new technology. In addition, librarianship has always attracted humanists who tend to have a limited interest in machinery, and it is unlikely that this kind of person will be sold quickly or easily on a change to a more automated library system.

Real leadership, however, should take an active form wherein solutions rather than resistance are the result of a library's effort. The economics of publishing, both traditional and electronic, depend on the library market. No one will buy the product of much of primary and secondary publishing in great enough quantity to justify its offering if libraries do not continue their massive acquisition programs.

Online services and gateways are good only if the data are deliverable in some usable form. The library, as an economic and intellectual force, remains the most likely distribution channel. According to ROSZAK (1986a, p. 176):

> The library is not only there as a socially owned and governed institution, a true people's information service; it is staffed by men

and women who maintain a high respect for intellectual values. Because they are also the traditional keepers of the books, the librarians have a healthy sense of the hierarchical relationship between data and ideas, facts and knowledge. They know what one goes to a data base to find and what one goes to a book to find. In their case, the computers might not only generate more information for the public, but information itself is more likely to stay in its properly subordinate place in the culture.

Yet librarians do not see themselves as having influence on the information market. It seems clear that the profession is well positioned to assume a critical and influential role in the emerging information society if it can find the will to grasp it. FAYEN (1986, p. 242) notes: "We have a chance now to redefine what we do, to provide what our users want, and to have our services valued at their true worth. We must not sell ourselves short. The future of libraries may depend on it."

THE DIRECT INFLUENCE ON LEARNING

One phrase by ROCHELL is troubling. He says that one role of libraries is to be "neutral providers." SPIGAI refers to "passive librarians." Why neutral? Why passive? Why not have a point of view about how learning and research should occur and about the requirement for and the application of information to both?

All good teaching is a function of point of view. A good history professor is not a neutral provider. He does not merely list all political dates and events from 1917 to 1939 to teach about why World War II occurred. Instead he assigns readings that describe, relate, and interpret events, and then he lectures on the subject, providing his own interpretation of the events and their interrelationships. His interpretation has been informed by years of study and consideration; the student is the beneficiary of his best judgment concerning a particular subject. Nor is this the end of the process. The student will engage with the professor, via tests, papers, and dialog in class, commenting on the professor's understanding of the event, questioning, sometimes challenging, until he has his own understanding, which can serve as his basis for further research.

Librarians and information professionals likewise have well-informed points of view, gleaned from decades of research and experience, as to how information should be organized and presented and, in many cases, interpreted and evaluated, but they have mostly stopped short of performing the latter. As FAYEN (1986, p. 240) says, "Librarians have traditionally defined the service they provide as directional; that is, they point users at appropriate sources, but in most circumstances, do not provide answers." She continues, "librarian's reluctance to provide real answers and to vouch for their correctness contributes to the low value that is placed on libraries and on the service librarians traditionally provide (p. 241)."

It is important that this reluctance not be reflected in the design, implementation, and institutionalization of online informing systems. Libraries have at their disposal both media and mechanisms to take advantage of prior effort, to point to alternate directions, to encourage comment from users, and to analyze, interpret, and report user experience. These perspectives are critical, and libraries should use all of them in their systems to ensure that their position as an active and influential participant in the learning process not be lost.

One need only examine some of the commercial brokering software to realize how important this is. The tendency is often to underestimate the users' intelligence and make no attempt to determine their understanding of a problem. The software usually requires users to interact with a prescribed and simplistic vocabulary of topics after which interaction choices are made for them. The goal is to fit the user to the system rather than the other way around. Any good librarian could give better directions in a few sentences, along with alternatives and suggestions for other paths.

It is this knowledge that librarians first must recognize has value and then must incorporate into the online environment. ROCHELL (p. 48) demands that librarians have "a perception of themselves as society's agents, uniquely charged with responsibility for the preservation of our intellectual heritage" and that they "surrender the anonymity of the past for the militancy of the future."

CONCLUSION

The literature of library automation is most preoccupied with decrying sins of the past or heralding wonders of the future. It is less effective in describing the substance of current effort or in predicting trends that are realistic outgrowths of that effort. It is optimistic in imagining the multiplicity of access that is becoming available but slow to exercise control over the design of such access and weak in understanding how that access can have a direct and dramatic effect on learning.

In short, the literature is not as good as it should be. It reports and describes and complains and enthuses, but it does not often take risks, probably because the profession itself does not often take risks. It largely reflects the notion that librarians are facilitators of others' efforts rather than generators of their own.

The profession is well positioned to take the lead in designing "systems that inform," with years of experience in interpretation, service, and retrieval and two decades of experience in designing local systems and using commercial retrieval technologies. DE GENNARO (p. 40) predicts the following scenario for future effort:

> In the next decade, the creative development work will no longer be limited to the small entrepreneurial vendors as it has been in recent years. They will be joined by the systems groups that are now being reconstituted and revitalized in many libraries after a decade of relative inactivity and decline. The work of this new generation of systems librarians will be augmented and supple-

mented by the efforts of a growing army of enthusiastic volunteers drawn from the rank and file of the professional staff.

The challenge is to ensure that the profession takes full advantage of its opportunity.

BIBLIOGRAPHY

ALLOWAY, CATHERINE SUYAK. 1986. Naisbitt's Megatrends: Some Implications for the Electronic Library. The Electronic Library. 1986; 4 (2): 114-118. ISSN: 0264-0473.

AMERICAN LIBRARIES. 1970-. Chicago, IL: American Library Association; 1970-. ISSN: 0002-9769.

ARRET, LINDA. 1985. Can Online Catalogs be too Easy? American Libraries. 1985 February; 16 (2): 118-120. ISSN: 0002-9769.

ASSOCIATION OF RESEARCH LIBRARIES, OFFICE OF MANAGEMENT STUDIES. 1984. Nonbibliographic Machine-Readable Data Bases in ARL Libraries. Washington, DC: Association of Research Libraries; 1984. 100p. (SPEC Kit 105). ISSN: 0160-3574.

ASSOCIATION OF RESEARCH LIBRARIES, OFFICE OF MANAGEMENT STUDIES. 1986. End-User Searching Services. Washington, DC: Association of Research Libraries; 1986. 112p. (SPEC Kit 122). ISSN: 0160-3582.

ATKINSON, HUGH C. 1972. The Ohio State On-line Circulation System. In: Lancaster, F. Wilfrid, ed. Proceedings of the 1972 Clinic on Library Applications of Data Processing; 1972 April 30–May 3; Urbana, IL. Urbana, IL: University of Illinois Graduate School of Library Science; 1972. 22-28. ISBN: 0-87845-035-1.

AVENEY, BRIAN; BUTLER, BRETT, eds. 1984. Online Catalogs, Online Reference: Proceedings of a Library and Information Technology Association Preconference Institute; 1983 June 23-24; Los Angeles, CA. Chicago, IL: American Library Association; 1984. 211p. (Library and Information Technology Series, no. 2). ISSN: 0743-7900; ISBN: 0-8389-3308-4.

AVRAM, HENRIETTE. 1986. Current Issues in Networking. The Journal of Academic Librarianship. 1986 September; 12 (4): 205-209. ISSN: 0099-1333.

BAKER, BETSY. 1986. A Conceptual Framework for Teaching Online Catalog Use. The Journal of Academic Librarianship. 1986 May; 12 (2): 90-96. ISSN: 0099-1333.

BATES, MARCIA J. 1986. Subject Access in Online Catalogs: A Design Model. Journal of the American Society for Information Science. 1986 November; 37 (6): 357-376. ISSN: 0002-8231.

BAWDEN, DAVID. 1986. Information Systems and the Stimulation of Creativity. Journal of Information Science (England). 1986; 12 (5): 203-216. ISSN: 0165-5515.

BEARMAN, TONI CARBO; GUYNUP, POLLY; MILEVSKI, SANDRA N. 1985. Information and Productivity. Journal of the American Society for Information Science. 1985 November; 36 (6): 369-375. ISSN: 0002-8231.

BELLARDO, TRUDI. 1985. What Do We Really Know About Online Searchers? Online Review (England). 1985 June; 9 (3): 223-239. ISSN: 0309-314X.

BELLARDO, TRUDI; STEPHENSON, JUDY. 1986. The Use of Online Numeric Databases in Academic Libraries: A Report of a Survey. The Journal of Academic Librarianship. 1986 July; 12 (3): 152–157. ISSN: 0099-1333.

BERRY, JOHN; WILLIAMS, DELMUS. 1986. The Second Generation: Planning for the Replacement of Automated Systems. Resource Sharing and Information Networks. 1985 Fall/1985–86 Winter; 3 (1): 39–49. ISSN: 0737-7797.

BINDER, MICHAEL B. 1985. Videotex and Teletext: New Online Resources for Libraries. Greenwich, CT: JAI Press Inc.; 1985. 160p. (Foundations in Library and Information Science, Volume 12). ISBN: 0-89232-612-3.

BORGMAN, CHRISTINE. 1986. Why are Online Catalogs Hard to Use? Lessons Learned from Information-Retrieval Studies. Journal of the American Society for Information Science. 1986 November; 37 (6): 387–400. ISSN: 0002-8231.

BROOKS, H. M.; DANIELS, P. J.; BELKIN, N. J. 1986. Research on Information Interaction and Intelligent Information Provision Mechanisms. Journal of Information Science (England). 1986; 12 (1/2): 37–44. ISSN: 0165-5515.

BROPHY, PETER. 1986. Management Information and Decision Support Systems in Libraries. Hants, England: Gower Publishing Company Limited; 1986. 158p. ISBN: 0-566-03551-0.

CARNEY, RICHARD D. 1986. InfoTrac vs. the Confounding of Technology and It's Applications. Database. 1986 June; 9 (3): 56–61. ISSN: 0162-4105.

CART, MICHAEL. 1987. Caveats, Qualms, and Quibbles: A Revisionist View of Library Automation. Library Journal. 1987 February 1; 112 (2): 38–41. ISSN: 0363-0277.

CAWKELL, A. E. 1986. The Real Information Society: Present Situation and Some Forecasts. Journal of Information Science (England). 1986; 12 (3): 87–95. ISSN: 0165-5515.

CHERWITZ, RICHARD A. 1986. Communication and Knowledge. Columbia, SC: University of South Carolina Press; 1986. 192p. ISBN: 0-87249-465-9.

COCHRANE, PAULINE ATHERTON. 1985. Redesign of Catalogs and Indexes for Improved Online Subject Access. Phoenix, AZ: The Oryx Press; 1985. 484p. ISBN: 0-89774-158-7.

CULKIN, PATRICIA; SHAW, WARD. 1985. The CARL System. Library Journal. 1985 February 1; 110 (2): 68–70. ISSN: 0363-0277.

DE GENNARO, RICHARD. 1985. Integrated Online Library Systems: Perspectives, Perceptions, & Practicalities. Library Journal. 1985 February 1; 110 (2): 37–40. ISSN: 0363-0277.

DENENBERG, RAY. 1985. Open Systems Interconnection. Library Hi Tech. 1985; 3 (1): 15–26. ISSN: 0737-8831.

DISKIN, GREGORY M.; MICHALAK, THOMAS J. 1985. Beyond the Online Catalog: Utilizing the OPAC for Library Information. Library Hi Tech. 1985; 3 (1): 7–13. ISSN: 0737-8831.

DOSZKOCS, TAMAS E. 1986. Natural Language Processing in Information Retrieval. Journal of the American Society for Information Science. 1986 July; 37 (4): 191–196. ISSN: 0002-8231.

DOUGHERTY, RICHARD M.; LOUGEE, WENDY P. 1985. What Will Survive? Library Journal. 1985 February 1; 110 (2): 41–44. ISSN: 0363-0277.

DOWLIN, KENNETH E. 1984. The Electronic Library. New York, NY: Neal-Schuman Publishers, Inc.; 1984. 199p. ISBN: 0-918212-75-8.

DRABENSTOTT, JON. 1985a. Automation Planning and Organizational Change: A Functional Model for Developing a Systems Plan. Library Hi Tech. 1985; 3 (3): 15–24. ISSN: 0737-8831.

DRABENSTOTT, JON, Forum ed. 1985b. The Consultants' Corner. Automating Libraries: The Major Mistakes Librarians are Likely to Make. Library Hi Tech. 1985; 3 (1): 93–99. ISSN: 0737-8831.

DRABENSTOTT, JON, Forum ed. 1985c. The Consultants' Corner. Automating Libraries: The Major Mistakes Vendors are Likely to Make. Library Hi Tech. 1985; 3 (2): 108–113. ISSN: 0737-8831.

DRABENSTOTT, JON, Forum ed. 1985d. The Consultants' Corner. What Lies Beyond the Online Catalog? Library Hi Tech. 1985; 3 (4): 105–114. ISSN: 0737-8831.

DUNN, KATHLEEN. 1986. Psychological Needs and Source Linkages in Undergraduate Information-seeking Behavior. In: Nitecki, Danuta A., ed. Energies for Transition: Proceedings of the 4th National Conference of the Association of College and Research Libraries (ACRL), 1986 April 9–12; Baltimore, MD. Chicago, IL: ACRL, A Division of the American Library Association, 1986. 172–178. ISBN: 0-8389-6976-3. Also published in: College & Research Libraries. 1986 September; 47 (5): 475–481. ISSN: 0010-0870.

ERNEST, DOUGLAS J.; MONATH, JENNIFER. 1986. User Reaction to a Computerized Periodical Index. College & Research Libraries News. 1986 May; 47 (5): 315–318. ISSN: 0099-0086.

FAYEN, EMILY GALLUP. 1983. The Online Catalog: Improving Public Access to Library Materials. White Plains, NY: Knowledge Industry Publications, Inc.; 1983. 148p. ISBN: 0-86729-053-6.

FAYEN, EMILY GALLUP. 1986. Beyond Technology: Rethinking "Librarian." American Libraries. 1986 April; 17 (4): 240–242. ISSN: 0002-9769.

FORD, BARBARA J. 1986. Reference Beyond (and Without) the Reference Desk. In: Nitecki, Danuta A., ed. Energies for Transition: Proceedings of the 4th National Conference of the Association of College and Research Libraries (ACRL), 1986 April 9–12; Baltimore, MD. Chicago, IL: ACRL, A Division of the American Library Association; 1986. 179–181. ISBN: 0-8389-6976-3. Also published in: College & Research Libraries. 1986 September; 47 (5): 491–494. ISSN: 0010-0870.

FORESTER, TOM. 1985. The Information Technology Revolution. Cambridge, MA: The MIT Press; 1985. 674p. ISBN: 0-262-56033-X.

FOX, CHRISTOPHER JOHN. 1983. Information and Misinformation. Westport, CT: Greenwood Press; 1983. 223p. (Contributions in Librarianship and Information Science, no. 45). ISSN: 0084-9243; ISBN: 0-313-23928-2.

FOX, CHRISTOPHER JOHN. 1986. Future Generation Information Systems. Journal of the American Society for Information Science. 1986 July; 37 (4): 215–219. ISSN: 0002-8231.

GORMAN, MICHAEL. 1984. Online Access and Organization and Administration of Libraries. In: Aveney, Brian; Butler, Brett, eds. Online Catalogs, Online Reference: Proceedings of a Library and Information Technology Association Preconference Institute; 1983 June 23–24; Los Angeles, CA. Chicago, IL: American Library Association; 1984. 153–

164. (Library and Information Technology Series, no. 2). ISSN: 0743-7900; ISBN: 0-8389-3308-4.

GORMAN, MICHAEL. 1986. Laying Siege to the "Fortress Library." American Libraries. 1986 May; 17 (5): 325–328. ISSN: 0002-9769.

GREEN, STEPHEN. 1986. RLIN: The View from the Reference Desk. Colorado Libraries. 1986 March; 12 (1): 13–16. ISSN: 0147-9733.

GURNSEY, JOHN. 1985. The Information Professions in the Electronic Age. London, England: Clive Bingley, 1985. 206p. (Looking Forward in Librarianship). ISBN: 0-85157-380-0.

HAAR, JOHN M. 1986. The Politics of Information: Libraries & Online Retrieval Systems. Library Journal. 1986 February 1; 111 (2): 40-43. ISSN: 0363-0277.

HARTER, STEPHEN P., PETERS, ANNE ROGERS. 1985. Heuristics for Online Information Retrieval: A Typology and Preliminary Listing. Online Review (England). 1985 October; 9 (5): 407-424. ISSN: 0309-314X.

HELAL AHMED H.; WEISS, JOACHIM W. 1984. Local Library Systems. Essen Symposium; 1984 September 24–27; Essen, Federal Republic of Germany. Essen, Federal Republic of Germany: Gesamthochschulbibliothek; 1984. 337p. (Publications of Essen University Library no. 7). ISSN: 0721-0469; ISBN: 3-922602-08-8.

HELAL, AHMED H., WEISS, JOACHIM W. 1986. Future of Online Catalogues. Essen Symposium; 1985 September 30–October 3; Essen, Federal Republic of Germany. Essen, Federal Republic of Germany: Gesamthochschulbibliothek; 1986. 443p. (Publications of Essen University Library no. 8). ISSN: 0721-0469; ISBN: 3-922602-09-6.

HILDRETH, CHARLES R. 1984. Pursuing the Ideal: Generations of Online Catalogs. In: Aveney, Brian; Butler, Brett, eds. Online Catalogs, Online Reference: Proceedings of a Library and Information Technology Association Preconference Institute; 1983 June 23–24; Los Angeles, CA. Chicago, IL: American Library Association; 1984. 31–56. (Library and Information Technology Series, no. 2). ISSN: 0743-7900; ISBN: 0-8389-3308-4.

HILDRETH, CHARLES R. 1985. Online Public Access Catalogs. In: Williams, Martha E., ed. Annual Review of Information Science and Technology: Volume 20. White Plains, NY: Knowledge Industry Publications for the American Society for Information Science; 1985. 233–285. ISSN: 0066-4200; ISBN: 0-86729-175-3.

HILDRETH, CHARLES R. 1986. Communicating with Online Catalogs and Other Retrieval Systems: The Need for a Standard Command Language. Library Hi Tech. 1986; 4 (1): 7–11. ISSN: 0737-8831.

JACOB, M. E. L., ed. 1986. Telecommunications Networks: Issues and Trends White Plains, NY: Knowledge Industry Publications for the American Society for Information Science; 1986. 179p. ISBN: 0-86729-166-4.

JARAMILLO, GEORGE R. 1986. CARL PAC: A Public Service Look. Colorado Libraries. 1986 March; 12 (1): 10–12. ISSN: 0147-9733.

KEHOE, CYNTHIA A. 1985. Interfaces and Expert Systems for Online Retrieval Online Review (England). 1985 December; 9 (6): 489–505. ISSN: 0309-314X.

KESNER RICHARD M., JONES, CLIFTON H. 1984. Microcomputer Applications in Libraries. Westport, CT: Greenwood Press, Inc.; 1984. 250p. (New Directions in Librarianship no. 5). ISSN: 0147-1090; ISBN: 0-313-22939-2.

KLAPP, ORRIN E 1986. Overload and Boredom. Westport, CT: Greenwood Press, Inc.; 1986. 174p. (Contributions in Sociology no. 57). ISSN: 0084-9278; ISBN: 0-313-25001-4.

KOENIG MICHAEL E. D. 1987. Information Systems Technology: On Entering Stage III. Library Journal. 1987 February 1; 112 (2): 49-54. ISSN: 0363-0277.

KRIZ HARRY M.; KOK, VICTORIA T. 1985. The Computerized Reference Department: Buying the Future. RQ: Reference Quarterly. 1985 Winter, 25 (2): 198-203. ISSN: 0033-7072.

LIBRARY JOURNAL. 1876-. New York, NY: R. R. Bowker; 1876-. ISSN: 0363-0277.

LIPOW, ANNE G.; ROSENTHAL, JOSEPH A. 1986. The Researcher and the Library: A Partnership in the Near Future. Library Journal. 1986 September 1; 111 (14): 154-156. ISSN: 0363-0277.

LOWRY, CHARLES B. 1985. Technology in Libraries: Six Rules for Management. Library Hi Tech. 1985; 3 (3): 27-29. ISSN: 0737-8831.

LUNDEEN, GERALD W.; DAVIS, CHARLES H. 1982. Library Automation. In: Williams, Martha E., ed. Annual Review of Information Science and Technology: Volume 17. White Plains, NY: Knowledge Industry Publications Inc. for the American Society for Information Science; 1982. 161-186. ISSN: 0066-4200; ISBN: 0-86729-032-3.

LYNCH, CLIFFORD A. 1986. Linked Systems Protocol: A Practical Perspective. In: Jacob, M. E. L., ed. Telecommunications Networks: Issues and Trends. White Plains, NY: Knowledge Industry Publications for the American Society for Information Science; 1986. 67-81. ISBN: 0-86729-166-4.

MARKEY, KAREN. 1984. Subject Searching in Library Catalogs. Dublin, OH: Online Computer Library Center, Inc.; 1984. 176p. (OCLC Library, Information, and Computer Science Series, no. 4). ISBN: 0-933418-54-X.

MATTHEWS, JOSEPH R. 1985. Public Access to Online Catalogs. 2nd edition. New York, NY: Neal-Schuman Publishers, Inc.; 1985. 497p. (Library Automation Planning Guides Series, no. 1). ISBN: 0-918212-89-8.

MATTHEWS, JOSEPH R. 1986. Growth & Consolidation: The 1985 Automated Library System Marketplace. Library Journal. 1986 April 1; 111 (6): 25-37. ISSN: 0363-0277.

MCCOY, RICHARD W. 1986. The Linked Systems Project: Progress, Promise, Realities Library Journal. 1986 October 1; 111 (16): 33-39. ISSN: 0363-0277.

MILLER CONNIE, TEGLER, PATRICIA. 1986. Online Searching and the Research Process. College & Research Libraries. 1986 July; 47 (4): 370-373. ISSN: 0010-0870.

MILLER, DAVID C. 1986a. Laser Disks at the Library Door: The Microsoft First International Conference on CD-ROM. Library Hi Tech. 1986; 4 (2): 55-68. ISSN: 0737-8831.

MILLER, DAVID C. 1986b. The New Optical Media Mid-1986: A Status Report. Benicia, CA: DCM Associates, 1986 August. 37p. (Special Report to Fred Meyer Charitable Trust). Available from: DCM Associates, P.O. Drawer 605, Benicia, CA 94510.

MILLER, DAVID C. 1986c. The New Optical Media in the Library and the Academy Tomorrow. Benicia, CA: DCM Associates; 1986 August. 29p. (Special Report to Fred Meyer Charitable Trust). Available from: DCM Associates P.O. Drawer 605, Benicia, CA 94510.

MISCHO, WILLIAM. 1987. End-User Searching of Bibliographic Databases. In: Williams, Martha E., ed. Annual Review of Information Science and Technology: Volume 22. New York, NY: Elsevier Science Publishers, B.V., 1987. ISSN: 0066-4200; ISBN: 0-444-70302-0.

MOLHOLT, PAT. 1985. On Converging Paths: The Computing Center and the Library. Journal of Academic Librarianship. 1985 November; 11 (5): 284-288. ISSN: 0099-1333.

MOLHOLT, PAT. 1986. The Information Machine. A New Challenge for Librarians. Library Journal. 1986 October 1; 111 (16): 47-52. ISSN: 0363-0277.

MORAN, BARBARA B. 1984. Academic Libraries. The Changing Knowledge Centers of Colleges and Universities. Washington, DC: Association for the Study of Higher Education; 1984. ASHE-ERIC Higher Education Research Report no. 8. ISSN: 0737-1292; ISBN: 0-913317-17-9.

MURPHY, BROWER. 1985. CD-ROM and Libraries. Library Hi Tech. 1985; 3 (2): 21-26. ISSN: 0737-8831.

NEFF, RAYMOND K. 1985. Merging Libraries and Computer Centers: Manifest Destiny or Manifestly Deranged? EDUCOM Bulletin. 1985 Winter, 20 (4): 8-12,16. ISSN: 0424-6258.

NEUFELD, M. LYNNE, CORNOG, MARTHA. 1986. Database History: From Dinosaurs to Compact Discs. Journal of the American Society for Information Science. 1986 July; 37 (4): 183-190. ISSN: 0002-8231.

NIELSEN, BRIAN. 1984. Online Reference and the "Great Change". In: Aveney, Brian; Butler, Brett, eds. Online Catalogs, Online Reference: Proceedings of a Library and Information Technology Association Preconference Institute; 1983 June 23-24; Los Angeles, CA. Chicago, IL: American Library Association; 1984. 75-88. (Library and Information Technology Series no. 2). ISSN: 0743-7900; ISBN: 0-8389-3308-4.

OCLC, INCORPORATED. OFFICES OF RESEARCH AND TECHNICAL PLANNING. 1986. Annual Review of OCLC Research. Dublin, OH: OCLC Online Computer Library Center, Inc.; 1986. 53p. OCLC No.: 14563869.

OJALA, MARYDEE. 1986. Views on End-User Searching. Journal of the American Society for Information Science. 1986 July, 37 (4): 197-203. ISSN: 0002-8231.

OSHERSON DANIEL, STOB MICHAEL, WEINSTEIN, SCOTT. 1986. Systems That Learn. Cambridge, MA: The MIT Press; 1986. 205p. (The MIT Press Series in Learning, Development, and Conceptual Change). ISBN: 0-262-15030-1.

POST WILLIAM; SESSIONS, JUDITH A. 1986. Academic Institutions and Information Services: The Position of the Library. Library Hi Tech News. 1986 July/August; Issue no. 9: 7-9. ISSN: 0741-9058.

POTTER, WILLIAM GRAY. 1986. Online Catalogues in North America. An Overview. Program. Automated Library and Information Systems (England). 1986 April; 20 (2): 120-130. ISSN: 0033-0337.

REYNOLDS, DENNIS. 1985. Library Automation. Issues and Applications. New York, NY: R. R. Bowker Company; 1985. 615p. ISBN: 0-8352-1489-3.

ROCHELL, CARLTON C. 1987. The Next Decade: Distributed Access to Information. Library Journal. 1987 February 1; 112 (2): 42-48. ISSN: 0363-0277.

ROOSE, TINA. 1987. Public Libraries Online: Free versus Fee Reexamined. Library Journal. 1987 January; 112 (1): 64–65. ISSN: 0363-0277.

ROSZAK, THEODORE. 1986a. The Cult of Information. New York, NY: Pantheon Books, 1986. 238p. ISBN: 0-394-54622-9.

ROSZAK, THEODORE. 1986b. Partners for Democracy: Public Libraries and Information Technology. Wilson Library Bulletin. 1986 February; 60 (6): 14–17. ISSN: 0043-5651.

SACK, JOHN R. 1986. Open Systems for Open Minds: Building the Library without Walls. College & Research Libraries. 1986 November; 47 (6): 535–544. ISSN: 0010-0870.

SANDORE, BETH, BAKER, BETSY. 1986. Attitudes Toward Automation: How They Affect the Services Libraries Provide. In: Hurd, Julie M., ed. ASIS '86: Proceedings of the American Society for Information Science 49th Annual Meeting: Volume 23; 1986 September 28–October 2; Chicago, IL. Medford, NJ: Learned Information, Inc.; 1986. 291–299. ISSN: 0044-7870; ISBN: 0-938734-14-8.

SANTOSUOSSO, JOE. 1985. The Library as a Gateway to Online Services. In: Parkhurst, Carol A., ed. ASIS '85: Proceedings of the American Society for Information Science 48th Annual Meeting: Volume 22; 1985 October 20–24; Las Vegas, NV. White Plains, NY: Knowledge Industry Publications Inc.; 1985. 106–110. ISSN: 0044-7870; ISBN: 0-86729-176-1.

SARACEVIC, TEFKO, KANTOR, PAUL, CHAMIS, ALICE, TRIVISON, DONNA. 1985. Experiments on the Cognitive Aspects of Information Seeking and Information Retrieving: Executive Summary. 1985 Final Report on NSF grant IST 85-05411. Available from: Tefko Saracevic, Rutgers School of Communication, Information, and Library Studies, 4 Huntington Street, New Brunswick, NJ 08903.

SCHAUB JOHN A. 1985. CD-ROM for Public Access Catalogs. Library Hi Tech. 1985, 3 (3): 7–13. ISSN: 0737-8831.

SHNEIDERMAN, BEN. 1986. Designing Menu Selection Systems. Journal of the American Society for Information Science. 1986 March; 37 (2): 57–70. ISSN: 0002-8231.

SITTS MAXINE K., ed. 1985. The Automation Inventory of Research Libraries. Washington, DC: Office of Management Studies, Association of Research Libraries (ARL), 1985. 1 volume (unpaged). Available from: Association of Research Libraries, Washington, DC.

SKOVIRA ROBERT J. 1986. Information. A Description and Analysis of the Debonian and Foxian Views. In: Hurd, Julie M., ed. ASIS '86: Proceedings of the American Society for Information Science 49th Annual Meeting: Volume 23; 1986 September 28- October 2; Chicago, IL. Medford, NJ: Learned Information, Inc.; 1986. 306–309. ISSN: 0044-7870; ISBN: 0-938734-14-8.

SMITH KAREN F. 1986. Robot at the Reference Desk? In: Nitecki, Danuta A., ed. Energies for Transition: Proceedings of the 4th National Conference of the Association of College and Research Libraries (ACRL); 1986 April 9–12; Baltimore, MD. Chicago, IL: ACRL, A Division of the American Library Association; 1986. 198–201. ISBN: 0-8389-6976-3. Also published in: College & Research Libraries. 1986 September; 47 (5). 486–490. ISSN: 0010-0870.

SPIGAI FRAN. 1984. Online Reference for Professions and Businesses. In: Aveney, Brian; Butler, Brett, eds. Online Catalogs, Online Reference:

Proceedings of a Library and Information Technology Association Pre-conference Institute; 1983 June 23–24; Los Angeles, CA. Chicago, IL: American Library Association; 1984. 89–104. (Library and Information Technology Series no. 2). ISSN: 0743-7900; ISBN: 0-8389-3308-4.

SUGNET, CHRIS ed. 1986a. The Vendors' Corner. Critical Issues in Providing Integrated Library Systems. Library Hi Tech. 1986; 4 (1): 105-109. ISSN: 0737-8831.

SUGNET, CHRIS, ed. 1986b. The Vendors' Corner. Where are We Headed? Library Hi Tech. 1986; 4 (2): 95–104. ISSN: 0737-8831.

SWANSON, DON R. 1986. Subjective versus Objective Relevance in Bibliographic Retrieval Systems. Library Quarterly. 1986 October; 56 (4): 389–398. ISSN: 0024-2519.

SYBROWSKY, PAUL; WILSON, KEITH. 1986. DYNIX Automated Library Systems. Library Hi Tech. 1986 Summer; 4 (2): 39–49. ISSN: 0737-8831.

TEGLER PATRICIA, MILLER CONNIE. 1986. Online Searching and the Research Process. In: Nitecki, Danuta A., ed. Energies for Transition: Proceedings of the 4th National Conference of the Association of College and Research Libraries (ACRL); 1986 April 9–12; Baltimore, MD. Chicago, IL: ACRL, A Division of the American Library Association; 1986. 202–205. ISBN: 0-8389-6976-3.

TESKEY, F. N. 1984. Information Retrieval Systems for the Future. Boston Spa, England: British Library; 1984. 72p. (Library and Information Research Report 26). ISSN: 0263-1709; ISBN: 0-7123-3037-2.

TURTLE, HOWARD. 1986. The Open Systems Interconnection (OSI) Reference Model. In: Jacob, M. E. L., ed. Telecommunications Networks: Issues and Trends. White Plains, NY: Knowledge Industry Publications for the American Society for Information Science; 1986. 41–65. ISBN: 0-86729-166-4.

VAN ARSDALE, WILLIAM. 1986. The Rush to Optical Discs. Library Journal. 1986 October 1; 111 (16): 53–55. ISSN: 0363-0277.

VIGIL, PETER J. 1986. The Software Interface. In: Williams, Martha E., ed. Annual Review of Information Science and Technology: Volume 21. White Plains, NY: Knowledge Industry Publications, Inc. for the American Society for Information Science; 1986. 63–86. ISSN: 0066-4200; ISBN: 0-86729-209-1.

WAGSCHAL, PETER H. 1985. Interactive Technologies in the Academic Library. Library Trends. 1985 Summer, 34 (1): 141-150. ISSN: 0024-2594.

WEISKEL, TIMOTHY C. 1986. Libraries as Life-Systems: Information, Entropy, and Coevolution on Campus. College & Research Libraries. 1986 November; 47 (6): 545–563. ISSN: 0010-0870.

WILLIAMS, MARTHA E. 1985. Electronic Databases. Science. 1985 April 26; 228 (4698): 445–456. ISSN: 0036-8075.

WILLIAMS, MARTHA E. 1986. Transparent Information Systems Through Gateways, Front Ends, Intermediaries, and Interfaces. Journal of the American Society for Information Science. 1986 July; 37 (4): 204–214. ISSN: 0002-8231.

WOODARD, BETH, GOLDEN GARY A. 1985. The Effect of the Online Catalogue on Reference: Uses, Services, and Personnel. Information Technology and Libraries. 1985 December; 4 (4): 338–345. ISSN: 0730-9295.

9 Agricultural Information Systems and Services

ROBYN C. FRANK
National Agricultural Library

American agriculture always has been the envy of the world for its basic research and practical application of new production and marketing techniques. A major ingredient in making this system work has been the collection and sharing of information among government and academic sources, private businesses and individual citizens.

LETT (p. 122)

INTRODUCTION

Agriculture is the biggest industry and largest employer in the United States (U. S. DEPARTMENT OF AGRICULTURE, 1986a). About 21 million people work in some phase of agriculture, from growing food and fiber to selling it at the supermarket. Agriculture requires the service of about 18.3 million people to store, transport, process, and merchandise the output of the nation's farms. Approximately one of every five jobs in private industry is involved in agriculture.

A recent study by the U.S. Office of Technology Assessment suggests that continuing, rapid advances in biotechnology and information technology promise to revolutionize agricultural production and to alter dramatically the structure of the U.S. agricultural sector from a system dominated by the moderate-size farm to one dominated by large and very large industrialized farms (U.S. CONGRESS. OFFICE OF TECHNOLOGY ASSESSMENT).

The author expresses her appreciation to Holly Irving, Susan Whitmore, John Forbes, Carolyn Costa, and Carol Nelson for their assistance in the preparation of this chapter. This chapter is in the public domain.

Annual Review of Information Science and Technology (ARIST), Volume 22, 1987
Martha E. Williams, Editor
Published for the American Society for Information Science (ASIS)
by Elsevier Science Publishers B.V.

U.S. agriculture is at a major turning point as it enters the information communication age. In fact, some agricultural scientists indicate that the computer is possibly the most significant technological invention in American agriculture today (EXTENSION COMMITTEE ON ORGANIZATION AND POLICY). The use of computer technology and telecommunications offers great potential for increasing agricultural productivity by improving the efficiency of information exchange between and among scientists, educators, producers, and consumers.

Because agriculture is the world's oldest industry and the cornerstone of every culture, a tremendous body of agricultural literature exists. This review focuses primarily on the agricultural information activities in the United States from 1980 through 1986. The literature reviewed for this chapter was identified through database searches and solicitation letters to agricultural librarians throughout the United States and to selected librarians throughout the world.

This is the first *ARIST* chapter devoted solely to agricultural information. VAUPEL & ELIAS wrote an *ARIST* review in 1981 on life sciences information that included agriculture as one of the life sciences. Significant changes have taken place during the past seven years in information science and, in particular, its use in agriculture. Before we look at recent developments, let us take a brief look at the definition and history of agricultural information in the United States.

Definition

Agriculture involves much more than growing food and raising livestock for consumption. It is a multidisciplinary subject that requires a broad knowledge base of the biological, physical, and social sciences as well as technology.

For the purposes of this chapter, the definition of food and agricultural sciences as cited in the Agriculture and Food Act of 1981 is used. It is perhaps the broadest and most encompassing. According to this act:

> the term "food and agricultural" sciences means basic, applied, and developmental research, extension, and teaching activities in the food, agricultural, renewable natural resources, forestry and physical and social sciences, in the broadest sense of these terms, including but not limited to, activities relating to:
>
> (A) agriculture, including soil and water conservation and use of organic waste materials to improve soil tilth and fertility, plant and animal production and protection, and plant and animal health;
>
> (B) the processing, distributing, marketing, and utilization of food and agricultural products;
>
> (C) forestry, including range management, production of forest and range products, multiple use of forest and rangelands, and urban forestry;
>
> (D) aquaculture;

(E) home economics, including consumer affairs, food and nutrition, clothing and textiles, housing and family well-being and financial management;
(F) rural community welfare and development;
(G) youth development, including 4-H clubs;
(H) domestic and export market expansion for United States agricultural products; and
(I) production inputs, such as energy, to improve productivity (U.S. CONGRESS. SENATE, p. 88).

Agricultural Information in the United States—A Review

In the early 17th century, European settlers brought to America the tools, seeds, and technology of their old world and combined them with the tools, seeds, and technology of the American Indian to achieve the best farming methods. Thus, some of the colonies in New England and Virginia became the first agricultural "experiment stations" in the new world (U.S. CONGRESS. HOUSE). In those days agricultural information was exchanged by word of mouth, hand-carried letters, and, by the late 18th century, by books such as *The New England Farmer* by Samuel Dean and *The Old Farmer's Almanac* published in Sterling, Mass. *The American Farmer* magazine began publication in 1819 in Baltimore.

In 1785 the first scientific society devoted entirely to agriculture, the Philadelphia Society for Promoting Agriculture, was established. Many local agricultural societies were later organized throughout the United States.

The early involvement of the government in scientific agriculture can be traced to the activities of the U.S. Patent Office in 1839 when Congress appropriated funds for collecting agricultural statistics, conducting agricultural investigations, and distributing seeds. In fact, those early library collections on agriculture are said to be the beginning of what is now the National Agricultural Library (NAL).

In 1862 the U.S. Congress created the Department of Agriculture (USDA) with the mission "to acquire and diffuse among the people of the United States useful information on subjects connected with agriculture in the most general and comprehensive sense of that word" (U.S. CONGRESS. 37th CONGRESS, p. 387-388). The Morrill Act was passed that same year, which provided public land to each state for a land-grant college to support agriculture. The Hatch Act of 1887 provided federal funds to establish state agricultural experiment stations. These three pieces of legislation recognized the need for an agricultural information delivery system.

The land-grant colleges and the agricultural experiment stations functioned as the foundation of scientific agriculture on the state level. Early research projects were reported in state experiment station annual reports. Some states started publication series directed towards scientists, others toward farmers and consumers. Agricultural publications, such as *The Breeder's Gazette* and the *American Agriculturalist*, began printing articles on agricultural research (OVERFIELD).

At the same time the federal government began to support agricultural research. USDA focused its research on regional and national problems and researched problems that were too risky or lengthy for industry to undertake profitably.

The Smith-Lever Act of 1914 created the Cooperative Extension Service. Thus, the formal "agricultural extension model" was established. ROGERS describes this information delivery system as a centralized diffusion system consisting of three main components: 1) a research base consisting of each of the state's agricultural experiment stations and the research conducted by the USDA; 2) the state extension specialists who link the researchers to the county agents; and 3) the county extension agents who act as "change agents" with farmers, consumers, and other rural people at the local level. The establishment of direct mail service to farm homes and the creation of the Rural Electrification Administration within USDA to make loans to ensure telephone service and electric power to rural households and businesses were other significant milestones.

Commercial computers began appearing at land grant universities around 1960, and the agricultural researchers and extension service workers were among their earliest users. Software programs and information systems and services were developed to meet the needs of the farmer. Extension agents exposed farmers to computers, which facilitated the early adoption of personal computers on the farm in the 1980s (U.S. CONGRESS. HOUSE).

For many years USDA has carried out educational and public awareness programs to provide a wide range of information concerning the results of research, regulatory actions, and public services to farmers, agribusiness researchers, educators, and consumers (LETT). USDA has published and distributed free of charge a vast number of publications and audiovisual materials. However, because of the Paperwork Reduction Act of 1980 (Public Law 96-511), all publications, new and old, were carefully reviewed, and many were eliminated. User fees were charged for many items. As a result, USDA became committed to ensuring that the data collected worldwide would be readily accessible to those who needed it by using alternative methods of information delivery. Cooperative efforts with the private sector appeared to be the best solution to this problem.

AGRICULTURAL LITERATURE

Key Reference Tools

The most commonly cited guide to the agricultural literature in the United States is *Guide to Sources for Agricultural and Biological Research* (BLANCHARD & FARRELL). Directed to librarians, researchers, and graduate students, this guide covers important reference tools, key monographs, periodicals, and bibliographic databases related to all research activities concerned with food production. Several peripheral areas are also covered. *Information Sources in Agriculture and Food Science* (G. P. LILLEY) and *Key Guide to Information Sources in Agricultural Engineering* (MORGAN), which covers the specific area of agricultural engineering, are also useful.

With the assistance of other agricultural librarians MATHEWS (1981; 1986) identified a core list of serials in agriculture for public and academic libraries geared to the needs of a broad spectrum of the general population. Serials concerning general sciences, engineering, and social sciences are not included since these basic disciplines should already be well represented in libraries serving agricultural research clientele.

In addition to the commercially published monographs, journals, and other media, there is a wealth of unconventional, fugitive, or "grey" literature on agriculture. It is often difficult to locate because it is not advertised through normal channels. Sources of unconventional literature include theses and dissertations, government publications (national and international), university publications, research institution publications, state publications, conference proceedings, technical bulletins, symposia, and patents (G. P. LILLEY).

Directories are important for identifying many of the issuing bodies of fugitive literature. A selected list of useful directories in agriculture includes: *Directory of Professional Workers in State Agricultural Experiment Stations and other Cooperating State Institutions* (U.S. DEPARTMENT OF AGRI-CULTURE. COOPERATIVE STATE RESEARCH SERVICE, 1983); *County Agents Directory* (CENTURY COMMUNICATIONS); *Directory of Food and Nutrition Information Services and Resources* (FRANK, 1984); *Directory of Aquaculture Information Resources* (U.S. DEPARTMENT OF AGRI-CULTURE. NATIONAL AGRICULTURAL LIBRARY, 1982a); *International Directory of Forestry and Forest Products Libraries* (U.S. DEPARTMENT OF AGRICULTURE. FOREST SERVICE); *Rural Resources Guide: A Directory of Public and Private Assistance for Small Communities* (U.S. DEPARTMENT OF AGRICULTURE. OFFICE OF RURAL DEVELOPMENT POLICY); *The Biotechnology Directory* (COOMBS); *Information Sources in Biotechnology* (CRAFTS-LIGHTY); *Agricultural Research Centres* (HARVEY); and *Commodities Futures Trading* (NICHOLAS).

INFORMATION NEEDS

Little has been published about the information needs of the U.S. agricultural community. This is disturbing because development of information products and services should be based on marketing research studies of current and potential users. On the other hand, many information needs studies or evaluations have been published in other parts of the world, especially in developing countries.

Just who comprises the agricultural community? Basically it includes a wide spectrum: researchers, educators, government personnel, agricultural associations, economists, exporters, journalists, bankers, consultants, agricultural librarians and information providers, nutritionists, home economists, persons involved in agribusiness, farmers, ranchers, and those living in rural areas, and, in fact, all consumers of agricultural products. Two approaches have been taken to identify their information needs. CHARTRAND identified ten types of information required in the daily personal operations of various agricultural audiences. He also identified other types of information such as recreational systems and community service information that add to the enrichment of their lives.

The second approach highlights the diverse needs of different segments of the agricultural community. These groups are not mutually exclusive. Findings from RUSSELL, PALMER, STORM, and M. G. JONES have been integrated in Table 1 to illustrate the various needs of each of these user groups.

Rural Information Needs

A neglected but important segment of the agricultural community is the rural sector. Although many people tend to equate rural with agriculture (POWERS & MOE), only one in nine rural residents is directly involved in agricultural production. The rural population increased 16% during the 1970s, but that rate has slowed considerably during the 1980s. The interests and backgrounds of new rural residents are very diverse. Many are young and well educated and grew up in metropolitan areas where they have learned to expect sophisticated information services (HAYCOCK & WILDE; VAVREK).

The NABRIN Report (U.S. NATIONAL ADVISORY BOARD ON RURAL INFORMATION NEEDS) addresses the diverse information needs of rural citizens including those concerned with managing small towns and communities. CARTER identified the specific needs of the farmer to be: 1) operations and production management, 2) financial management, 3) community interaction, and 4) personal and family concerns. Information technologies can play a major role in meeting these needs, but Carter believes that the government must take the lead in encouraging the private sector to design programs and services specifically to respond to the wants and needs of the agricultural/rural community.

At this time the full benefits of new technology have not been realized in rural areas because: 1) the communities are too isolated, 2) there are few other resources from which to draw, and 3) few of the local people are trained in the use of the technologies. Likewise, the library situation is dismal. Eighty-two percent of U.S. public libraries are located in communities of 25,000 or fewer citizens (VAVREK). However, most rural libraries are characterized by small collections, low financial support, and lack of professional staff (only 25% of the librarians have a M.L.S.). Many libraries are run by volunteers. In 1982, a 12-year-old boy testified at a congressional hearing that out of frustration he set up and operated a local library because there were no services available to him (U.S. DEPARTMENT OF AGRICULTURE; NATIONAL COMMISSION ON LIBRARIES AND INFORMATION SCIENCE).

Only recently have rural sociologists addressed the role of information technologies (SOPHAR). DILLMAN, a rural sociologist, addresses five kinds of information technology that will play a role in the rural information age: 1) expanding telecommunications capability; 2) hardware for using telecommunications; 3) information technologies embedded in tools and materials; 4) a rapid delivery system of goods and services; and 5) the capability of persons to effectively use those technologies. Dillman predicts the growth of electronic cottage industries located in rural areas as a result of expanded tele-

TABLE 1
Agricultural Information Needs of Various Users

User Population	Information Needs	Sources
Policy makers/ administrators (government and private industry).	Production levels; use of resources, market outlook; state and national outlook; (perishable information).	Analyses prepared by support staff; press clippings; management information systems; unconventional literature.
Research scientists; agricultural librarians and information providers; all segments of agriculture.	Research—past, present, and future; rapid access to the latest findings.	Scientific serial literature; conference proceedings; informal communication with colleagues; traditional library services; online search services.
Diagnostic, analytical, and industrial scientist; economists.	Immediate access to details on new standards, techniques and procedures; patents and product details; trade information, market intelligence and outlook statements.	Scientific literature, manuals, reference texts, technical reports, brief technical notes, trade journals and product specifications; online search services; unconventional literature.
Specialist advisors (State Cooperative Extension Specialist).	Similar to needs of researcher; new developments and who is conducting relevant work in their specialist area.	Wide range of scientific, technical, and industry sources; research information systems, directories and specialist newsletters; subject-oriented current awareness; industry or subject reviews; outlook information; production monitoring statistical reports; direct reports from colleagues; state-of-the-art reports.
General advisors: County Cooperative Extension personnel, home economists, journalists.	Practical information; factual information; current practices and happenings throughout the agricultural sector; need for up-to-date information.	Personal contact with specialist advisors and other colleagues; review articles, trade publications, product specifications; materials produced by other advisors, fact sheets; potential heavy user of computerized information systems; Extension Service publications; press releases; exhibitors; project reports.
Educators and students (all levels, formal and informal).	Current practices and issues; computer literacy and experience in accessing databases.	Reviews; current serial literature and conference proceedings; textbooks (diminishing in value due to lack of currency), Extension Service publications; computer searches.

TABLE 1. Continued

User Population	Information Needs	Sources
Agricultural service industries: banks, feed and fertilizer suppliers, agricultural associations, produce brokers, chemical companies, exporters, accountants, economists, and other farm consultants, agribusiness, and service groups.	Market trends, production estimates and prospects for agricultural industries, research results, new practices and government policy; rapid access to new information is critical; data analysis and interpretation.	Government agencies; information utilities providing perishable information (i.e., statistics, economic indicators, and sociological data).
Consumers: farmers/ranchers and rural residents; general public.	Integrated technical and economic information for making decisions on production, marketing, and consumption; information to help them manage their lives successfully, cope with everyday problems, and realize opportunities.	Colleagues, friends, community leaders, other farmers, service agents, extension, rural and popular press; direct use of computerized and electronic services; videotex; libraries; radio and television presentations; demonstrations; exhibitions.

communication capabilities. This telecommunications revolution might also bring more people back to rural areas.

HAYCOCK & WILDE describe the rural community as a microcosm of an urban community. Most of the needed services are the same, but fewer professional and information resources are available. They note two current major facts related to the adoption of information technology in rural areas: slow adoption of the technology and lack of training.

A survey by the Intermountain Community Learning and Information Service (ICLIS) of current and potential users found that over 58% of the respondents had sought information from community, state, or federal sources during 1981-1982, yet nearly 50% had to obtain it from outside the community. The community library was identified as the favored resource center.

Under its charter, the National Commission on Libraries and Information Science (NCLIS) is responsible for studying the information needs of the United States and making recommendations to the appropriate government agencies for meeting these needs. Early in 1984, a planning committee was formed to consider the creation of a national advisory board on rural information needs. The NABRIN Report recommended the establishment of a NABRIN Advisory Committee under the auspices of USDA. Unfortunately, this report was issued when the U.S. Office of Management and Budget was unfavorable to the establishment of advisory boards. Also neither additional funding nor staffing was provided to support this information activity within the NAL, the designated agency.

Meanwhile, however, in 1986, the multistate (Utah, Colorado, Wyoming, and Montana) ICLIS project received $4.1 million in grants to help provide better lifelong educational opportunities for rural residents. The funds are providing new courses, programs, and information services for rural citizens and making them available through their community libraries. A major component in this project is the use of modern telecommunications and technologies, including computers, electronic blackboards, and closed-circuit television. This project also joins together in a cooperative partnership the various information providers in rural communities including the Cooperative Extension Service, public libraries, state and local governments, and public and private agencies. Much attention will be paid to this project to see if its approach is a viable prototype.

AGRICULTURAL DATABASES

In the *Agricultural Databases Directory*, WILLIAMS & ROBINS cite 428 databases dealing with topics related to agriculture. Of these, 302 are word oriented including bibliographic, referral, or natural language text. One hundred twenty-six are numeric databases that provide statistical data and modeling systems- e.g., toxicology data or commodity prices.

Bibliographic Databases

Three major databases cover agriculture comprehensively: AGRICOLA, produced by USDA's NAL; CAB, produced by U.K.'s Commonwealth Agricultural Bureaux; and AGRIS, produced by the U.N. Food and Agriculture Organization (FAO). Notable changes have occurred over recent years to each of these databases.

AGRICOLA. AGRICOLA (Agricultural OnLine Access), formerly known as CAIN, contains over 2 million citations on a wide range of agriculturally related topics. The scope of coverage is reflected in the definition provided earlier in this chapter. Approximately 90% of AGRICOLA consists of indexing records covering over 4,500 journal titles that are regularly scanned. The remaining 10% of the database is composed of cataloging records, including books, technical reports, booklets, audiovisuals, and microcomputer software.

AGRICOLA corresponds to several printed products including the *BIBLIOGRAPHY OF AGRICULTURE*, which covers journal articles and other analytic records, and the *NATIONAL AGRICULTURAL LIBRARY CATALOG*, which covers the monographic literature. Book catalogs of the food and nutrition subfile include the *FOOD AND NUTRITION QUARTERLY INDEX* (formerly the *Food and Nutrition Bibliography*), and the *Audiovisual Resources in Food and Nutrition* (U.S. DEPARTMENT OF AGRICULTURE. NATIONAL AGRICULTURAL LIBRARY. FOOD AND NUTRITION INFORMATION AND EDUCATION RESOURCES CENTER).

Traditionally AGRICOLA has had a heavy emphasis on the international literature. In 1980, only 60% of the citations were in English with the remaining 40% consisting predominantly of Russian, German, and French. PETERS points out that one of AGRICOLA's strongest assets is its nonjournal coverage, which includes many state and federal publications not indexed elsewhere.

One of AGRICOLA's past weaknesses was that it did not use a controlled vocabulary to index journal articles, but in 1985, NAL adopted the *CAB Thesaurus* as the standard vocabulary for indexing records (THOMAS, 1985). Identifiers were also incorporated into the record when appropriate. However, all cataloging records (e.g., monographs, audiovisuals) use the Library of Congress (LC) subject headings, so it is important to use both vocabularies when searching AGRICOLA.

Until 1986 all of the citations represented exclusively the holdings of NAL in Beltsville, Md. NAL is currently establishing a network of libraries to participate in a cooperative cataloging project of agricultural monographs using OCLC, Inc. as the host bibliographic utility. This means that AGRICOLA will include not only NAL holdings but also records of other libraries (HOWARD).

In an effort to avoid some of the duplication found in AGRICOLA and AGRIS, an evaluation study was conducted early in 1986 of foreign (non-U.S.) journal indexing (EDWARDS). As a result, NAL's objective is to focus primarily on U.S. publications and publications not indexed elsewhere. Other foreign literature will be covered on a more selective basis. The new NAL indexing priorities call for USDA, state experiment station, State Cooperative Extension Service and U.S. publications to be indexed before non-U.S. literature.

An exception concerns aquaculture. By agreement with Aquatic Sciences and Fisheries Information System (ASFIS), NAL will no longer index aquaculture with the exception of USDA and state experiment station publications and literature that has agricultural applications but will cooperate in trying to ensure that everything published will be covered by the Aquatic Sciences and Fisheries Abstracts (ASFA) database. NAL, in turn, will receive copies of selected indexed materials and make them available via interlibrary loan. (*AGRICULTURAL LIBRARIES INFORMATION NOTES*, 1986a, p.3)

NAL is also working with the land grant libraries to ensure that all publications of the state experiment stations and cooperative extension service are entered into AGRICOLA (THOMAS, 1986). NAL and the land grant libraries are sharing the responsibility of collecting, cataloging, indexing, preserving, and delivering documents.

CAB.   CAB (the Commonwealth Agricultural Bureaux) is currently the world's largest agricultural database, providing citations with abstracts from publications in nearly 40 languages. The major areas of coverage include: crop science and production; animal science and production; crop protection (pest control); plant and animal breeding; veterinary medicine; human nutrition; agricultural machinery and buildings; and economics and social services. CAB does not include post-harvest technology because that subject is covered by Food Science and Technology Abstracts (FSTA). There are 26 main indexing journals and 19 specialist journals that correspond to all the records in the database. Approximately 8,500 journals are reviewed regularly for articles. CAB is noted for its scientific integrity and for its information content.

The first *CAB Thesaurus* (TIDBURY) was published in 1983 and contains more than 48,000 terms. A second edition, to be published in 1987, will include many "Americanisms" (i.e., phrases, spellings, indigenous terms) as well as terminology for topics outside CAB's scope, such as in home economics and social science (see AGRICOLA). The revision will also attempt to bring the *CAB Thesaurus* and FAO's multilingual thesaurus as closely in line as possible.

During 1987-1988 all the CAB information services will be centralized into one location rather than dispersed among multiple information services at each research station. The new name of the service, CABI, reflects the growing international nature of the CAB database. It is hoped that this change will avoid duplication of effort, improve coordination, and be more responsive to users' needs (CAB INTERNATIONAL). CAB provides document delivery through a network of special libraries throughout the United Kingdom and overseas that acquire and supply photocopies of most source documents that have been abstracted.

AGRIS. AGRIS was established in 1975 to serve the agricultural needs of both developed and developing countries. It is an international database with participation from 120 countries and 14 international centers and is coordinated by the AGRIS Coordinating Center (ACC) of the FAO (UNITED NATIONS. FOOD AND AGRICULTURE ORGANIZATION, 1984). Each national center submits bibliographic data on all its agricultural publications, conventional and unconventional. Extension publications intended for farmers and consumers are not included. The major weakness of AGRIS is that it depends entirely on what the participating organizations supply.

The bibliographic citations are made available monthly in the form of a book catalog called *AGRINDEX* and on AGRIS magnetic tapes. The multilingual thesaurus *AGROVOC* (LEATHERDALE ET AL.) was published in 1982 to aid in the indexing and retrieval of relevant bibliographic citations. The first edition in English, French, Spanish, and German contained approximately 9,000 descriptor terms. The full incorporation of *AGROVOC* into AGRIS will make the database a multilingual international information system for agriculture (UNITED NATIONS. FOOD AND AGRICULTURE ORGANIZATION, 1986).

AGRIS became available online in the United States on DIALOG in 1986. However, the AGRIS records on DIALOG do not contain the U.S. records so a U.S. searcher would have to search AGRICOLA to find U.S. publications.

Cooperative efforts. Over the past 12 years, AGRIS has grown in size but has also encountered several obstacles. One major barrier has been to get all the nations of the world to cooperate. The U.S. position toward AGRIS has vacillated over the years according to the attitude of NAL's administration (SIMMONS). The lack of a U.S. or USDA policy on participation in an international information system has contributed to this vacillation, but the current administration is very supportive of international cooperation (HOWARD; THOMAS, 1985).

When AGRIS was developing its format and vocabulary, NAL chose not to be involved and as a result finds the AGRIS format and vocabulary incompatible with those used by NAL. Since the AGRIS format differs from that of CAB and AGRICOLA, for example, data must be converted in order to be used by AGRIS. For several years, CAB has not entered any records into AGRIS because it lacks the resources to make this conversion.

Currently, both CAB and AGRIS have separate and incompatible thesauri, but efforts are under way to bring these vocabularies more in line with one another (MANN). As previously discussed, NAL recently decided to adopt the *CAB Thesaurus* over *AGRIVOC* for its database. SIMMONS criticizes NAL and FAO for not cooperating at the early stages of AGRIS's development when present problems could have been avoided.

Comparison studies. Expansion within the past 12 years in the size of AGRIS has made earlier comparisons among AGRICOLA, CAB, and AGRIS, such as the one by LONGO & MACHADO, irrelevant (LEBOWITZ). The 1983 study by DESELAERS of AGRICOLA, AGRIS, and CAB shows substantial duplication in citations among all three databases. The smallest overlap was between AGRICOLA and AGRIS (34%), and the greatest was 46% duplication in the number of citations appearing in AGRIS that also appeared in both CAB and AGRICOLA. Looking at a total of 1,082 key international agricultural scientific journals covered by the three databases, Deselaers discovered that 23% of the articles recorded in AGRICOLA were also analyzed in the other two databases. The corresponding percentages were 24% for CAB and 31% for AGRIS.

English was found to be the predominant language in all three databases (AGRICOLA-71%, CAB-68%, and AGRIS-60%). In terms of regional emphasis, AGRICOLA's strength is North America and eastern Europe. CAB's emphasis is in western and eastern Europe as well as Asia. AGRIS covers western Europe and Asia relatively well.

Deselaers makes a case for having one integrated agricultural information system, citing cost effectiveness and efficient use of time in searching. He estimates that $2.5 million in searching costs could be saved each year. Because many organizations cannot afford to search all three databases online, they are missing much of the literature.

MANN suggests that one way to avoid the overlap problem would be to define more clearly areas that each database should cover. He suggests that: 1) NAL be responsible for all U.S. agricultural literature both conventional and unconventional; 2) AGRIS be responsible for all unconventional and some conventional literature produced in individual countries or regions; and 3) CAB be responsible for monitoring most of the world's conventional literature (excluding the U.S. literature and some of the national and regional literature covered by AGRIS). Like Deselaers, Mann recommends an overall, single, coordinated, world food and agricultural information system. Deselaers submits, however, that cooperation among the "big three" will take place only if it results in advantages for all concerned.

This proposal sounds appealing, but some hard facts prevent its realization. After all, NAL and CAB are not working with the same agenda as AGRIS. For example, AGRIS is mission oriented rather than subject oriented. Its mission is broadly defined to include "any subject or discipline which will help in the attempt to free mankind from the scourge of hunger and ensure a better life for rural people everywhere" (LEBOWITZ). While CAB has the goal to provide a worldwide information service for agriculture and allied sciences, it is also required to be self-supporting and therefore needs to be managed more like a business than a government agency that receives a subsidy (CAB INTERNATIONAL). Royalties gained from use of the database are important considerations. NAL has the mission "to facilitate access to and utilization of needed information in any medium by agricultural researchers, regulators, educators, and extension personnel; those employed in agriculture; those living in rural areas and communities; consumers of agricultural products; and the public in large, insofar as they need agricultural information" (U.S. DEPARTMENT OF AGRICULTURE. OFFICE OF THE SECRETARY). Like AGRIS, AGRICOLA is not working in a financially motivated environment.

Discipline Online Searching

To gain a better appreciation for the large scope of agriculture and its multidisciplinary nature, let us look at the databases that serve some specific areas of agriculture. Note that it is often necessary to search more than one database to get the most useful citations.

Veterinary medicine. MACNEIL, NIELSEN, and MACK contend that the most useful databases for veterinarians are MEDLINE, BIOSIS, AGRICOLA, and CAB. MEDLINE contains many citations to clinical and research information in veterinary medicine. BIOSIS is most useful in the areas of parasitology, public health, pathology, nutrition, and toxicology. AGRICOLA's strongest point is its coverage of government publications and reports because these are not covered well in other databases (MACNEIL). A major advantage of CAB is that it contains clinical veterinary articles on patient management and patient care. However, MacNeil suggests searching CAB after MEDLINE and BIOSIS because of CAB's cost, lack of timeliness, and the availability of journals cited. DIFT (Drug Information FullText), prepared by the American Society of Hospital Pharmacists, is a full-text database of two useful publications: the *American Hospital Formulary Service Drug Information 84* and the third edition of the *Handbook on Injectable Drugs.* Additional databases of interest to veterinarians include TOXLINE, DISSERTATION ABSTRACTS ONLINE, CA SEARCH, INTERNATIONAL PHARMACEUTICAL ABSTRACTS, and TOXICOLOGY DATA BANK. For further information on bibliographic databases as well as epidemiologic, laboratory and clinical, and research-in-progress data banks, see USDA's *International Directory of Animal Health and Disease Data Banks* (U.S. DEPARTMENT OF AGRICULTURE. NATIONAL AGRICULTURAL LIBRARY, 1982b).

Agronomy. HALL ET AL. found that CAB, BIOSIS, and AGRICOLA were the most comprehensive databases for agronomic studies. CA SEARCH and SCISEARCH are also helpful.

Agricultural economics. SIESS & BRADEN cite 38 bibliographic and 14 nonbibliographic databases that are relevant to agricultural economics. They report on a study conducted at the University of Illinois at Urbana-Champaign on agricultural economics; the three databases searched most often were

AGRICOLA, CAB, AND CRIS (Current Research Information System) (AGRIS was not yet available).

Forestry. BROOKS compared four databases- AGRICOLA, BIOSIS, CAB, and SCISEARCH—in terms of their coverage of the literature on agriculture and forestry. Subject coverage in AGRICOLA, while comprehensive, primarily emphasizes USDA's research interests. Thus, some areas generally associated with forestry, such as recreation and archeology, which is the responsibility of the U.S. Department of the Interior, are not emphasized in AGRICOLA.

Brooks states that it is not possible to identify one best database in agriculture and forestry. She reports an overlap of 20% between AGRICOLA and CAB. Because of its comprehensiveness and low searching cost, AGRICOLA is the best place to start. The abstracts in CAB make it a valuable database, but, again, cost is an important consideration. BIOSIS is best when one is looking for information on ecologically oriented topics. SCISEARCH is useful primarily for its cited references.

Additional information on forestry databases can be found in the *Directory of Selected Forestry-Related Bibliographic Databases* (EVANS) which lists 117 databases maintained by scientists of USDA's Forest Service. Most of the databases cited are bibliographic, but several provide raw data or referral information.

Food technology. For questions dealing with food chemistry, analytical methods, and biological aspects of food science, SZE identifies the best databases as CA SEARCH, FSTA, and BIOSIS. Sze also recommends searching AGRICOLA, CAB, CRIS, NTIS (National Technical Information Service), COMPENDEX, FOODS ADLIBRA, and the Federal Register database for other food research and development topics. TCHOBANOFF recommends FSTA as the first database to search for food technology, followed by CA SEARCH, AGRICOLA, and NTIS. SCISEARCH is useful because of its timeliness and its citation indexing.

Both Sze and Tchobanoff recommend searching PIRA (produced by the Research Association of the Paper and Board, Printing and Packaging Industries, U.K.) and PAPERCHEM (produced by the Institute of Paper Chemistry, U.S.) for information on food packaging. RAPRA (produced by the Rubber and Plastics Research Association, U.K.) might also be useful.

Patent information for the food industry is found in CLAIMS (U.S. patents), WPI (wordwide patents), CA SEARCH, FSTA, and FOODS ADLIBRA. Marketing information can be found in FOODS ADLIBRA, FSTA, and Predicasts Terminal System (PTS). Food engineering is covered in FSTA and COMPENDEX. Toxicology issues can be found in TOXLINE and TOXBACK as well as FSTA, CA SEARCH, and BIOSIS.

Human nutrition. FRANK (1982) identified 24 major databases (bibliographic, referral, and numeric) representing some of the most useful and generally recognized resources on food and human nutrition. Those bibliographic databases with a major emphasis on food and nutrition include AGRICOLA (particularly the food and nutrition subfile that contains abstracts), CAB, FSTA, and FOODS ADLIBRA. Databases with in-depth coverage of selected food and nutrition areas include BIOSIS, CA SEARCH, COMPENDEX, EXCERPTA MEDICA, and MEDLINE. Two useful referral databases are CRIS and HNRIMS (both discussed in the next section), which provide information on research in progress.

Tchobanoff recommends MEDLINE, BIOSIS, and CAB for nutrition and food microbiology questions. Sze recommends CA SEARCH, CAB, BIOSIS, FSTA, EXCERPTA MEDICA, and MEDLINE for nutritional, toxicological, and medical problems. AGRICOLA is conspicuously missing from their recommendations since it has a strong emphasis on human nutrition.

Agricultural Research Project Referral Databases

In addition to bibliographic information, researchers and practitioners may use databases to obtain valuable information on current research activities.

CRIS. Current Research Information System (CRIS) is USDA's computer-based documentation and reporting system for agricultural and forestry research. CRIS contains over 27,000 descriptions of current, publicly supported agricultural and forestry research projects of USDA agencies, state agricultural experiment stations, and other cooperating state institutions. The research projects cited are typically problem oriented, such as farm and forest product development, protection of crops from insects, diseases, or pests, and human nutrition (U.S. DEPARTMENT OF AGRICULTURE. COOPERATIVE STATE RESEARCH SERVICE, n.d.).

HNRIMS. Similar to CRIS is the Human Nutrition Research and Information Management System (HNRIMS) database that provides access to information on human nutrition research and research training activities supported wholly or partly by the federal government. Contributors to this database include USDA, U.S. Department of Health and Human Services (DHHS), the Veteran's Administration (VA), the Agency for International Development (AID), the Department of Defense (DOD), and the National Oceanic and Atmospheric Administration (NOAA) of the Department of Commerce. HNRIMS is maintained as a private file on DIALOG and is also available through the National Institutes of Health's (NIH) in-house computer system.

TEKTRAN. Technology Transfer Automated Retrieval System (TEKTRAN) provides ready access to over 6,000 brief, easy-to-read summaries of the latest research results that have been peer-reviewed and cleared by USDA's Agricultural Research Service (ARS). About 300 new findings are entered into the database each month. TEKTRAN is accessible through state cooperative extension offices as well as through USDA (KINNEY).

CARIS. Current Agricultural Research Information System (CARIS), coordinated by FAO, is a database of agricultural research projects currently under way in or related to developing countries. Like AGRIS, CARIS is an international cooperative network, with each participating country contributing information on projects in progress within its boundaries. CARIS provides a mechanism whereby developing and developed countries can exchange information on their current agricultural research activities (UNITED NATIONS. FOOD AND AGRICULTURE ORGANIZATION, 1985).

Numeric Databases

The U.S. government supplies much of the information found in agricultural numeric databases. The USDA, U.S. Geological Survey (USGS), U.S. Census Bureau, and the National Weather Service are a few of the major government sources. Other sources of data include international organizations and agencies, private companies, commodity exchanges, surveys, and the literature. A more complete list of government-generated statistical databases appears in the *Handbook of Agricultural Statistical Data* (GARKEY & CHERN).

In 1986 USDA made its perishable information, such as agricultural market reports, available in electronic form. Several services have purchased this information and made it available through various databases- e.g., the AGRIBUSINESS USA database that is available on DIALOG. AGRIBUSINESS USA provides access to the full text of USDA statistical reports published since 1986, to numeric data, and to indexes and abstracts of more than 300 business and government publications. Two other important databanks maintained by USDA include the Feed Composition Data Bank and the Nutrient Data Bank.

Feed composition data. The Feed Composition Data Bank (FCDB) was originally maintained by Utah State University, but in 1986, this responsibility was shifted to USDA's NAL with support from ARS. This data bank is one of the most complete compilations of data on feedstuffs in the world and contains information on more than 23,000 feeds and feed ingredients (U.S. DEPARTMENT OF AGRICULTURE. NATIONAL AGRICULTURAL LIBRARY, n.d.). FCDB exchanges feed composition information internationally within the International Network of Feed Information Centers (INFIC).

FCDB contains data on plant tissue and proximate analysis, vitamins, minerals, protein constituents, antinutrients and toxicants, and energy use by animal species. Many printed products are available from the National Technical Information Service (NTIS) on international tables of feed composition.

Nutrient Data. USDA's premier nutrient database, the USDA Nutrient Data Base for Standard Reference, provides the most complete and up-to-date food composition data on American foods. Nutrient data in the form of reference tables for more than 60 food components in thousands of American foods are generated, expanded, and updated continually (HEPBURN). These tables are available in printed form (U.S. DEPARTMENT OF AGRICULTURE. AGRICULTURAL RESEARCH SERVICE) from the U.S. Government Printing Office (GPO) and in machine-readable form from NTIS (U.S. DEPARTMENT OF AGRICULTURE. HUMAN NUTRITION INFORMATION SERVICE). Annual conferences bring together professionals from industry, academia, the medical sciences, and the government to discuss all aspects of nutrient data bank technology (PERLOFF & GRAY).

USDA and DHHS are developing the National Nutrition Monitoring System (NNMS) to track the dietary and nutritional status of the American population (PETERKIN & RIZEK). The system draws its information mainly from the Nationwide Food Consumption Survey (NFCS) and the DHHS's National Health and Nutrition Examination Survey (NHANES). Among its many uses, NNMS provides information for evaluating changes in agricultural policy related to food production, processing and distribution of the U.S. food supply (WELSH).

Another recent movement in the nutrient data arena is the development of an international organization called INFOODS (International Network of Food Data Systems), which is concerned with all aspects of food composition data including assessing data quality, nomenclature and classification, systems design, examining content and form of databases, accessing the needs and uses of food composition data and maintaining an international inventory of food composition data (BUTRUM & YOUNG).

USE OF COMPUTERS IN AGRICULTURAL INFORMATION SYSTEMS

"The computer is possibly the most significant technological innovation during the 1980's in American agriculture" (EXTENSION COMMITTEE ON ORGANIZATION AND POLICY, p. 1). This sentiment is reflected in the hundreds of articles on how computer technology has been applied to agriculture. Journals, such as *COMPUTER APPLICATIONS IN THE BIOSCIENCES*, have appeared to assist researchers and practitioners in determining the appropriate use of information technology in their work.

Minicomputer and Mainframe Systems

Many of the articles written on agricultural information systems deal with programs using a mainframe or a minicomputer system; in some cases the system is not indicated. The wide use of computers includes the production of a nursery and seed catalog database from existing OCLC records as a resource for gardeners, plant systematists, librarians, and so forth (FEIDT); a food-animal residue-avoidance data bank to provide electronic access to agrichemical information to veterinarians, livestock producers, and extension specialists (SUNDLOF ET AL.); a cotton ginning systems model to optimize producer returns (ANTHONY); a standardized, centrally located germplasm resource system to facilitate information access and retrieval (LARSON); a food service operations matrix system to assist with planning and decision-making (MATTHEWS); surveillance systems to aid in the management of crop diseases (TENG); a forest pest-management system that uses geographic data (PENCE ET AL.); plant information network for convenient storage and retrieval of data to assist rare plant management (DITTBERNER ET AL.); a program to determine material requirements for furniture manufacturing (ARAMAN); and a system to analyze a variety of survey data with the purpose of providing pest information (JOHNSON).

Microcomputer Systems

The acceptance of microcomputers in agriculture has been rapid. In 1985 it was estimated that 6-8% of all farms presently have microcomputers and that these farms account for a large portion of the value of farm sales (U.S. DEPARTMENT OF AGRICULTURE. EXTENSION SERVICE). AUDIRAC & BEAULIEU have reviewed several computer adoption studies that show a strong correlation between farmers who use computers and high levels of formal education, farm income, and sophisticated farm management practices. They have proposed a model and framework for studying the distribution and adoption of microcomputer technology by farmers.

Cooperative Extension Service Involvement. USDA's Cooperative Extension Service has moved ahead rapidly in the acquisition of microcomputers. By 1985, 29 states had microcomputers in 100% of the county offices while only 9 states had them in less than 50% (EXTENSION COMMITTEE ON ORGANIZATION AND POLICY).

. Regional computer organizations working with the Extension Service have been established in the north central, northeast, western, and southern U.S. regions. The North Central Computer Institute (NCCI), located at the University of Wisconsin at Madison, has sponsored many useful workshops, seminars, and training sessions. NCCI facilitates the sharing of computer software throughout the region.

Farmers and ranchers have been vocal and persistent in their desire for land grant institutions (research and Cooperative Extension Service) to develop on-farm or on-ranch software (MCGRANN). Both farmers and agricultural producers value the objectivity, accountability, and educational support they expect from these institutions. However, software development by Cooperative Extension Service personnel is controversial. Where does one draw the line between the responsibilities of public and private sectors? Coordination between the public and private sectors is necessary to reduce duplication. The land grant institution has the knowledge base for the software whereas the private sector has expertise in repackaging, marketing, and delivery.

Another major concern to Extension Service administrators is the cost of hardware and software, which represent a major investment. Decisions regarding their purchase should be coordinated so that systems are compatible and are in sufficient supply throughout the Extension Service (MCCLELLAND).

The New Agricultural Computer Professional

AUDIRAC & BEAULIEU note the arrival of a new professional—the farm computer consultant who will act as a bridge between the computer salesman and the farmer. In some U.S. areas, farm computer user groups and computer cooperatives have been established. These groups sometimes take the place of the computer consultant and share expertise to reduce the costs and risks of adopting computer technology. Unfortunately little is known about how many of these groups exist in the farming community.

Audirac and Beaulieu also point out some of the consequences of adopting microcomputer technology. They believe that while this technology offers definite opportunities and advantages to traditional approaches, the need for managerial knowledge and scientific/technical expertise will increase. At the same time certain routine types of farm labor will be replaced by the technology. They propose that computer-aided farm managers will be almost indistinguishable from other industry managers due to their increased skills in finance, marketing, and management. LEVINS states that agricultural economists have a major role to play in introducing microcomputers to farmers and to agribusiness.

LA FERNEY discusses the impact and opportunities of microcomputers on teaching, research, and Cooperative Extension Service work in the land grant institutions. The challenge in teaching is to be certain that all graduates are computer literate. Researchers need to provide opportunities for students to learn how to conduct computer-aided research. Extension service personnel will be called on by homemakers and farmers for advice on purchasing hardware and software as well as instruction on how to use these devices and packages. As a result, they must be well trained in the use of microcomputers so that they can educate others.

Both MCGRANN and La Ferney point out that one reason why researchers do not spend more time in developing software is the lack of professional recognition. If the software does not count toward promotion, as a traditional publication would, he suggests that a peer-review process be established so that this creative activity would be fully recognized. LEVINS, on the other hand, believes that administrators who are looking for recognition in the information age will welcome software as the format of publication. After all, software requires published documentation.

INFORMATION UTILITIES

An information utility is a home or office computer-based information system. It can be interactive or noninteractive. Information utilities usually offer more than a typical computer-based news service such as Associated Press or Reuters. The service is usually conveyed directly to the end user, and the interactive systems also allow for the use of educational programs based on self-instruction or tutorial assistance (WHITING).

Videotex is the transmission of unlimited numbers of words, numbers, and pictures either via television transmitter or by central computer over telephone lines (also known as videotex or viewdata) (SIGEL). In 1978, the British PRESTEL system initiated the first major effort to provide perishable information to the British public via a videotext system (HOEY). Other countries, such as France, Denmark, Finland, Japan, and Australia, have developed similar systems (JUDGE). Each has met with varying degrees of success and, in some cases, failure. SIGEL proposes that one reason why videotext has not been more popular with consumers is that the information is not presented in an entertaining format. On the other hand, farmers know only too well the importance of timely information on market situations and are thus a prime target audience for interactive retrieval of commodity price data.

General Utilities

Several general information utilities provide agricultural services. Two examples are The Source and DIALCOM.

The Source provides access to news wire services, electronic mail, stock prices, travel services, desktop shopping, computerized conferencing, and an educational software library among other sources. DIALCOM is being used by USDA to disseminate all nationally significant press releases issued from Washington, D.C., as well as a weekly newsletter. In 1986, the Food and Nutrition Information Service was made available as a part of USDA Online on DIALCOM. This service provides consumer-oriented food and nutrition information including current news items, bibliographies, publications lists, the full text of selected USDA food and nutrition publications charts, and lists, and brief descriptions of USDA agencies responsible for food and nutrition (VINCENT).

Agricultural Utilities

Green Thumb. In 1980, the first major attempt at a videotex information system in the U.S. was launched and was called the Green Thumb project. It was a cooperative project among USDA, the University of Kentucky's Cooperative Extension Service, and the National Weather Service. Its goal was to develop field tests and evaluate a computerized system for disseminating weather, market, and other information on agricultural production and management on a daily basis to farmers.

Green Thumb allowed farmers to selectively retrieve computer-based information by using the telephone system to transmit the data and a home television set to display it. Even though Green Thumb had its share of technical and organizational problems, the project leaders believed that there was a real future for this type of information delivery system that would disseminate research, farming information, and educational information on a timely basis to farmers at an affordable price (WARNER & CLEARFIELD). This very successful pilot project ended in 1982 when funding was discontinued. Kentucky modified the Green Thumb system to form AgText in 1985. AgText broadcasts weather, market, and agricultural advisory information through the local public broadcasting station to television sets equipped with decoders.

AGNET. The Agriculture Computer Network (AGNET) is the largest agricultural computing network developed at a land grant university. It has been in existence since 1975 at the University of Nebraska at Lincoln, and it services clients in 47 states and 10 foreign countries. HUMPHREY credits AGNET's success to the fact that it is "the friendliest of the user friendly agricultural computer networks."

According to MURRAY, AGNET's client base has changed over the past 12 years. In 1975 the primary audience was educational institutions—universities, community colleges, and high schools. Today, more than half of its clients are producers and agribusinesses, with education, government, credit institutions, farm management consultants, computer vendors, and nonagricultural organizations comprising the rest.

AGNET presently offers three types of services: 1) management models, 2) current information, and 3) national and international electronic communications. It is most used and best known for its "what if" games as management models. They are kept current by subject-matter experts at the various land grant universities.

Programs help users make decisions in diet analysis, local livestock feed data, financial management, crop and livestock production and marketing, and home economics. Using AGNET, clients can calculate the cost per acre to produce a crop, determine the appropriate fungicide treatment for a plant disease, and predict the growth rate, costs, and returns for cattle, hog, or poultry feeding operations.

When using AGNET one can download the "input form" so that it can be filled in offline on the client's personal computer and then uploaded into the system. This procedure saves significant costs in connect-time and telephone charges.

One of the primary sources of information for AGNET's current information service is USDA. Government data include commodity marketing and outlook and situation reports, livestock shipping requirements, press releases, foreign trade leads and attaché reports from USDA's Economic Research Service, Statistical Reporting Service, Foreign Agricultural Service, the Animal and Plant Health Inspection Service, and the Office of Information. Other sources of perishable information include specialists at universities, state departments of agriculture, and private agribusiness concerns. AGNET does not edit any of the data.

Infotext.   Infotext represents the first step toward electronic delivery of Cooperative Extension Service information directly into the home in Wisconsin. It is a teletext, alphanumeric news and information service provided by WHA-TV, the five Wisconsin Educational Television Network stations, and WMVS-TV in Milwaukee. This one-way communication uses television decoders that display captioning similar to that used for the hearing impaired.

Users of Infotext have access to a 15-minute cycle of bulletins from the University of Wisconsin Cooperative Extension Service, agricultural weather information, and market reports. VEDRO believes that this type of communications system has great potential for providing timely and useful local agricultural information.

COMNET.   COMNET provides centralized office automation for Michigan State University's campus offices and an electronic information-dissemination network that links campus offices with county extension service offices (MARTIN). The system also uses USDA's electronic mail system, DIALCOM. COMNET uses both broadcast and "bulletin board" functions.

Telplan.   Telplan, another service offered by Michigan State University, provides computerized programs for educational purposes, such as university instruction, extension work, and commercial agricultural operations. More than 100 agricultural programs are available (RENEAU & PATTERSON).

FACTS.   Through FACTS, the Fast Agricultural Communications Terminal System, the Indiana Cooperative Extension Service provides information and programs to the Indiana agricultural community (FREDERICKS). Electronic mail, full text of Extension Service newsletters, and more than 50 different applications programs (e.g., Cropland Lease Comparisons, Retired Couples' Budget, and Home Vegetable Garden Planner) can be run in each field Cooperative Extension office.

Agridata. Agridata, formerly AgriStar, provides perishable information of interest to the agricultural community. Agridata is accessible via any computer with communications capability. It contains a large database of continuously updated information on business, finances, marketing, prices, weather, livestock, and crops along with analysis and recommendations from reporters, market analysts, economists, meteorologists, and researchers throughout the United States and other countries.

Grassroots. Telidon is Canada's generic name for videotex technology. Canada's first major attempt to provide a videotex agricultural information service called Grassroots was provided commercially by Infomart in Winnipeg. This service was later made available to U.S. customers in selected states.

Grassroots provides market farm prices, calendar of events for the agricultural community, agricultural extension courses, interactive farm management programs, research reports, summaries, weather forecasts, agribusiness product information, and stock exchange trading information.

A 1982 study done by the University of Guelph to introduce Grassroots to Ontario farmers evaluated both the agricultural information and its education system. While the study reaffirmed that Grassroots is an easy-to-use home service for inexperienced computer users, the cost of the system both in hardware and telecommunications charges and service to rural areas appears to be the biggest obstacle to its widespread use in the farm community (BECKMAN).

REMOTE SENSING

Geographic information is used by natural resource decision makers in each state. In the past, it has been difficult to get the necessary information because agreement and coordination among the various federal and state agencies were needed.

The University of Missouri developed the Geographic Information System (GIS), which gathers, stores, analyzes, and disseminates information on natural resources and socioeconomic data (JOHANNSEN). Heavy emphasis is placed on remote sensing and special analysis. Some of the data include: vegetation types, land ownership, watershed boundaries, land use, surface water resources, geology, soil types, underground water resources, slope, aspect, and water and air quality. Statistical data are also available on precipitation records, water, oil and gas well locations, mineral locations, and the location of distinctive natural features. This system has prodded agencies to be more cooperative in sharing their data. GIS has already been used for a variety of projects: an inventory or assessment of forest cover, agricultural crops, and wildlife habitat; the monitoring of forest stand conversions and harvesting; water resources or suburban development; and the analysis of land use impacts and of mined land reclamation.

MICROCOMPUTER SOFTWARE

In recent years there has been a proliferation of agriculturally related microcomputer software developed by public and private organizations. STRAIN & SIMMONS cite approximately 800 microcomputer software programs developed by state Cooperative Extension Service offices. Their inventory of computer programs is available in printed form as well as on Virginia Polytechnic Institute and State University's CMN (Computerized Management Network) system in Virginia. The inventory lists programs used by personal computers, programmable calculators, and mini- and mainframe computer programs. Topics range from business management, crop or livestock management, veterinary medicine, and natural resources to community development, home and family, and youth programs. Research and teaching applications as well as general administrative programs are also included.

Several journals regularly review and evaluate agriculturally related software: *FARM COMPUTER NEWS*, *AGRICULTURAL COMPUTING*, and *AGRICOMP*. *BIOTECHNOLOGY SOFTWARE* reviews microcomputer applications in biotechnology. Those interested in software programs on home economics and nutrition should consult the *JOURNAL OF DIETETIC SOFTWARE* and *THE HOME ECONOMIST'S COMPUTER NEWSLETTER.*

Feed composition databases and nutrient databases have been used in microcomputer programs. In Australia microcomputer software has been developed, drawing from the database of the Australian Feeds Information Centre (AFIC). Subunits of the database can be downloaded via an interactive computer program prepared and supplied by AFIC (M. W. JONES; OSTROWSKI-MEISSNER ET AL.).

Food and Nutrition Software

The existing body of food and nutrition-related microcomputer software can be classified into three general groups: diet analysis, nutrition education, and food service management. The diet analysis programs determine the nutrient content of meals, individual foods, recipes, menus, and so forth. Data from one or several sources may be used to formulate the nutrient analysis. Supporting information (recommended dietary allowances, Metropolitan Life Insurance Co. height-weight tables, formulas to determine basal metabolic rates, etc.) can be used in conjunction with the nutrient analysis data to allow the user to make decisions about diet quality and/or adequacy. These programs have been implemented in several clinical, research, and educational settings including the National School Lunch Program (NSLP) (FULTON & DAVIS; KIRBY & SCHILLING).

NSLP has also used features of food service management software, such as production and menu planning, participation and cost reporting, warehouse and inventory control, and forecasting to reduce costs, simplify procedures, and increase accuracy and production (HIEMSTRA & VANEGMOND-PANNELL). Food service management software is used in institutional as well as private organizations.

Nutrition education software is a tool for group or self-directed nutrition education. It is available for different audience levels, including school-aged children, adolescents, and adults. Graphic and game formats are used widely (PLUMMER). Programs may present one or many food or nutrition concepts and frequently include diet analysis exercises as an educational tool. Several noteworthy resources provide listings and/or critical reviews of food and nutrition microcomputer software (BYRD-BREDBENNER ET AL;*JOURNAL OF NUTRITION EDUCATION;* HAMILTON; HOOVER; PLUMMER).

Farm Application Software

Software for farmers tends to fall into two types: that which can be used for production control and that which can be used to enhance managerial tasks (AUDIRAC & BEAULIEU). Many of the crop record programs are available as stand-alone systems or are integrated with an accounting system. Integrated systems can merge crop production data (e.g., yields, quantities of inputs used) to provide information on production costs; data are keyed only once and used for both purposes.

WILSDORF (1985a) provides guidelines on how a farmer can set up his own crop management database. Programs are available to provide fertility and management information to crop producers on a field-by-field basis. One such program, "CropConsultanT," can manipulate data to provide recommendations for crop management, such as seeding rates, timing of soil tests, fertilization, and so forth, for crop management. It is designed to work with corn for grain, corn silage, soybeans, sorghum for grain, sorghum silage, wheat, oats, barley, rye, and flax. With additional manipulation the program can also be used for sweet corn, white corn, popcorn, seed corn, and seed beans. Related crop management programs provide spreadsheet analyses of all of the current crop insurance options. They offer farmers a quick method for identifying their most effective insurance plan (LAY). With such a program it may be necessary for the farmer to update numbers and formulas to reflect changes in federal crop insurance requirements, rules, and calculations.

PONTIUS ET AL. report on a software program developed as a model for managing the European corn borer in yellow field corn. Extension personnel and agronomic management researchers are the primary users.

HERDMATE, an interesting example of herd breeding management software, was developed at Texas A&M University (B. LILLEY). This innovative program provides a dating service for dairy bulls and cows. By comparing data on females and bulls, dairymen can select service sires that will maximize their investment dollars.

Natural resource management has been the subject of many software programs. State Soil Conservation Service offices are working cooperatively with State Agricultural Stabilization and Conservation Service offices to develop databases using microcomputers to show how much soil erosion is occurring and which conservation practices and systems can best control it (GARLITZ). CUNNINGHAM ET AL. of Pennsylvania State University developed a microcomputer program for delivering soil information through digitized soil maps and interpretive records. With this program, soil map data are displayed as color-interpretive maps. THATCHER, URBANO, and FOREST report on microcomputer programs for nurserymen, landscapers, and farm business management in general, respectively.

AGRICULTURAL MICROCOMPUTER SOFTWARE COLLECTIONS IN LIBRARIES

DEMAS ET AL. emphasize that a library must ask four critical questions when establishing a machine-readable collection: 1) will the computerized data be treated as print materials are; 2) what are the hardware limitations; 3) who are the patrons; and 4) how will the collection be used.

Cornell University's Albert R. Mann Library of the College of Agriculture and Life Sciences has taken the lead in the United States by establishing a microcomputer center in an academic library and developing library collections of information in machine-readable formats. The Mann Library took a systematic approach in collecting, processing, and providing instruction in the use of machine-readable files. A major objective was to integrate these materials into the general collection as much as possible.

The Food and Nutrition Information Center (FNIC), a unit of NAL, houses a collection of food and nutrition-related microcomputer software programs (FRANK, 1987). Visitors to the center can gain "hands-on" experience with over 130 privately and government-developed packages. Initially the software was purchased; however, all newer items have been donated to the center. Software is not loaned but can be used on site. FNIC staff members, each with a strong educational background in nutrition and knowledge of computer technology, provide individual assistance. NAL, as a whole, has adopted FNIC's approach and is developing a demonstration center for all other types of agricultural microcomputer software. All agricultural software housed at NAL is being included in the AGRICOLA database.

EMERGING TECHNOLOGIES

Recent developments in microcomputer software and laser disk technologies have been used to improve the quality and facilitate the distribution of agricultural information. These innovative information systems offer faster, wider, and more reliable access to data.

One form of laser disk (also called videodisk) contains both still and moving images (including graphics and text), which have been imprinted onto well-protected metal disks. The laser disk has high storage capacity (54,000 to 108,000 frames) and is "read" optically by a laser beam. At one time it was marketed actively to the public as an alternative to the videocassette recorder. Now, in concert with videodisk players, microcomputers, and interactive software, videodisks are being used in instructional settings. NAL has created a videodisk orientation to the library and has acquired privately produced disks for its collection (WATERS ET AL). An interactive course of instruction for searching the AGRICOLA database is currently being developed and evaluated by NAL in conjunction with the University of Maryland's Center for Instructional Development and Evaluation (BUTLER). The Cooperative

Extension Service is examining the use of videodisks as "point of purchase" educational and decision-making tools for consumers (DIK & TRAVIESO). A second application of laser disk technology uses text that has been digitized. An example of this technique is a project that integrates digitized text with graphics in analog form to reproduce the *Pork Industry Handbook* (ANDRE). These data combined with a database management system allow for full-text searching, rapid data retrieval, and storage capacity for up to 800 million characters on one disk.

CD-ROM (compact disk, read-only memory) is another form of laser disk technology. Using the same premise as the popular music compact disk, several numeric and bibliographic databases, including AGRICOLA and CAB, have been made available in CD-ROM. Those with a CD player, microcomputer, and appropriate software, can search databases without paying telecommunications or printing fees. Because the CD can only be "read," information loaded onto databases more recently than the publication date of the disks must be obtained online.

Expert systems are a form of artificial intelligence software that solve a problem according to the user's responses to a series of questions; the questions guide the system to the pertinent information in the expert system's database. The system can be knowledge-based to provide factual information and/or advisory-based to refer users to sources of data. With some programs, external systems can be added such as telecommunications software, CD-ROM, word processing, database management, spreadsheets, and graphics to expand the system's capabilities. Several agricultural expert systems have been developed by the Cooperative Extension Service (BECK ET AL.; MCGRANN & FREDERICKS), and the medical community has adopted expert system technology for diagnosis (SMITH). NAL has developed samples of advisory-based expert systems to assist library reference staff (WATERS). More powerful expert systems recognize key words and identify the relevant information in the system's database, side-stepping the necessity for intermediate questions. In the future, expert system software may use a voice-recognition feature, eliminating the need for keying responses.

AGRICULTURAL LIBRARIES AND INFORMATION CENTERS

BLANCHARD & FARRELL trace the history and development of agricultural libraries throughout the world. To coordinate and encourage bibliographic work in agriculture, the International Association of Agricultural Librarians and Documentalists (IAALD) was founded in 1955. It meets every five years and publishes a quarterly bulletin (*QUARTERLY BULLETIN OF THE IAALD*).

Forums for sharing information among agricultural libraries and information centers include the Special Libraries Association's (SLA) Food, Agriculture, and Nutrition Division; the American Society for Information Science's (ASIS) Special Interest Group on Rural Information Services (SIGRIS); and the Science and Technology Section of the Association of College and Research Libraries of the American Library Association (ALA).

Veterinary medical librarians participate in the Veterinary Medical Section of the Medical Library Association (MLA) (JETTE).

Libraries

Agricultural libraries fall into three major categories: government, academic, and private. Most have a strong research orientation. There are, however, several international libraries that may fall into more than one group. Most of the principal agricultural libraries in the world are located in Europe, the United Kingdom, the Soviet Union, and North America.

In the United States, the largest agricultural library is USDA's National Agricultural Library (NAL) located in Beltsville, Md. NAL's collection now numbers over 1.9 million volumes, and is particularly strong in chemistry, botany, zoology, and other sciences related to agriculture (BLANCHARD & FARRELL). As mentioned earlier, NAL maintains the AGRICOLA database that is available online as well as in printed form.

USDA also supports field libraries for its research centers throughout the United States. USDA's Forest Service has an elaborate network of field libraries to serve it.

Academic agricultural libraries in the United States received a major boost in 1862 with the passage of the Federal Land Grant Act (the Morrill Act) which created the land-grant colleges and universities throughout the United States. Each state and U.S. territory land-grant university has an agricultural collection located in a separate agricultural library or as a part of the graduate or undergraduate library on campus. Along with serving the students and faculty, land-grant libraries also serve State Experiment Stations and personnel of the Cooperative Extension Service. Several of these libraries, such as Cornell's Mann Library and the Agricultural Library at the University of Minnesota, have opened their services to the general public and business community for a fee.

Hundreds of libraries and specialized information centers in the United States serve the agricultural community although they are located in private industry, professional and trade associations, research institutes and laboratories. Many of them contain valuable special collections and provide extensive service. While much of the nonproprietary material is available to others via interlibrary loan, in general, the libraries' service is limited to its own organization. Such libraries are found in the Food Marketing Institute, General Mills, Inc., Sunkist Growers, Inc., American Farm Bureau Federation, and Cargill, Inc.

Information Centers

NAL has recently established specialized information centers (PISA) to be more responsive to the Library's current clientele and to reach new user groups. The centers also take a proactive role and work with support groups in industry, federal and state agencies, and the scientific community.

The subject areas for information centers are determined by congressional mandate, priorities within USDA, availability of subject expertise on NAL staff, and financial support and interest by agricultural organizations. The information centers cover: agricultural marketing and trade; alternative farming systems; animal welfare; aquaculture; biotechnology; critical agricultural materials; family; fibers and textiles; food and nutrition; food irradiation; and horticulture. Each information center is at a different stage of development and is taking a unique approach to reaching its clientele.

The U.S. Agency for International Development (AID) also sponsors specialized library services to developing countries. Two examples, the Postharvest Institute for Perishables Information Center at the University of Idaho in Moscow, Idaho, and the Postharvest Documentation service at Kansas State University in Manhattan, Kansas, maintain and develop collections in addition to offering current awareness and document delivery. The need for these services will diminish as developing countries adopt microcomputer technology and use telecommunications to obtain information (SCHENCK-HAMLIN & GEORGE).

Information Analysis Centers

The Consultative Group on International Agricultural Research (CGIAR), an international consortium sponsored by the World Bank, the United Nations Development Program, and FAO, sponsors 13 research institutes. Because of the combined support of their libraries and scientific facilities, these institutes can establish specialized information analysis centers. CGIAR has established CGNET, a data transfer network, to use computer-based messaging, bulletin-board services, and conferencing systems (BALSON).

Canada's International Development Research Centre provides grants to support specialized information analysis centers dealing with international agricultural development as well as health, environmental sanitation, geotechnical engineering, and human settlements (BRANDRETH).

CONCLUSION

Since the 18th century the U.S. agricultural community has been involved in gathering and disseminating information to support its own activities. This process has evolved in scope, size, and sophistication into a market of both privately and publicly produced databases and bibliographies, and, most recently, a variety of innovations in the use of computer technologies.

Among the important issues facing the agricultural community, there exists a newly recognized and pressing need for the delivery of useful information to rural communities. The impact of recent changes such as slowed rural population growth, the "graying of America," rapid advances in technology, and shifts in the world economy, has jeopardized the survival of U.S. rural communities. Revitalization of rural America will demand careful analysis of the information needs of this population, along with changes in the presentation and nature of access to agricultural and rural information. The effectiveness of information providers is a key to the success of this effort.

Information providers, whether they are librarians, extension specialists, or government agencies, need to fully understand their clientele's information-seeking behavior, their use of and satisfaction with existing products, and their impressions of desirable improvements. Sensitivity to trends is absolutely critical. As new, expanded, or repackaged information products and services are developed to meet the information needs of all segments of the agricultural community, information providers must be aware of these developments and use the most appropriate technologies to reach their clientele.

The activities of information providers must change as ease of access to data changes. Researchers are doing their own online database searching, libraries have begun to include microcomputer software, and farmers have computers and information utilities in their homes. Survival, not only in rural communities but throughout the world, depends increasingly on an individual's or an organization's ability to make decisions based on vast amounts of data. Therefore, a more intimate knowledge of the methods and mechanics of information gathering has developed among end users. Availability of data no longer depends on access to a few specialists (librarians or researchers), and information does not need to trickle down through a highly structured and traditional network of teachers. For example, an extension agent can develop a microcomputer expert system that allows a farmer to make farm management decisions based on the same raw data that the agents themselves would use.

Agricultural information covers much more than bibliographic databases. As shown in this chapter, many other information sources such as the extension agent are crucial to the everyday functioning of various segments of the agricultural community. Sweeping advances in information technology provide challenges and opportunities in the methods used in research, education, and information dissemination. It is hoped that the successful adaptation and utilization of information technology will play an important role in the revitalization of the agricultural and rural community.

BIBLIOGRAPHY

AGNET. 1983. Information for Modern Agriculture: Symposium Proceedings; 1983 September 7-9 Lincoln, NE. Lincoln, NE: AGNET; 1983. 40p. Available from: Central AGNET, University of Nebraska-Lincoln, Lincoln, NE 68583.

AGRICOMP. 1982-. 103 Outdoors Building, Columbia, MO. ISSN: 0738-5978.

AGRICULTURAL COMPUTING. 11701 Borman Drive, St. Louis, MO: Doane Publishing. ISSN: 0882-9284.

AGRICULTURAL LIBRARIES INFORMATION NOTES. 1986a. Information Centers: In Touch with Agriculture. Agricultural Libraries Information Notes. 1986 January; 12(1): 1-4. ISSN: 0095-2699.

AGRICULTURAL LIBRARIES INFORMATION NOTES. 1986b. New Programs Offer Help For Microcomputer Software Users. Agricultural Libraries Information Notes. 1986 September; 12(9): 4. ISSN: 0095-2699.

AGRINDEX. 1975-. Rome, Italy: Food and Agriculture Organization of the United Nations. ISSN: 0254-8801.

ANDRE, PAMELA Q. J. 1986. Full-Text-Access and Laser Videodiscs: The

National Agricultural Library System. Library Hi Tech. 1986 Spring; 4(1): 13-21. ISSN: 0737-8831.

ANTHONY, W. S. 1985. Evaluation of an Optimization Model of Cotton Ginning Systems. Transactions of the ASAE. 1985; 28(2): 411-414. ISSN: 0001-2351.

ARAMAN, P. A. 1983. BLANKS: A Computer Program for Analyzing Furniture Rough-Part Needs in Standard-Size Blanks. Broomall, PA: U.S. Department of Agriculture, Forest Service, Northeastern Forest Experiment Station; 1983. 8p. (Research Paper NE-521).

AUDIRAC, IVONNE; BEAULIEU, LIONEL J. 1986. Microcomputers in Agriculture: A Proposed Model to Study Their Diffusion/Adoption. Rural Sociology. 1986 Spring; 51(1): 60-77. ISSN: 0036-0112.

BALSON, DAVID A. 1985. CGNET: A Data Transfer Network for the CGIAR. Paper presented at: International Association of Agricultural Librarians and Documentalists (IAALD) 7th World Congress; 1985 June 2-6; Ottawa, Canada. 19p. Available from: Author, Telecommunication Systems, Information Sciences Division, International Development Research Centre, Ottawa, Canada, K1G 3HP.

BECK, HOWARD W.; STIMAC, JERRY L.; JOHNSON, FRED A. 1984. Florida Agricultural Information Retrieval System (FAIRS): Design of a Computerized Consultation System for Agricultural Extension. Gainesville, FL: University of Florida, Institute of Food and Agricultural Sciences; 1984 June 20. 12p. OCLC: 11477745.

BECKMAN, MARGARET. 1985. Updates and Innovations- Session: Information Services at the Farm Gate. Paper presented at: International Association of Agricultural Librarians and Documentalists (IAALD) 7th World Congress; 1985 June 2-6; Ottawa, Canada. 10p. Available from: Author, Executive Director for Information Technology, University of Guelph, Guelph, Ontario, Canada N2G 2W1.

BIBLIOGRAPHY OF AGRICULTURE. 1940-. Phoenix, AZ: The Oryx Press. ISSN: 0006-1530.

BIOTECHNOLOGY SOFTWARE. 1984-. New York, NY: Mary Ann Liebert, Inc. ISSN: 0749-0372.

BLANCHARD, J. RICHARD; FARRELL, LOIS, eds. 1981. Guide to Sources for Agricultural and Biological Research. Berkeley, CA: University of California Press; 1981. 735p. ISBN: 0-520-03226-8.

BOYLE, P. J. 1981. Commonwealth Agricultural Bureaux. A Survey of the Agricultural Literature. Paper presented at: CAB 4th Annual Training Course; 1981 September; St. Anne's College, Oxford. 19p. Available from: The author, Commonwealth Bureau of Pastures and Field Crops, Hurley, Maidenhead, Berks, UK SL6 5LR.

BRANDRETH, MIKE. 1982. Specialized Information Analysis Centers in International Development. Montebello, Quebec, Canada: International Development Research Centre; 1982 October 4-8. 60p. OCLC: 11251493.

BROOKS, KRISTINA. 1980. A Comparison of the Coverage of Agricultural and Forestry Literature on AGRICOLA, BIOSIS, CAB, and SCISEARCH. Database. 1980 March; 3: 38-49. ISSN: 0162-4105; CODEN: DTBSDQ.

BUTLER, ROBERT W. 1987. New Technologies: Computerizing the Future. In: Crowley, John J., ed. Research for Tomorrow. Washington, DC: U.S. Department of Agriculture; 1987. 294-300. (1986 Yearbook of Agriculture). Available from: Government Printing Office, Washington, DC 20402.

BUTRUM, RITVA; YOUNG, VERNON R. 1984. Development of a Nutrient Data System for International Use: INFOODS (International Network of Food Data Systems). Journal of the National Cancer Institute. 1984 December; 73(6): 1409–1413. ISSN: 0027-8874; CODEN: JNCIAM.

BYRD-BREDBENNER, CAROL; SAMPSON, DEBORAH; FISHMAN, LORRI; MARTIN, TANYA. 1986. Computer Software for Nutrition & Food Education. University Park, PA: Pennsylvania State Nutrition Center, Pennsylvania State University; 1986. 73p. Available from: Pennsylvania State Nutrition Center, Pennsylvania State University, Benedict House, University Park, PA 16802.

CAB INTERNATIONAL. 1986. Report for April 1985–March 1986. Farnham House, Farnham Royal, Slough SL2 3BN, United Kingdom: CAB International; 1986. 42p. ISBN: 0-85198-576-9.

CARTER, DANIEL H. 1982. Testimony of Daniel H. Carter. In: U.S. Department of Agriculture; National Commission on Libraries and Information Science. Joint Congressional Hearings on the Changing Information Needs of Rural America: The Role of Libraries and Information Technology. Washington, DC: Government Printing Office; 1982 July 21. 34–38.

CENTURY COMMUNICATIONS. 1986. County Agents: The Reference Directory for Agricultural Extension Workers. 70th edition. Skokie, IL: Century Communications Inc.; 1986. 81p. ISSN: 0739-4330.

CHARTRAND, ROBERT L. 1983. Keynote Address: In: Information for Modern Agriculture: Symposium Proceedings; 1983 September 7–9; Lincoln, NE. 1983. 3–6. Available from: Central AGNET, University of Nebraska-Lincoln, Lincoln, NE 68583.

CHRONOLOG. 1986. Announcing AGRIBUSINESS U.S.A. Chronolog. 1986 May; 14(5): 83–85.

COMPUTER APPLICATIONS IN THE BIOSCIENCES. 1985-. Oxford, England and Washington, DC: IRL Press. ISSN: 0266-7061; CODEN: COABER.

COOMBS, J. 1985. The Biotechnology Directory. New York, NY: Nature Press; 1985. 494p. ISBN: 0-943818-06-0.

CRAFTS-LIGHTY, A. 1983. Information Sources in Biotechnology. New York, NY: Nature Press; 1983. 306p. ISBN: 0-943818-04-4.

CUNNINGHAM, R. L; PETERSEN, G. W.; SACKSTEDER, C. J. 1984. Microcomputer Delivery of Soil Survey Information. Journal of Soil and Water Conservation. 1984 July/August; 39(4): 241–243. ISSN: 0022-4561.

DEMAS, SAMUEL; CHIANG, KATHERINE S.; OCHS, MARY A.; CURTIS, HOWARD. 1985. Developing and Organizing Collections of Computer-Readable Information in an Agriculture Library. Paper presented at: International Association of Agricultural Librarians and Documentalists (IAALD) 7th World Congress; 1985 June 2–6; Ottawa, Canada. 15p. Available from: The authors, Albert R. Mann Library, Cornell University, Ithaca, NY 14853.

DESELAERS, NORBERT. 1986. The Necessity for Closer Cooperation Among Secondary Agricultural Information Services: An Analysis of AGRICOLA, AGRIS and CAB. Quarterly Bulletin of the IAALD. 1986; 31(1): 19–26. ISSN: 0020-5966.

DIK, DAVID W.; TRAVIESO, CHARLOTTE B. 1987. Future Role of Electronic Technology in Agricultural Research and Extension. In: Crowley, John J., ed. Research For Tomorrow. Washington, DC: U.S. Department of Agriculture; 1987. 272-281. (1986 Yearbook of Agriculture). Available from: Government Printing Office, Washington, DC 20402.

DILLMAN, DON A. 1984. The Social Impacts of Information Technologies in Rural North America. Paper presented at: Rural Sociological Society 47th Annual Meeting; 1984 August 23; Texas A&M University, College Station, TX. 38p. Available from: Author, Department of Sociology and Research, Washington State University, Pullman, WA 99164.

DITTBERNER, P. L.; BRYANT, G.; VORIES, K. C. 1981. The Use of the Plant Information Network (PIN) in Rare Plant Conservation. In: Morse, Larry E.; Henijin, Mary Sue, eds. Rare Plant Conservation: Geographical Data Organization. Bronx, NY: The New York Botanical Garden; 1981. 149-165. ISBN: 0-89327-23-X.

EDWARDS, SHIRLEY. 1986. NAL Conducts Journal Evaluation. Agricultural Libraries Information Notes. 1986 March; 12(3): 7. ISSN: 0095-2699.

ENDRUD, G. 1983. Review: AgriStar. AgriComp. 1983 July/August; 2(1): 18-20, 46. ISSN: 0738-5978.

EVANS, PETER A., comp. 1979. Directory of Selected Forestry-related Bibliographic Data Bases. Berkeley, CA: U.S. Department of Agriculture, Forest Service, Pacific Southwest Forest and Range Experiment Station; 1979 September. 42p. Available from: Government Printing Office, Washington, DC 20402.

EXTENSION COMMITTEE ON ORGANIZATION AND POLICY. 1982. The Computer Management Power for Modern Agriculture. West Lafayette, IN: Purdue University; 1982 July. 19p. Available from: H.G. Diesslin, Director, Cooperative Extension Service, Agricultural Administration Building, Purdue University, West Lafayette, IN.

FARM COMPUTER NEWS. 1716 Locust, Des Moines, IA: Successful Farming Magazine. ISSN: 0736-8263.

FEIDT, WILLIAM B. 1986. Nursery and Seed Trade Catalog: Creation of a Local Database from OCLC Cataloging Data. In: Proceedings of the 7th National Online Meeting; 1986 May 6-8; New York, NY. Medford, NJ: Learned Information; 1986. 119-128. OCLC: 14473538.

FELLER, IRWIN; KALTREIDER, LYNNE; MADDEN, PATRICK; MOORE, DAN; SIMS, LAURA. 1984. The Agricultural Technology Delivery System: A Study of the Transfer of Agricultural and Food-Related Technologies. University Park, PA: The Pennsylvania State University, Institute for Policy Research and Evaluation; 1984 December. 5 volumes. (discontinuous paging). OCLC: 13333323.

FOOD AND NUTRITION QUARTERLY INDEX. 1985-. Phoenix, AZ: The Oryx Press. ISSN: 0887-0535.

FOREST, WAYNE. 1985. Review: System ABC. AgriComp. 1985 March/April; 3(5): 30-34. ISSN: 0738-5978.

FRANK, ROBYN C. 1982. Information Resources for Food and Human Nutrition. Journal of the American Dietetic Association. 1982 April; 80(4): 344-350. ISSN: 0002-8223; CODEN: JADAAE.

FRANK, ROBYN C., ed. 1984. Directory of Food and Nutrition Information Services and Resources. Phoenix, AZ: Oryx Press; 1984. 287p. ISBN: 0-89774-078-5.

FRANK, ROBYN C. 1987. Food and Nutrition Microcomputer Demonstration Center. In: Lehmann, Klaus-Dieter; Strohl-Goebel, Hilde, eds. The Application of Microcomputers in Information, Documentation and Libraries. Amsterdam, The Netherlands: Elsevier Science Publishers B.V.; 1987. 672-674. (Contemporary Topics in Information Transfer: volume 4). ISBN: 0-444-70135-4.
FREDERICKS, ELDON E. 1983. Facts. In: North Central Computer Institute (NCCI) Workshop: The Use of Computers in Agricultural Information; 1983 May 2-5; Chicago, IL. Madison, WI: NCCI; 1983. 15-16. Available from: North Central Computer Institute, 667 WARF Office Building, 610 Walnut Street, Madison, WI 53705.
FULTON, LOIS H.; DAVIS, CAROLE A. 1985. Microcomputer System for Assessing the Nutrient Content of School Lunches. Paper presented at: National Nutrient Data Bank 10th Annual Conference; 1985 July 22-24; San Francisco, CA. 32p. Available from: Authors, Nutrition Education Division, Human Nutrition Information Service, USDA, Hyattsville, MD 20782.
GARKEY, JANET; CHERN, WEN S. 1986. Handbook of Agricultural Statistical Data. Washington, DC: Economic Statistics Committee, American Agricultural Economics Association; 1986. 139p. Available from: Government Printing Office.
GARLITZ, NANCY M. 1985. Computer Data Base to Improve Conservation Planning. Soil and Water Conservation News. 1985 June; 6(3): 3-4. ISSN: 0199-9060.
HALL, JULIE; DAVIS, MARTHA; WOLF, DUANE. 1984. Printed and Computer Sources for Retrieval and Publication of Agronomic Literature. Journal of Agronomic Education. 1984 Spring; 13: 57-61. ISSN: 0094-2391; CODEN: JAEDDV.
HAMILTON, CHARLENE. 1986. Software for Computer Assisted Nutrition Analysis: A Comparative Study. Framingham, MA; Framingham State College; 1986. 139p. Available from: Charlene Hamilton, Department of Home Economics, Framingham State College, 100 State Street, Framingham, MA 01701.
HARVEY, NIGEL, ed. 1986. Agricultural Research Centres: A World Directory of Organizations and Programmes. 8th ed. Harlow, Essex, U.K.: Longman; 1986. 1500p. ISBN: 0-582-90014-X.
HAYCOCK, RICHARD C.; WILDE, GLENN R. 1984. Information/Educational Needs in Rural Intermountain Communities. Bulletin of the American Society for Information Science. 1984 October; 11(1): 20-23. ISSN: 0095-4403; CODEN: BASICR.
HEPBURN, F. N. 1986. Report from USDA's Nutrient Data Research Branch. Paper presented at: National Nutrient Data Bank 11th Annual Conference; 1986 June 27-July 2; Athens, GA. 5p. Available from: Author, Human Nutrition Information Service, USDA, Hyattsville, MD 20782.
HIEMSTRA, STEPHEN J.; VANEGMOND-PANNELL, DOROTHY. 1984. Computer Applications in School Food Service. School Food Service Research Review. 1984 Fall; 8(2): 86-96. ISBN: 0-85198-522-X.
HINTZ, THOMAS R. 1986. The IFAS Computer Network: The Florida Experience. In: Proceedings of International Conference on Computers in Agricultural Extension Programs; 1986 February 5-6; Lake Buena Vista, FL. Gainesville, FL: Florida Cooperative Extension Service, Insti-

tute of Food and Agricultural Sciences (IFAS), University of Florida; 1986. 12p. Available from: Author at IFAS.

HNRIMS DATABASE. n.d. Washington, DC: U.S. Department of Agriculture. 1p. (Information Sheet). Available from: HNRIMS/USDA System Coordinator, Current Research Information System, Cooperative State Research Service, NAL Bldg., 5th floor, Beltsville, MD 20705.

HOEY, P. O'N. 1985. Use of Information Technology in UK Agricultural Information Work: A Review. Journal of Information Science. 1985; 11(1): 9-17. ISSN: 0165-5515.

HOME ECONOMIST'S COMPUTER NEWSLETTER, THE. Elaine Muller, ed. Denville, NJ: E. M. Enterprises. Available from: E.M. Enterprises, 3 Hickory Road, Denville, NJ 07834.

HOOVER, LORETTA W., ed. 1986. Nutrient Data Bank Directory. 5th edition. Columbia, MO: Curators of the University of Missouri; 1986. 73p. Available from: Department of Human Nutrition, Foods and Food Systems Management, College of Home Economics, 217 Gwynn Hall, University of Missouri-Columbia, Columbia, MO 65211.

HOWARD, JOSEPH H. 1985. The National Agricultural Library as a Source of Agricultural Information. Paper presented at: International Association of Agricultural Librarians and Documentalists (IAALD) 7th World Congress; 1985 June 2-6; Ottawa, Canada. 13p. Available from: Author, National Agricultural Library, Beltsville, MD 20705.

HUMPHREY, CHUCK. 1983. AGNET: A Management Tool for Agriculture. In: North Central Computer Institute (NCCI) Workshop, The Use of Computers in Agricultural Information; 1983 May 2-5; Chicago, IL. Madison, WI: NCCI; 1983. 17-21. Available from: North Central Computer Institute, 667 WARF Office Building, 610 Walnut Street, Madison, WI 53705.

JETTE, JEAN-PAUL. 1982. The Veterinary Medical Libraries in North America. Bulletin of the Japan Association of Agricultural Librarians and Documentalists. 1982 February; 47: 1-10. OCLC: 2246434.

JOHANNSEN, CHRIS J. 1983. Use of Geographic Data Bases in Missouri. In: North Central Computer Institute (NCCI) Workshop, The Use of Computers in Agricultural Information; 1983 May 2-5; Chicago, IL. Madison, WI: NCCI; 1983. 67-72. Available from: NCCI, 667 WARF Office Building, 610 Walnut Street, Madison, WI 53705.

JOHNSON, RONALD L. 1986. Plant Protection and a National Biological Survey. In: Kim, Ke Chung; Knutson, Lloyd. Foundations for a National Biological Survey. Lawrence, KS: Association of Systematics Collections in Cooperation with the Holcomb Research Institute and Illinois Natural History Survey; 1986. 77-84. ISBN: 0-94292-413-4.

JONES, M. W. 1983. Australian Feeds Information Centre– A Microcomputer Database. In: Robards, G. E.; Packham, R. G., eds. Feed Information and Animal Production: Proceedings of the International Network of Feed Information Centres 2nd Symposium; 1983; Farnham Royal, Slough, UK. Blacktown, Australia: Commonwealth Agricultural Bureaux and International Network of Feed Information Centres; 1983. 101-104. ISBN: 0-85198-522-X.

JONES, MARY GARDINER. 1983. The Challenge of the New Information Technologies: The Need to Respond to Citizens' Information Needs. The Information Society. 1983; 2(2): 145-156. ISSN: 0197-2243; CODEN: INSCD8.

JOURNAL OF DIETETIC SOFTWARE. 1983-. Sara A. Gill, ed. P.O. Box 2565, Norman, OK. ISSN: 0742-826X.

JOURNAL OF NUTRITION EDUCATION. 1984. Computers in Nutrition Education. Journal of Nutrition Education. 1984 June; 16(2): 1-123. ISSN: 0022-3182.

JUDGE, P. J. 1983. The International Feed Information Centres System in Relation to Other Databases. In: Robards, G. E.; Packham, R. G., eds. Feed Information and Animal Production: Proceedings of the International Network of Feed Information Centres 2nd Symposium; 1983; Farnham Royal, Slough, U. K. Blacktown, Australia: Commonwealth Agricultural Bureaux and International Network of Feed Information Centres; 1983. 79-100. ISBN: 0-85198-522-X.

KINNEY, TERRY B., JR. 1986. Reporting Science: Whose Responsibility? Paper presented at: American Chemical Society, Midwest Regional Meeting; 1986 June 3; Bowling Green State University, Bowling Green, OH. 12p. Available from: Author, Agricultural Research Service, USDA, Washington, DC 20250.

KIRBY, ALICE L.; SCHILLING, PRENTISS E. 1980. Computers Lend a Hand. School Food Service Journal. 1980 August; 34(7): 86. ISSN: 0036-6641.

LA FERNEY, PRESTON E. 1983. An Administrative Perspective on Microcomputers for Agricultural Research and Education. Southern Journal of Agricultural Economics. 1983 July; 15(1): 53-55. ISSN: 0081-3052.

LARSON, JEAN. 1986. "GRIN" and Share It. Agricultural Libraries Information Notes. 1986 October; 12(10): 2-3. ISSN: 0095-2699.

LAY, WILLIAM D. 1985. Crop Insurance Analyzer: A Spreadsheet View of Federal Crop Insurance. AgriComp. 1985 January/February; 3(4): 36-40. ISSN: 0738-5978.

LEATHERDALE, DONALD; TIDBURY, G. ERIC; MACK, ROY. 1982. AGROVOC: A Multilingual Thesaurus of Agricultural Terminology. Luxembourg and Rome, Italy: Commission of the European Communities; Food and Agriculture Organization; 1982. 530p. ISBN: 88-7643-001-6.

LEBOWITZ, ABRAHAM I. 1985. AGRIS and North America. Paper presented at: American Society for Information Science (ASIS) Annual Meeting; 1985 October 20-25; Las Vegas, NV. 23p.

LETT, RAYMOND D. 1983. Computers: The Newest Technology for American Farmers. The Information Society. 1983; 2(2): 121-129. ISSN: 0197-2243.

LEVINS, RICHARD A. 1983. Implications of Microcomputer Technology for Agricultural Economics Programs. Southern Journal of Agricultural Economics. 1983 July; 15(1): 27-29. ISSN: 0081-3052.

LILLEY, BETH. 1986. Computers Now Find Perfect Mates for Herd. Enid Morning News. 1986 June 15; C: 1, 3; 1.

LILLEY, G. P. 1981. Information Sources in Agriculture and Food Science. London, England: Butterworths; 1981. 603p. (Butterworths Guides to Information Sources). ISBN: 0-408-10612-3.

LONGO, ROSE MARY JULIANO; MACHADO, UBALDINO DANTAS. 1981. Characterization of Databases in the Agricultural Sciences. Journal of the American Society for Information Science. 1981 March; 32(2): 83-91. ISSN: 0002-8231; CODEN: AISJB6.

LUCHSINGER, ARLENE E. 1987. Agriculture. In: Shapiro, Beth; Whaley, John, eds. Selection of Library Materials in Applied and Inter-

disciplinary Fields. Chicago, IL: Collection Management and Development Committee, Resources and Technical Services Division, American
Library Association; 1987. 1-9. ISBN: 0-83890-466-1.

MACK, ROY. 1980. Veterinary Information Within the European Communities. EUROVET Bulletin. 1980 Summer; 2: 10-13. OCLC: 7727234.
Available from: Secretary to Eurovet. British Veterinary Association,
7 Mansfield Street, London W1M OAT. UK.

MACNEIL, KATHARINE. 1986. Veterinary Computing: On-Line Medical
Databases. Modern Veterinary Practice. 1986 January; 67(1): 68-72.
ISSN: 0362-8140.

MANN, ERNEST J. 1986. Past, Present, and Future Developments in the
Transfer and Dissemination of Agricultural Information: The Case for a
Single, Coordinated, World Agricultural Information System. Quarterly
Bulletin of the IAALD. 1986; 31(1): 5-9. ISSN: 0020-5966.

MARTIN, FRANK E. 1983. COMNET: A Computer Based Communications Network. In: North Central Computer Institute (NCCI) Workshop:
The Use of Computers in Agricultural Information; 1983 May 2-5;
Chicago, IL. Madison, WI: NCCI; 1983. 7-13. Available from: NCCI,
667 WARF Office Building, 610 Walnut Street, Madison, WI 53705.

MATHEWS, ELEANOR. 1981. When Tillage Begins Other Arts Follow. . .
A Core List of Agriculture Serials. Serials Review. 1981 July-September;
7(3): 9-50. ISSN: 0098-7913.

MATHEWS, ELEANOR. 1986. Agriculture. In: Katz, Bill; Katz, Linda
Sternberg. Magazines for Libraries. 5th ed. New York, NY: R. R.
Bowker Company; 1986. 69-75. ISBN: 0-83521-495-8.

MATTHEWS, M. EILEEN. 1983. A Foodservice Information Framework
for Decision Making. Food Technology. 1983 December; 39(12): 46-
49. ISSN: 0015-6639.

MCCLELLAND, JERRY. 1986. Reaping the Benefits of Computers.
Journal of Extension. 1986 Summer; 24: 11-13. ISSN: 0022-0140.

MCGRANN, JAMES M. 1983. Microcomputer Software Development and
Distribution. Southern Journal of Agricultural Economics. 1983 July;
15(1): 21-25. ISSN: 0081-3052.

MCGRANN, JAMES M.; FREDERICKS, ELDON E. 1986. Land-Grant
Universities' Artificial Intelligence—Expert Systems Development
Activities. College Station, TX: Department of Agricultural Economics,
Texas Agricultural Experiment Station, Texas A&M University; 1986
October. 69p. (Staff Paper Series: SP-7; Department Information
Report: DIR 86-1).

MORGAN, BRYAN. 1985. Key Guide to Information Sources in Agricultural Engineering. London, England: Mansell Publishing Limited; 1985.
209p. ISBN: 0-7201-1720-8.

MURRAY, PAMELA J. 1985. AGNET: After a Decade. . .Different Clientele
and Enhanced Services. Online. 1985 November; 9(6): 109-117. ISSN:
0146-5422.

NATIONAL ADVISORY BOARD ON RURAL INFORMATION NEEDS
(NABRIN). PLANNING COMMITTEE. 1985. The NABRIN Report.
Washingtion, DC: National Commission on Libraries and Information
Science; 1985. 30p. Available from: NCLIS, General Services Administra-
tion Building, 7th & D Streets, S.W., Suite 3122, Washington, DC 20024.

NATIONAL AGRICULTURAL LIBRARY CATALOG. 1966-. Totowa,
NJ: Rowman and Littlefield; 1966. ISSN: 0027-8505.

NICHOLAS, DAVID. 1985. Commodities Futures Trading: A Guide to Information Sources and Computerized Services. London, England: Mansell Publishing Ltd.; 1985. 144p. ISBN: 0-7201-1703-8.

NIELSEN, J. N. 1985. Searching the Veterinary Literature via Computer. Journal of the American Veterinary Medical Association. 1985 May 15; 186(10): 1058-1061. ISSN: 0003-1488; CODEN: JAVMA4.

OSTROWSKI-MEISSNER, HENRY T.; JACKSON, NEVILLE; WESTWOOD, NEVIL H. 1987. National Feed Database in Australia- I. Computerization of Data Collection System: Micro-Computer Version. II. System for Feed Database Interrogation and Data Retrieval. In: Lehmann, Klaus-Dieter; Strohl-Goebel, Hilde; eds. The Application of Micro-Computers in Information, Documentation and Libraries. Amsterdam, The Netherlands: Elsevier Science Publishers B.V.; 1987. 665-671. (Contemporary Topics in Information Transfer, volume 4). ISBN: 0-444-70135-4.

OVERFIELD, RICHARD A. 1981. State Agricultural Experiment Stations and the Development of the West, 1887-1920: A Look at the Sources. Government Publications Review. 1981; 8A: 463-472. ISSN: 0196-335X.

PALMER, JUDITH. 1986. Information Services in a Beleaguered Agricultural Industry. Aslib Proceedings. 1986 April; 38(4): 101-113. ISSN: 0001-253X; CODEN: ASLPAO.

PENCE, R.; CIESLA, W. M.; HUNTER, D. O. 1983. Geographic Information System—A Computer Assisted Approach to Managing Forest Pest Data. Fort Collins, CO: U.S. Department of Agriculture, Forest Service, Pest Management, Methods Application Group; 1983 October. 19p. (Report No. 84-1). Available from: Government Printing Office.

PERLOFF, BETTY; GRAY, BRUCY. 1986. Using USDA Data Tapes. Hyattsville, MD: U.S. Department of Agriculture, Human Nutrition Information Service; 1986. 2p. Available from: Authors, Human Nutrition Information Service, USDA, Hyattsville, MD 20782.

PETERKIN, BETTY B.; RIZEK, ROBERT L. 1984. National Nutrition Monitoring System. Family Economics Review. 1984; 4: 15-19. ISSN: 0425-676X.

PETERS, JEFFREY R. 1981. AGRICOLA. Database. 1981 March; 1: 13-27. ISSN: 0162-4105; CODEN: DTBSDQ.

PISA, MARIA G. 1987. Information Centers: From Irradiation to Biotechnology and Beyond. In: Crowley, John J., ed. Research for Tomorrow. Washington, DC: U.S. Department of Agriculture; 1987. 285-293. (1986 Yearbook of Agriculture). Available from: Government Printing Office, Washington, DC 20402.

PLUMMER, PATRICIA, n.d. An Evaluation of Nutrition Education Software. Framingham, MA: Framingham State College; n.d. 36p. Available from: Charlene Hamilton, Department of Home Economics, Framingham State College, 100 State Street, Framingham, MA 01701.

PONTIUS, JEFFREY S.; CALVIN, DENNIS D.; WELCH, STEPHEN M.; POSTON, FREDDIE L. 1984. Software for a European Corn Borer Management Model in Yellow Field Corn. North Central Computer Institute Software Journal. 1984 November; 1(1): 1-105. OCLC: 12175801. Available from: North Central Computer Institute, 610 Walnut St., Madison, WI 53705.

POWERS, RONALD C.; MOE, EDWARD C. 1982. The Policy Context for Rural-Oriented Research. In: Dillman, Don A.; Hobbs, Daryl J., eds.

Rural Society in the U.S.: Issues for the 1980s. Boulder, CO: Westview Press; 1982. 10-20. ISBN: 0-86531-100-5.

QUARTERLY BULLETIN OF THE IAALD. 1956-. S. C. Harris, ed. The Hague, The Netherlands. International Association of Agricultural Librarians and Documentalists. ISSN: 0020-5966; CODEN· QBALAE. Available from: ISNAR, P.O. Box 93375, 2509 AJ The Hague, The Netherlands.

RENEAU, FRED; PATTERSON, RICHARD. 1984. Comparison of Online Agricultural Information Services. Online Review. 1984 August; 8(4): 313-322. ISSN: 0309-314X.

ROGERS, EVERETT M. 1983. Diffusion of Innovations. 3rd edition. New York, NY: The Free Press; 1983. 453p. LC: 82-70998.

RUSSELL, H. M. 1983. Agricultural User Populations and their Information Needs in the Industrialized World. Quarterly Bulletin of the IAALD. 1983; 28(2): 40-52. ISSN: 0020-5966.

SARETTE, M.; TETTE, J. P.; BARNARD, J. 1981. SCAMP—A Computer Information Delivery System for Cooperative Extension. New York's Food and Life Sciences Bulletin. 1981; 90: 1-8. ISSN: 0362-0069.

SCHENCK-HAMLIN, DONNA; GEORGE, PAULETTE FOSS. 1986. Using Libraries to Interface with Developing Country Clientele. Special Libraries. 1986 Spring; 77(4): 80-89. ISSN: 0038-6723; CODEN: SPLBAN.

SIESS, JUDITH A.; BRADEN, JOHN B. 1982. Online Databases Relevant to Agricultural Economics. American Journal of Agricultural Economics. 1982 November; 64(4): 761-767. ISSN: 0002-9092.

SIGEL, EFREM. 1983. The Future of Videotext. White Plains, NY: Knowledge Industry Publications, Inc.; 1983. 197p. ISBN: 0-86729-025-0.

SIMMONS, WENDY. 1986. The Development of AGRIS: A Review of the United States Response. Quarterly Bulletin of the IAALD. 1986; 31(1): 11-18. ISSN: 0020-5966.

SMITH, KENT A. 1986. Medical Information Systems. Bulletin of the American Society for Information Science. 1986 April/May; 12(4): 17-18. ISSN: 0095-4403; CODEN: BASICR.

SOPHAR, GERALD J. 1984. Rural Information Services in North America-and the World. Bulletin of the American Society for Information Science. 1984 August; 10(6): 26. ISSN: 0095-4403; CODEN: BASICR.

STORM, BONNIE L. 1982. Information and Rural Development: From Bangladesh to Guatemala, Audio and Video Tapes Deliver Important Messages. Bulletin of the American Society for Information Science. 1982 April; 8(4): 25-29. ISSN: 0095-4403; CODEN: BASICR.

STRAIN, J. ROBERT; SIMMONS, STEPHANIE. 1984. The Cooperative Extension Service Updated Inventory of Computer Programs. Gainesville, FL: Cooperative Extension Service, University of Florida, Institute of Food and Agricultural Sciences; 1984 April. 2 volumes. 646p. (Circular no. 531-A and 531-C). Available from: C. M. Hinten, Publications Distribution Center, IFAS Building 664, University of Florida, Gainesville, FL 32611.

SUNDLOF, STEPHEN F.; RIVIERE, J.; EDMOND; CRAIGMILL, ARTHUR L.; BUCK, WILLIAM B. 1986. Computerized Food-Animal Residue-Avoidance Data Bank for Veterinarians. Journal of the American Veterinary Medical Association. 1986 January 1; 188(1): 73-76. ISSN: 0003-1488.

SZE, MELANIE C.   1980.   Computer-Based Information Retrieval for the Food Industry.   Food Technology.   1980 June; 34(6): 64–66, 68, 70. ISSN: 0015-6639; CODEN: FOTEAO.

TCHOBANOFF, JAMES B.   1980.   The Databases of Food: A Survey of What Works Best. . . and When.   Online.   1980 January; 1: 20–25. ISSN: 0146-5422; CODEN: ONLION.

TENG, P.S.   1984.   Surveillance Systems in Disease Management.   Plant Protection Bulletin.   1984; 32(2): 51–60.   ISSN: 0014-5637; CODEN: PPRBA.

THATCHER, RICHARD H.   1983.   Computer Use at Lucky Peak Nursery. In: Proceedings, Southern Nursery Conference (Eastern Session); 1982 July 12–15; Savannah, GA.   U.S. Forest Service, Southern Region; 1983 August.   66–76.   (Technical Publication R8-TP4).   OCLC: 10625283.

THOMAS, SARAH E.   1985.   Use of the CAB Thesaurus at the National Agricultural Library.   Quarterly Bulletin of the IAALD.   1985; 30(3): 61–65. ISSN: 0020-5966.

THOMAS, SARAH E.   1986.   More Than You Ever Wanted to Know About State Agricultural Publications.   Agricultural Libraries Information Notes.   1986 February; 12(2): 1–4.   ISSN: 0095-2699.

TIDBURY, ERIC, comp.   1983.   CAB Thesaurus.   Farnham Royal, Slough, UK: Commonwealth Agricultural Bureaux; 1983.   1230p.   ISBN: 0-85198-540-8.

U.S. CONGRESS. 37TH CONGRESS, 2ND SESSION.   1862.   An Act to Establish a Department of Agriculture.   United States Statutes at Large. 1862 May 15; 72: 387–388.

U.S. CONGRESS. HOUSE. COMMITTEE ON AGRICULTURE. 97TH CONGRESS, 2ND SESSION.   1983.   Information Technology for Agricultural America.   Prepared by Library of Congress Congressional Research Service.   Washington, DC: Government Printing Office; 1983.   358p.

U.S. CONGRESS. OFFICE OF TECHNOLOGY ASSESSMENT. 99TH CONGRESS.   1985.   Technology, Public Policy, and the Changing Structure of American Agriculture: A Special Report for the 1985 Farm Bill.   Washington, DC: Government Printing Office; 1985 March.   91p.   (OTA-F-272).

U.S. CONGRESS. SENATE. 97TH CONGRESS, 1ST SESSION.   1981.   Agriculture and Food Act of 1981: Senate Report no. 97-290. 97th Congress, 1st Session. Washington, DC: Government Printing Office; 1981.   278p. OCLC: 8099901.

U.S. DEPARTMENT OF AGRICULTURE.   1986a.   Fact Book of U.S. Agriculture.   Washington, DC: U.S. Department of Agriculture; 1986.   131p. (Miscellaneous Publication Number 1063).   Available from: Government Printing Office, Washington, DC 20402.

U.S. DEPARTMENT OF AGRICULTURE.   1986b.   High-Tech Financial Help.   USDA News Feature.   1986 January 2; 1–2.   Available from: News Division, Room 406-A, U.S. Department of Agriculture, Washington, DC 20250.

U.S. DEPARTMENT OF AGRICULTURE. AGRICULTURAL RESEARCH SERVICE.   1976-.   Composition of Foods.   Washingtion, DC: U.S. Department of Agriculture; 1976.   (Agriculture Handbook no. 8).   Available from: Government Printing Office, Washington, DC 20402.

U.S. DEPARTMENT OF AGRICULTURE. COOPERATIVE STATE RESEARCH SERVICE.   n.d.   Guide to CRIS Services.   Washington, DC: U.S. Department of Agriculture.   16p.   Available from: CRIS, 5th floor, The National Agricultural Library, Beltsville, MD 20705.

U.S. DEPARTMENT OF AGRICULTURE. COOPERATIVE STATE RE-
SEARCH SERVICE. 1987. 1986–87 Directory of Professional Workers
in State Agricultural Experiment Stations and Other Cooperating State
Institutions. Washington, DC: U.S. Department of Agriculture; 1984.
239p. (Agriculture Handbook number 305). Available from: Government Printing Office, Washington, DC 20402.

U.S. DEPARTMENT OF AGRICULTURE. EXTENSION SERVICE. 1985.
Cooperative Extension and Electronic Technology– New Initiatives in
Information Delivery, Educational Delivery, and Problem Solving. Wash-
ington, DC: U.S. Department of Agriculture, Extension Service; 1985
September. 19p. OCLC: 12634467.

U.S. DEPARTMENT OF AGRICULTURE. FOREST SERVICE. PACIFIC
SOUTHWEST FOREST AND RANGE EXPERIMENT STATION. 1983.
International Directory of Forestry and Forest Products Libraries.
Berkeley, CA: U.S. Department of Agriculture; 1983. 90p. Available
from: Government Printing Office, Washington, DC.

U.S. DEPARTMENT OF AGRICULTURE. HUMAN NUTRITION INFOR-
MATION SERVICE. NUTRITION MONITORING DIVISION. SURVEY
STATISTICS BRANCH. 1986. Machine-Readable Data Sets on Com-
position of Foods and Results from Food Consumption Surveys. Hyatts-
ville, MD: U.S. Department of Agriculture; 1986 May. 39p. (Administra-
tive Report no. 378). Available from: Human Nutrition Information
Service, U.S. Department of Agriculture, Nutrition Monitoring Division,
Rm. 304, Federal Building, Hyattsville, MD 20782.

U.S. DEPARTMENT OF AGRICULTURE. NATIONAL AGRICULTURAL
LIBRARY. n.d. Feed Composition Data Bank at the National Agri-
cultural Library; n.d. 3p. Available from: The Feed Composition Data
Bank, National Agricultural Library, 5th Floor, Beltsville, MD 20705.

U.S. DEPARTMENT OF AGRICULTURE. NATIONAL AGRICULTURAL
LIBRARY. 1982a. Directory of Aquaculture Information Resources.
Washington, DC: U.S. Department of Agriculture; 1982 December. 58p.
(Bibliographies and Literature of Agriculture number 25). Available
from: Government Printing Office, Washington, DC 20402.

U.S. DEPARTMENT OF AGRICULTURE. NATIONAL AGRICULTURAL
LIBRARY. 1982b. International Directory of Animal Health and Disease
Data Banks. Washington, DC: U.S. Department of Agriculture; 1982
November. 92p. Available from: National Agricultural Library, Beltsville,
MD 20705.

U.S. DEPARTMENT OF AGRICULTURE. NATIONAL AGRICULTURAL
LIBRARY. FOOD AND NUTRITION INFORMATION CENTER. 1982.
Audiovisual Resources in Food and Nutrition: Volume 2. Phoenix, AZ:
The Oryx Press; 1982. 232p. ISBN: 0-89774-105-6.

U.S. DEPARTMENT OF AGRICULTURE. OFFICE OF RURAL DEVELOP-
MENT POLICY. 1985. Rural Resources Guide: A Directory of Public
and Private Assistance for Small Communities. Washington, DC: U.S.
Department of Agriculture; 1985. 1 volume. Available from: U.S.
Government Printing Office, Washington, DC 20402.

U.S. DEPARTMENT OF AGRICULTURE. OFFICE OF THE SECRETARY.
1986. Mission Statement for the National Agricultural Library. Washington, DC: U.S. Department of Agriculture; 1986 June 9. 2p. (Secretary's Memorandum 1020-25).

U.S. DEPARTMENT OF AGRICULTURE; NATIONAL COMMISSION ON
LIBRARIES AND INFORMATION SCIENCE. 1982. Joint Con-

gressional Hearings on the Changing Information Needs of Rural America: The Role of Libraries and Information Technology. Washington, DC: Government Printing Office; 1982 July 21. 83p.

U.S. NATIONAL ADVISORY BOARD ON RURAL INFORMATION NEEDS (NABRIN). PLANNING COMMITTEE. 1985. The NABRIN Report. Washington, DC: National Commission on Libraries and Information Science; 1985. 30p. Available from: NCLIS, General Services Administration Building, 7th & D Streets, S.W., Suite 3122. Washingtion, DC 20024.

UNITED NATIONS. FOOD AND AGRICULTURE ORGANIZATION (FAO). 1984. AGRIS Introduction. Rome, Italy: FAO; 1984. 33p. Available from: AGRIS Coordinating Center, Food and Agriculture Organization of the United Nations, Via delle Terme di Caracalla, 00100 Rome, Italy.

UNITED NATIONS. FOOD AND AGRICULTURE ORGANIZATION (FAO). 1985. CARIS Introduction: Current Agricultural Research Information System. Rome, Italy: FAO; 1985. 28p. Available from: CARIS Coordinating Centre, FAO, Via delle Terme di Caracalla, 00100, Rome, Italy.

UNITED NATIONS. FOOD AND AGRICULTURE ORGANIZATION (FAO). 1986. Fifth Technical Consultation of AGRIS Participating Centres: Progress Report. Rome, Italy: FAO; 1986 March. 8p. (GIL-AGRIS/TC/5/2). Available from: AGRIS Coordinating Center, FAO. Via della Terme di Caracalla, 00100 Rome, Italy.

URBANO, CYNTHIA CHAMPNEY. 1984. Slice: A Computer System for Landscapers. American Nurseryman. 1984 September 1; 160(5): 92–111 (16p. not consecutive). ISSN: 0003-0198.

UTAH STATE UNIVERSITY NEWS. 1986. Kellogg Foundation Awards $4.1 Million for Rural Education, Information Services. Logan, UT: Utah State University; 1986 March 17. 4p. (Press release). Available from: Information Services, Utah State University, Logan, UT 84322.

VAUPEL, NANCY G.; ELIAS, A. W. 1981. Information Systems and Services in the Life Sciences. In: Williams, M. E., ed. Annual Review of Information Science and Technology: Volume 16. White Plains, NY: Knowledge Industry Publications for the American Society for Information Science; 1986. 267–288. ISSN: 0066-4200; CODEN: ARISBC.

VAVREK, BERNARD. 1982. Appendix 2: Reference Services in Rural Public Libraries in Communities of 25,000 or Fewer People. In: U.S. Department of Agriculture; National Commission on Libraries and Information Science. Joint Congressional Hearings on the Changing Information Needs of Rural America: The Role of Libraries and Information Technology. Washington, DC: Government Printing Office; 1982 July 21. 46–70.

VEDRO, STEVEN R. 1983. Infotext: An Agricultural Electronic Text Service From WHA-TV. In: North Central Computer Institute (NCCI) Workshop: The Use of Computers in Agricultural Information; 1983 May 2-5; Chicago, IL. Madison, WI: NCCI; 1983. 37–41.

VINCENT, LILLIE. 1986. Online With Food and Nutrition Information. Agricultural Libraries Information Notes. 1986 August; 12(8): 1-3. ISSN: 0095-2699.

WARNER, PAUL; CLEARFIELD, FRANK. 1982. An Evaluation of a Computer-Based Videotext Information Delivery System for Farmers: The Green Thumb Project: Executive Summary. Lexington, KY: Department of Sociology, University of Kentucky; 1982 January. 32p.

WATERS, SAMUEL T. 1986. Answerman, the Expert Information Specialist: An Expert System for Retrieval of Information from Library Reference Books. Information Technology and Libraries. 1986 September;
5(3): 204-212. ISSN: 0730-9295; CODEN: ITLBDC.
WATERS, SAMUEL T.; BUTLER, ROBERT W.; JACOBS, MARILYN M.
1985. Interactive Educational Videodisk: An Evaluation Study. Agricultural Libraries Information Notes. 1985 May; 11(5): 1-4. ISBN:
0095-2699.
WELSH, SUSAN. 1986. The Joint Nutrition Monitoring Evaluation Committee. In: What Is America Eating: Proceedings of a Symposium of the
Food and Nutrition Board, Commission on Life Sciences, National Re-
search Council. Washington, DC: National Academy Press; 1986. 7-20.
ISBN: 0-30903-635-6.
WHITING, LARRY R. 1981. Communications Technology in the Land
Grant University Setting: A Focus on Computer-Based Innovations for
Information Dissemination to External Audiences. Ames, IA: Center
for Agricultural and Rural Development, Department of Journalism and
Mass Communications, Iowa State University; 1981. 210p. (Miscel-
laneous Report). Available from: Dr. Paul Yarbrough, Department of
Journalism and Mass Communication, Press Building, Iowa State Univer-
sity, Ames, Iowa 50011.
WILLIAMS, MARTHA E.; ROBINS, CAROLYN G., eds. 1985. Agricultural
Databases Directory. Washington, DC: US Department of Agriculture,
National Agricultural Library; 1985 October. 249p. (Bibliographies and
Literature of Agriculture, number 42). Available from: Government
Printing Office: 529-999-40-010.
WILSDORF, MARK. 1985a. Crop Records on Computer: Techniques for
Building a Worthwhile Management Database. AgriComp. 1985
September/October; 4(2): 24-25. ISSN: 0738-5978.
WILSDORF, MARK. 1985b. Review: Crop Consultant. AgriComp. 1985
September/October; 4(2): 19-23. ISSN: 0738-5978.

Introduction to the Index

Index entries have been made for names of individuals, corporate bodies, subjects, geographic locations, and author names that have been included in the text pages and for author and conference names found in the bibliography pages. The page numbers in the index referring to bibliography pages are set in italics and they are listed after the page numbers relating to the text pages. Thus, the user can readily distinguish references to bibliographic materials from references to text.

Postings to acronyms are listed either under the acronym or under the fully spelled-out form, depending on which form is more commonly used and known. In either case a cross reference to the alternate form is provided. Postings associated with PRECIS, for example, would be listed under PRECIS as readers are generally less familiar with the full name "Preserved Context Index System." In a few cases, such as the names of programs, systems and programming languages, there is no spelled-out form either because there is none or because the meaning has been changed or is no longer used.

The index is arranged on a word-by-word basis. The sort sequence employed sort on special characters, first, followed by alpha characters and then numeric characters. Thus, O'Neill precedes Oakman and 3M Company follows the Zs. Government organizations are generally listed under country name, with *see* references provided from names of departments, agencies, and other subdivisions. While index entries do correspond precisely in spelling and format, they do not necessarily follow the typographical conventions used in the text. Author names, which are all upper case in the text, and both programming languages and software packages (such as expert system shells), which are in small caps in the text, are in upper and lower case or normal upper case in the index.

Subject indexing is by concepts rather than by words. When authors have used different words or different forms of the same words to express the same or overlapping concepts, the terminology has been standardized. An effort was made to use the form of index entries for concepts that had previously appeared in *ARIST* indexes. Cross references have been used freely to provide broad access to subject concepts. *See also* references are used for overlapping or related (but not synonymous) concepts; *see* references are used to send the readers to the accepted form of a term used in the index.

The index was prepared by Debora Shaw, using the MACREX Indexing Program developed by Hilary and Drusilla Calvert and distributed in the U.S. by Bayside Indexing. The overall direction and coordination of the index were provided by Martha E. Williams. Comments and suggestions are welcome and should be directed to the Editor.

Index*

---

*Italicized page numbers refer to Bibliography pages.

Introduction to the Keyword
and Author Index

The following section is an author and keyword index to *ARIST* chapters for Volumes 1 through 22. It has been produced to assist users in locating specific topics, chapters, and author names for all *ARIST* volumes to date. The index terms are sorted alphabetically and include all author names and content words from titles (a stop-word list of articles, conjunctions, and other non-content words was used). The sort word is followed by the author(s) name(s) and the *ARIST* citation.

Keyword and Author Index
of *ARIST* Titles
for Volumes 1-22

413

Eastman, Caroline M.
  Eastman, Caroline M. Database Management Systems. **20**, p91
Eckert, Philip F.
  Brandhorst, Wesley T. and Eckert, Philip F. Document Retrieval and Dissemination Systems. **7**, p379
Economics
  Cooper, Michael D. **8**, p5; Hindle, Anthony and Raper, Diane. **11**, p27; Lamberton, Donald M. **19**, p3; Repo, Aaatto J. **22**, p3; Wilson, John H., Jr. **7**, p39
Economist's
  Spence, A. Michael. **9**, p57
Education
  Harmon, Glynn. **11**, p347; Jahoda, Gerald. **8**, p321; Silberman, Harry F. and Filep, Robert T. **3**, p357; Vinsonhaler, John F. and Moon, Robert D. **8**, p277; Wanger, Judith. **14**, p219
Eisenmann, Laura M.
  Schwartz, Candy and Eisenmann, Laura M. Subject Analysis. **21**, p37
Elias, Arthur
  Vaupel, Nancy and Elias, Arthur. Information Systems and Services in the Life Sciences. **16**, p267
Electronic
  Hjerppe, Roland. **21**, p123; Lunin, Lois F. **22**, p179
Emerging
  Shaw, Ward and Culkin, Patricia B. **22**, p265
Empirical
  Zunde, Pranas and Gehl, John. **14**, p67
End-user
  Mischo, William H. and Lee, Jounghyoun. **22**, p227
Energy
  Coyne, Joseph H., Carroll, Bonnie C., and Redford, Julia. **18**, p231
Engineering
  Grattidge, Walter and Creps, John E., Jr. **13**, p297
Entry
  Turtle, Howard, Penniman, W. David, and Hickey, Thomas. **16**, p55
Environmental
  Freeman, Robert R. and Smith, Mona F. **21**, p241
European
  Tomberg, Alex. **12**, p219
Evaluation
  Bourne, Charles P. **1**, p171; Cleverdon, Cyril W. **6**, p41; Debons, Anthony and Montgomery, K. Leon. **9**, p25; Kantor, Paul B. **17**, p99; Katter, Robert V. **4**, p31; King, Donald W. **3**, p61; Lancaster, F. Wilfrid and Gillespie, Constantine J. **5**, p33; Rees, Alan M. **2**, p63; Stern, Barrie T. **12**, p3; Swanson, Rowena Weiss. **10**, p43; Yang, Chung-Shu. **13**, p125
Evans, Glyn T.
  Evans, Glyn T. Library Networks. **16**, p211
Experimental
  McGill, Michael J. and Huitfeldt, Jennifer. **14**, p93
Expert
  Sowizral, Henry. **20**, p179

Intelligence

Interaction

Interactive

Interface

About the Editor . . .

Professor Martha E. Williams assumed the Editorship of the *ANNUAL REVIEW OF INFORMATION SCIENCE AND TECHNOLOGY* with Volume 11 and has produced a series of books that provide unparalleled insights into, and overviews of, the multifaceted discipline of information science.

Professor Williams holds the positions of Director of the Information Retrieval Research Laboratory and Professor of Information Science in the Coordinated Science Laboratory (CSL) as well as affiliate of the Computer Science Department at the University of Illinois, Urbana-Champaign, Illinois. As a chemist and information scientist Professor Williams has brought to the Editorship a breadth of knowledge and experience in information science and technology.

She serves as: Trustee, Director and the Chairman of the Board of Engineering Information, Inc.; founding editor of *COMPUTER-READABLE DATABASES: A DIRECTORY AND DATA SOURCEBOOK*; Editor of *ONLINE REVIEW* (Learned Information, Ltd., Oxford, England); and Program Chairman for the National Online Meetings which are sponsored by *ONLINE REVIEW*. She was appointed by the Secretary of Health, Education and Welfare, Joseph Califano, to be a member of the Board of Regents of the National Library of Medicine (NLM) in 1978 and has served as Chairman of the Board. She has been a member of the Numerical Data Advisory Board of the National Research Council (NRC), National Academy of Sciences (NAS). She was a member of the Science Information Activities task force of the National Science Foundation (NSF), was chairman of the Large Database subcommittee of the NAS/NRC Committee on Chemical Information, and was Chairman of the Gordon Research Conference on Scientific Information Problems in Research in 1980.

Professor Williams is a Fellow of the American Association for the Advancement of Science, Honorary Fellow of the Institute of Information Scientists in England, and recipient of the 1984 Award of Merit of the American Society for Information Science. She is a member of, has held offices in, and/or is actively involved in various committees of the American Association for the Advancement of Science (AAAS), the American Chemical Society (ACS), the Association for Computing Machinery (ACM), the American Society for Information Science (ASIS), and the Association of Information and Dissemination Centers (ASIDIC). She has published numerous books and papers and serves on the editorial boards of several journals. She is the founder and President of Information Market Indicators, Inc., and consults for many governmental and commercial organizations.

DATE DUE

			Printed in USA

HIGHSMITH #45230